Principles of Applied Statistics

An Integrated Approach using MINITAB and Excel

Principles of Management Series

Editor

Joseph G Nellis

Professor of International Management Economics
Cranfield School of Management, Cranfield University

Books in the series

Principles of Applied Statistics
Second Edition
Michael C Fleming and Joseph G Nellis

Principles of Operations Management
Second Edition
R Les Galloway

Principles of Human Resource Management
David Goss

Principles of Marketing
Geoffrey Randall

Principles of Law for Managers
Anne Ruff

Principles of Accounting and Finance
Peter Sneyd

Principles of Information Systems Management
John M Ward

For further information about any of these titles, or to order inspection copies or purchase books, please visit http://www.cengage.co.uk

Principles of Applied Statistics

An Integrated Approach using MINITAB and Excel

Michael C. Fleming
Professor Emeritus of Economics
Loughborough University

and

Joseph G. Nellis
Professor of International Management Economics
Cranfield School of Management
Cranfield University

SOUTH-WESTERN
CENGAGE Learning™

Australia • Brazil • Japan • Korea • Mexico • Singapore • Spain • United Kingdom • United States

**Principles of Applied Statistics,
2nd Edition**
Michael C. Fleming and Joseph G. Nellis

Publishing Director: John Yates

Commissioning Editor: Thomas Rennie

Manufacturing Manager: Helen Mason

Senior Production Controller: Maeve Healy

Marketing Manager: Angela Lewis

Typesetter: Techset Composition Ltd,
Salisbury, Wiltshire

Cover design: Acrobat

Text design: Design Deluxe, Bath

Microsoft® Windows® are registered
trademarks and Microsoft® Excel® is a
trademark of Microsoft Corporation in the
United States and / or other countries.

Screen shots reprinted by permission of the
Microsoft Corporation.

MINITAB? is a trademark on Minitab, Inc.
and is used herein with the owner's
permission.

Portions of MINITAB Statistical Software
input and output contained in this book are
printed with permission of Minitab, Inc. For
further information contact:
http://www.minitab.com

For product information and technology assistance,
contact **emea.info@cengage.com**.

For permission to use material from this text or product,
and for permission queries,
email **clsuk.permissions@cengage.com**

British Library Cataloguing-in-Publication Data
A catalogue record for this book is available from the
British Library.

ISBN: 978-1-86152-586-4

Cengage Learning EMEA
High Holborn House, 50-51 Bedford Row
London WC1R 4LR

Cengage Learning products are represented in Canada
by Nelson Education Ltd.

For your lifelong learning solutions, visit
www.cengage.co.uk

Purchase e-books or e-chapters at:
http://estore.bized.co.uk

Printed by TJ International, Padstow,
Cornwall
5 6 7 8 9 10 – 10 09 08

Contents

Figures in main text

Tables in main text

Worked examples

Glossary of symbols

(Symbols using Greek letters are listed separately first)

α	(Alpha) Significance level
$1 - \alpha$	Confidence level
Δ	Difference
μ	(Mu) Population mean
$\overline{\mu}$	Confidence interval for the population mean
$\mu_{\overline{D}}$	Mean of paired differences for the population
π	(Pi) Population proportion
$\overline{\pi}$	Confidence interval for population proportion
$\hat{\pi}$	('Pi hat') Pooled sample proportion for two samples (estimated common value of π for two population proportions)
v	(Nu) Degrees of freedom in the F distribution
ρ	(Rho) Population correlation coefficient
σ	(Lower case sigma) Population standard deviation
σ^2	Population variance
σ_b	Standard deviation of the sampling distribution of b – the standard error of b
$\sigma_{\overline{D}}$	Standard deviation of the sampling distribution of paired differences, D
σ_p	Standard deviation of the sampling distribution of the sample proportion p – the standard error of p
$\sigma_{p_1 - p_2}$	Standard deviation of the sampling distribution of the difference between sample proportions – the standard error of the difference between proportions
$\sigma_{\overline{x}}$	Standard deviation of the sampling distribution of $\overline{x}$ – the standard error of the mean
$\sigma_{\overline{x}_1 - \overline{x}_2}$	Standard deviation of the sampling distribution of the difference between means of two populations – the standard error of the difference between means
$\sum$	(Upper case sigma) Summation sign
χ^2	(χ is Greek letter chi) Chi-square statistic and distribution
A	y intercept (constant) in the population regression model
a	y intercept (constant) in the regression model estimated from sample data
B	Regression coefficient of the population regression line
b	Regression coefficient, measuring slope (gradient) of the regression line, estimated from sample data

C	Cyclical variation in time series analysis
D	(1) Difference between two paired values; (2) Decile
$\overline{D}$	Mean of the paired differences in a sample of paired values
d	Durbin–Watson statistic
df	Number of degrees of freedom
d_L	Lower critical value of the Durbin–Watson statistic
d_U	Upper critical value of the Durbin–Watson statistic
E	Expected frequency
e	(1) Random error (residual) $y - y_c$ in sample regression model; (2) Mathematical constant ($= 2.71828$) used in Poisson probability distribution
F	F statistic and distribution
f	Frequency of a class or category
g	Geometric mean
H_0	Null hypothesis
H_1	Alternative hypothesis
i	Subscript used to denote specific observations in a data set
k	(1) Number of independent variables in multiple regression; (2) Number of categories in a goodness-of-fit test; (3) Number of samples or columns in ANOVA
Md	Median
Mo	Mode
N	Population size
n	(1) Sample size; (2) Current period in index number formulae; (3) Number of time periods in compound rate of change formula
n!	n factorial
^nC_r	Number of combinations of n items selected r at a time
^nP_r	Number of permutations of n items selected r at a time
O	Observed frequency
o	Base period in index number formulae
$P(A)$	Probability that event A will occur
$P(A|B)$	Conditional probability that event A will occur given that event B has occurred
PI	Price index
p	(1) Sample proportion; (2) Probability of success in binomial experiments; (3) Prices in index number formulae
Q_1	First (lower) quartile
Q_3	Third (upper) quartile
QI	Quantity index
q	(1) Probability of failure in binomial experiments ($= 1 - p$); (2) Quantities in index number formulae
R	(1) Multiple correlation coefficient; (2) Random variation in time series analysis
R^2	Multiple coefficient of determination
$\overline{R}^2$	Adjusted multiple coefficient of determination
r	(1) Simple correlation coefficient estimated from sample data – estimator of ρ; (2) Number of rows in a contingency table; (3) Rate of increase per time period
r^2	Simple coefficient of determination
r_s	Spearman's rank correlation coefficient

S	Seasonal factor in time series analysis
s	Sample standard deviation
s^2	Sample variance
s_b	Estimator of σ_b
s_D	Standard deviation of paired differences in a sample of paired values
$S_{\overline{D}}$	Estimator of $\sigma_{\overline{D}}$
SEE	Standard error of estimate
s_p	Estimator of σ_p
$s_{p_1-p_2}$	Estimator of $\sigma_{p_1-p_2}$
$s_{\bar{x}}$	Estimator of $\sigma_{\bar{x}}$
$s_{\bar{x}_1-\bar{x}_2}$	Estimator of $\sigma_{\bar{x}_1-\bar{x}_2}$
s_{xy}	Covariance between x and y
T	Trend in time series analysis
t	(1) t statistic and distribution; (2) a time period
t_α	Value of t at the α level of significance
W	Weights in index number formulae
x or X	(1) Random variable; (2) Independent variable in a regression model
$\bar{x}$	Sample mean
y or Y	(1) Variable; (2) Dependent variable in a regression model
y_c	Computed (predicted) value of y from regression model
y^f	Extrapolated (forecast) value of y
Z	Standard normal variable
Z_α	Value of Z at the α level of significance

This book is dedicated to Lucy Grace Buckley who arrived just in time!

A judicious man looks at Statistics not to get knowledge but to save himself from having ignorance foisted on him.

Thomas Carlyle

Preface

Introduction

The aim in writing this book has been to introduce the **Principles of Statistics** and their applications in a user-friendly manner. With this aim in mind, we have sought to eliminate as much unnecessary jargon and mathematical proofs as possible. Emphasis throughout the book is placed on practical illustrations of the techniques and the application of computers to practical problems.

The essence of the book is on 'learning by doing' as well as 'learning by seeing'. To this end a large number of **Worked Examples** are included in each chapter along with **Self-study exercises**. The Worked examples show in detail how manual calculations may be carried out, the aim being not only to solve problems but also to support and strengthen the learning process.

At the same time, the use of computers to take the tedium and headache out of statistics is a key feature of this book. The use of two of the most popular computer packages – **MINITAB** and **Microsoft Excel** – is prominent throughout. The intention is to give students and instructors a choice of programs alongside the traditional manual approach to statistical analysis. These computer packages provide tools to solve a wide range of statistical problems and to create graphs and charts. Both of these programs receive a fully integrated treatment.

While the book has been written with the needs of a wide range of users in mind, it will be especially attractive to three groups in particular:

- to students taking formal courses in statistics in universities and colleges as part of degree and other programmes;
- to those studying for professional and vocational qualifications such as banking and accountancy examinations;
- to people following general post-graduate courses such as the Master of Business Administration (MBA) and Diploma in Management Studies (DMS), as well as practising managers and others following continuing studies programmes.

To serve the needs of different user groups, the book has been broken down into a large number of relatively small chapters so that they can be bundled together to satisfy the requirements of different examining bodies. For many users, for whom the use of statistics is not a regular need, the book will serve as a concise reference and *aide-mémoire*.

Key features of the book

The book is set out in a standard format with the material in each chapter consistently arranged under the following set of headings:

- Aims
- Learning outcomes
- Principles
- Applications and Worked Examples
- Key terms and concepts
- Self-study exercises (with answers at the back of the book)

Full explanations are given on how to use the two computer packages:

- MINITAB
- Microsoft Excel

Appendix B describes the MINITAB program and explains its application. Excel is a spreadsheet program, which is in widespread use as part of the Microsoft 'Office' suite of programs, and now has an extensive range of built-in statistical tools. Appendix C describes the Excel program and explains its application.

New to the second edition

The major new features of this edition are the following:

- Clear and concise statements of the aims of each chapter
- Descriptions of the learning outcomes from each chapter
- Summary statements of the key terms and concepts used
- Inclusion of 'Schematic Overviews' preceding each of the five 'Parts' of the book. These show at a glance the content and structure of each chapter and how each topic relates to other topics
- Inclusion of a completely new chapter on analysis of variance (ANOVA)
- An extension to the treatment of regression analysis to include the problem of autocorrelation
- Additional worked examples and self-study exercises where appropriate
- A greatly enlarged treatment of the use of the MINITAB statistics package operating under the Windows environment
- Inclusion of the use of the statistical facilities in the Microsoft Excel program
- Inclusion of a full glossary of terms at the end of the book
- Summary list of statistical terms and symbols
- An updated and enlarged list of references for further reading

A particularly important new feature of the book is the production of supplementary material for the use of instructors, details of which are given below.

To the instructor

Supplementary material is available to all instructors who adopt the book for their courses through the Cengage Learning website:

- www.cengage.co.uk

This material is designed to support the work of instructors and to facilitate student learning. There are three main categories of material:

- **Solutions manual**. This provides fully-worked solutions to all of the self-study exercises which are given at the end of each chapter.
- **Overhead transparency masters**. These provide, for each chapter, a full set of visual aids covering:
  - The learning outcomes
  - Schematic overviews
  - Key formulae and concepts
  - Tables and charts
- **Data bank**. This consists of a large number of data sets, including many which are too large to use conveniently in the book itself. They are meant to provide the basis for additional problem-solving and especially for use in practising the use of the computer programs. All the data sets may be downloaded into MINITAB or Excel. Descriptive details of the data sets, and their potential uses, is posted on the website. It may be noted too that the MINITAB package itself includes a range of data sets drawn from different fields of study.

To the student

Consideration of the needs of students has been at the forefront of our minds in writing this book, especially those who are new to the subject and those who see statistics as a daunting area of study. To facilitate learning, we have set out to focus only on the principles and application of statistics rather than the theoretical underpinnings. Extensive use is made of illustrative Worked examples solved manually as well as using MINITAB and Microsoft Excel.

The authors would be pleased to receive, through the publishers, any feedback from readers concerning areas for future improvement and development of the text.

Acknowledgements

We would like to acknowledge formally the advice and comments received from a number of people in helping to improve the content and style of the book. In particular, we thank our colleagues (Séan Rickard and John Mapes) and former students of the Cranfield School of Management as well as many former colleagues and students from the Department of Economics at Loughborough University. We would also like to express our thanks to those who have contacted us directly from around the world with many encouraging comments and suggestions. We are also grateful to the academic reviewers employed by the publishers to advise on the content and structure of the book. Their remarks have been extremely helpful in shaping the final product. Our thanks also go to the staff of Cengage Learning for managing the editorial and production process. Naturally, any errors or other shortcomings remain the responsibility of the authors alone.

As always we are grateful to Chris Williams for her ever-willing help and cheerfulness during the preparation of the manuscript. Finally, we would like to thank our families and ask for their forgiveness for breaking our promise never to write another book!

M.C.Fleming
J.G.Nellis

| Chapter 1 | # Introduction |

Aims

The purpose of this book is to set out the fundamental principles of statistical methods which will be useful to all individuals regardless of their fields of specialization. It is intended to equip you with the skills to analyse a wide range of statistical problems which arise in many areas of activity, in business, academic work and all walks of life generally. The book will also serve as a primer for further studies in selected areas involving quantitative analysis.

At the outset, it may be helpful to explain the meaning of the word *statistics*.

The word *statistics* has two meanings. First, it is commonly used as a collective noun to refer to sets of data relating to a wide range of topics such as the size of populations, production activity, retail prices, incomes, rainfall, etc. Statistical information of this kind is regularly produced by the government and other bodies and is part and parcel of every-day life. Indeed, the word itself can be traced back to the Latin words *status*, meaning state, and *statista*, meaning statesman. Second, the word statistics also refers to the theory and methods used for the collection, description, analysis and interpretation of numerical data. Nowadays, statistical techniques are widely employed in all areas of activity where the analysis and presentation of numerical data are useful. This may range, of course, from high-level research in the physical sciences, through applications involving the regular monitoring of production processes in industry, economic analysis and forecasting, to the simple tabulation and presentation of data on common subjects of interest such as football scores, examination results, etc.

1

Learning Outcomes   After working through the book, you will be able to:

- understand the meanings and uses of the word *statistics*
- apply statistical techniques to problems involving subject areas such as finance, economics and other social sciences, marketing, operations management etc.
- organize statistical information and data in a way which is meaningful and understandable for the purposes of report writing and other presentations
- generate charts of all types and assess their use for succinctly summarizing a diverse quantity of information
- calculate many summary measures for capturing the key characteristics of data
- analyse trends in time series data and make forecasts
- derive index numbers to describe changes in economic, business and social indicators
- examine relationships between variables involving techniques such as regression and correlation etc.
- appreciate the principles of probability as the basis of decision-making
- understand the meaning and application of the main probability distributions
- use the main statistical probability tables to solve problems
- appreciate what is meant by statistical hypotheses
- consider and make significance tests of statistical hypotheses
- use sample information to make reliable inferences about the populations from which the data have been obtained including fields such as market research and quality control
- use the well-known statistical software packages, MINITAB and Excel, to make statistical analyses
- interpret the results of statistical analyses and draw appropriate conclusions
- understand the limitations and pitfalls involved in the use of statistical data and techniques.

Structure and content

The book has been designed to meet the needs of the wide range of potential users referred to above. As far as possible, it has been written in a largely non-technical form, avoiding the unnecessary use of formal mathematics and mathematical proofs. Naturally, it is not possible to avoid the use of mathematics altogether, but the objective has been to make the book accessible to as many readers as possible. To this end, care has been taken to set out the meaning and use of any mathematical notation in a straightforward fashion. The only mathematical background needed for an understanding of the entire book is basic arithmetic and the elements of algebra.

The book is divided into five parts as follows:

Part I   descriptive statistics

Part II   probability and probability distributions

Part III   statistical inference

Part IV   relationships between variables

Part V   time series and index numbers

We comment on the structure and content of each of these parts below.

Part I   Descriptive statistics

This section of the book is concerned with two aspects of description: (a) the presentation of data in tables, charts and graphs and (b) the use of numerical measures to summarize data.

Part II   Probability and probability distributions

A large part of statistical theory involves the drawing of inferences on the basis of samples of data, rather than full data sets, referred to by statisticians as *statistical populations or statistical universes*, because it is often impractical or too costly to carry out *censuses* of full data sets. Such inferences are statistical inferences and analysis in this field is based on the concepts of probability. These concepts lie at the heart of statistical inference. This part of the book, therefore, sets out the basic principles of probability and of three particular probability distributions (binomial, Poisson and normal).

Part III   Statistical inference

This part of the book is concerned with the application of statistical inference to two basic areas:

1. **estimation:** the estimation of unknown summary measures for a statistical population using sample data;

2. **hypothesis testing:** the testing of hypotheses about statistical populations on the basis of sample information.

Part IV   Relationships between variables

The idea of a relationship between statistical variables is a familiar one in many fields of interest, e.g. the relationship between the price of a good and the quantity sold, between rainfall and crop yields, etc. Such relationships may be examined using some of the graphical methods considered in Part I of this book. In Part IV we are concerned with techniques for measuring such relationships.

Part V   Time series and index numbers

This part of the book covers analytical techniques of particular importance for the analysis and interpretation of economic and business data. Time series analysis is concerned with the measurement of trends as well as seasonal and cyclical fluctuations in data recorded over time. Index numbers provide another means of summarizing data but, unlike the summary measures discussed in Part I, they provide composite measures for groups of variables (frequently relating to rates of change over time or, for example, price changes for groups of commodities).

Format of chapters

With only a few exceptions, each chapter is arranged according to a common format. Each is divided into six sections as follows:

1. aims
2. learning outcomes
3. principles
4. applications and worked examples
5. key terms and concepts
6. self-study exercises

The first section provides a summary statement of the aims of each chapter, while the second, learning outcomes, section states in more detail how the aims are to be achieved by specifying the particular topics covered. The explanation and discussion of principles which underlie the statistical techniques are placed in a reasonably self-contained section – the third section. This is intended to make the book especially useful for student revision purposes. Throughout the book, emphasis is placed on the practical uses and applications of the techniques and methods of analysis. For this purpose the relevant material, covering both worked and computer-based examples and solutions, is brought together in the fourth section of each chapter. This is followed by a compilation of the key terms and concepts used in each chapter. Finally, the last section provides a number of graded self-study exercises. The answers to these exercises are provided at the back of the book.

Computer-based applications

A special feature of this book is the emphasis placed on the use of computer software to solve problems. Nowadays, a large number of statistical software packages are available which are programmed to process large volumes of data very quickly and to carry out a wide range of statistical analyses as well as draw graphs and charts of various kinds. For this purpose, we show computer applications, wherever appropriate, based on the MINITAB statistical package and the Microsoft Excel spreadsheet program – both operating under the Microsoft Windows environment. These are well-known and easy-to-use packages that run on both mainframe and personal computers (PCs) and are in widespread use throughout the world. Descriptions of these programs and their applications are given in two appendices (Appendix B covering MINITAB and Appendix C covering Excel).

Solutions using MINITAB and Excel show clearly the succession of steps required to solve any particular problem, including 'authors' comment' which explain each step as necessary.

Many other computer packages are also available which provide a similar range of facilities as MINITAB and Excel. We would emphasize that while computer solutions are provided in most chapters, they are not essential to an understanding of the content of each chapter. The book is designed to be used equally well as a conventional text.

Basic statistical concepts

In this introductory chapter, a number of basic terms and concepts used in statistics are introduced. These and a number of others provide a basic vocabulary used throughout the remainder of the book and we therefore give a summary of the essential terms and concepts below. These are arranged in three parts:

- terminology
- types of variables
- scales of measurement

Terminology

Primary data	Information originally collected at source (such as that obtained in questionnaire surveys) – often referred to as raw data.
Secondary data	Summary information based upon the raw data for the purposes of reporting and publication.
Census	The collection of information about every item or object in the statistical population.
Sample	A subset of the population, selected so as to represent the whole population, from which information is collected.
Random sample	A sample in which each element has the same chance of being selected for inclusion in the sample set.
Parameters	Numerical summary measures used to describe a statistical population (such as the population average).
Sample statistics	Numerical summary measures used to describe a sample (such as the sample average) whose values vary according to the random sample collected.
Descriptive statistics	Tabular, graphical and numerical methods for organizing and summarizing information clearly and effectively relating to either population or sample data.

Inferential statistics	Methods of drawing and measuring the reliability of conclusions about a statistical population based on information obtained from a sample data set.

Types of variables

Variable	A characteristic or phenomenon of a population or sample that can take *more than one value* (such as household income). A variable can be either quantitative or qualitative in nature.
Quantitative variable	A variable which can take different *numerical* values, and which is measured in discrete or continuous form (see below).
Qualitative variable	A variable which describes characteristics which may be categorized rather than measured (e.g. sex, hair colour, race).
Discrete variable	A quantitative variable in which the scale of measurement varies in discrete steps – typically associated with counts (e.g. the number of firms, number of goals scored).
Continuous variable	A quantitative variable which may be measured on a continuous scale and can take any value within an interval (e.g. temperature, height, weight).

Scales of measurement

Quite apart from the measurement of data in discrete or continuous forms as defined above, it is important to appreciate that different levels or scales of measurement are used for different types of variables. This is important because the choice of techniques of statistical analysis depends partly on the nature of the data available and the scale of measurement that applies. There are four such scales: nominal, ordinal, interval and ratio.

Nominal scale	A scale of measurement for a variable is *nominal* when the observations simply define categories. These categories may be given a *non-numeric* label (such as colour, country of origin, etc.) or, alternatively, a *numeric* code (e.g. house numbers, identification numbers for people or spare parts.) The essential point to note is that the basic arithmetic operations of addition, subtraction, multiplication and division do not make sense in the case of nominal data. It is not possible, therefore, to calculate an average value for such data.
Ordinal scale	A scale of measurement for a variable is *ordinal* when the data have the properties of nominal data but the variable can now be used to arrange the data in a meaningful order (e.g. subjective judgements of quality or performance as in competitions). Again, as with nominal data, even if the observations are given numeric codes denoting rank order, it is important to note that arithmetic operations do not make sense in the case of ordinal data.

Interval scale

A scale of measurement for a variable is *interval* when the data have the properties of ordinal data but the interval between observations can be expressed in terms of a fixed unit of measurement. A good example of such a variable is temperature, in which the fixed unit of measurement is a degree. So records of temperature can be ranked in order from hot to cold and the interval between observations can be expressed in terms of the fixed unit of measurement (number of degrees). An important point to note is that, in contrast to a ratio scale (discussed below), the value of zero on an interval scale is not meaningful in the sense of indicating that nothing exists at that point: $0°C$ is not the same as $0°F$ and the idea of 'no' temperature is meaningless. Hence, the ratio of values measured on an interval scale is also meaningless. For example, while we can state that $40°C$ is $20°C$ hotter than $20°C$, we cannot claim that it is twice as hot since there is no absolute benchmark for comparisons. This will be readily appreciated by comparing these temperatures in Fahrenheit: $20°C = 77°F$, $40°C = 104°F$. Such data are always numeric and the basic arithmetic operations *do* make sense.

Ratio scale

A scale of measurement for a variable is *ratio* when the data have all the properties of interval data but the ratio of two observations is now meaningful. In this case, the value of zero is meaningful and indicates that nothing exists for the variable at that point. Hence, we can compare heights, weights, distances, times, etc. and compare ratios of measurements. When data are collected on the cost of something, a ratio scale of measurement is, of course, implicit in that any value is measured relative to a zero cost (i.e. free). As a result, if one item costs £10 and another £5, then we can state on a ratio scale that the first one is twice as expensive as the second one. Since the ratio data have all the properties of interval data, ratio data are *always numeric* and the basic arithmetic operations *do* make sense.

Figure 1.1 provides a useful summary of the various scale classifications for numerical (i.e. quantitative) data. Note that discrete data can have any of the four scales of measurement while continuous data can only be assigned to an interval or ratio scale. Table 1.1 summarizes the differences between the four scales of measurement bringing together the key points noted above for both qualitative and quantitative data.

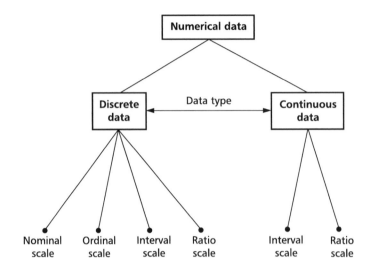

Figure 1.1 Classifications of numerical data

Table 1.1 Scales of measurement – a summary

Scale of measurement	Non-numerical data	Numerical data	Arithmetic operations
Nominal	Observations placed in categories only	Numerical codes assigned to categories	Not applicable
Ordinal	Observations placed in categories and may be arranged in rank order	Numerical values permit ranking or ordering of data	Not applicable
Interval	Not applicable	Numerical values expressed on a scale with fixed units so that the interval between them is meaningful but ratios are not	Applicable
Ratio	Not applicable	Numerical values for which ratios are meaningful	Applicable

The users and uses of statistics

The purpose of this chapter has been to lay out some of the fundamentals which apply to the collection, description, analysis and interpretation of numerical data and to indicate the aims and scope of the book in presenting the principles of statistics.

The use and presentation of statistical information permeates our everyday lives in almost every respect. The information explosion has affected everybody. This is increasingly the case as society becomes more and more skilled in the use of information technology. The correct use, understanding and interpretation of statistical information demands an appreciation of statistical theory and methods. The range of uses is extremely wide and the introduction of computers, especially PCs, has brought data analysis within the reach of many more people in all walks of life.

Statistical techniques are to be frequently encountered in a very wide range of uses covering government, business and commerce, market and academic research as well as for private purposes by individuals. Some examples of uses by each of these categories of user are given in Table 1.2.

Not all users of statistics, of course, will wish or need to study this book from beginning to end. For some, one or two parts, particularly Parts I and V, will be sufficient to give them a basic understanding of the principles and applications. To serve the needs of others the book has been sub-divided as much as possible so that topics may be combined as appropriate for various courses of study. We have also had in mind the different needs of such users by adopting a standard format, as described above, in which the basic principles, applications and self-study exercises are sub-divided so that the book may be used as a standard text, as a revision tool and for self-study purposes.

Finally, just as important as being able to use statistical methods correctly is the need to be able to recognize their misuse. Unfortunately, whether by accident or design, their misuse is all too common. It is hoped that this book will equip the reader with the necessary skills and knowledge to recognize and avoid the pitfalls and contribute to putting to rest once and for all the old quip of 'lies, damned lies and statistics'.

Table 1.2 Users and uses of statistics

User	Examples of uses
Government	Monitoring economic and social trends
	Forecasting
	Policy-making
Business and commerce	Monitoring performance
	Market appraisal
	Financial analysis and planning
	Process and quality control
	Employee and other records
	Forecasting
Market and academic research	Testing of hypotheses
	Development of new theories
	Consultancy services to government and business
Individuals	Personal information and decision-making
	Leisure activities
	Community work
	Personal finances
	Gambling

Part I | Descriptive statistics

Figures often beguile me, particularly when I have the arranging of them myself.

MARK TWAIN

The following diagram provides an overview of the whole of this part of the book.

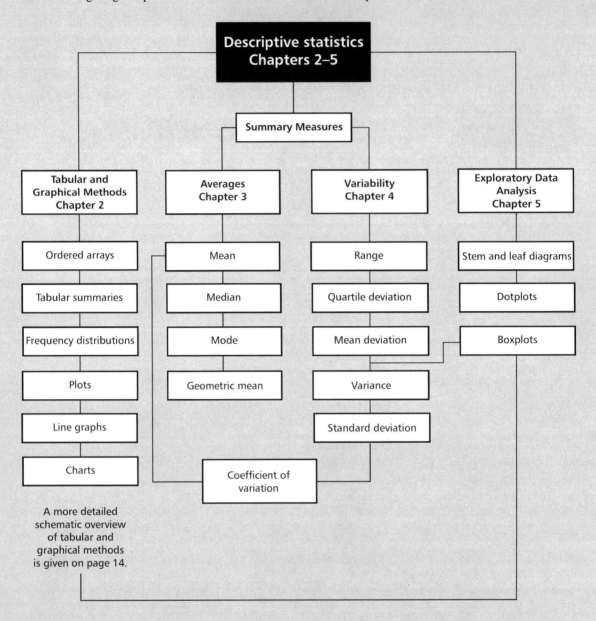

**Descriptive statistics
Chapters 2–5**

Summary Measures

| Tabular and Graphical Methods Chapter 2 | Averages Chapter 3 | Variability Chapter 4 | Exploratory Data Analysis Chapter 5 |

Ordered arrays — Mean — Range — Stem and leaf diagrams

Tabular summaries — Median — Quartile deviation — Dotplots

Frequency distributions — Mode — Mean deviation — Boxplots

Plots — Geometric mean — Variance

Line graphs — Standard deviation

Charts — Coefficient of variation

A more detailed schematic overview of tabular and graphical methods is given on page 14.

Chapter 2	# Describing data: tables, charts and graphs

Aims

In this chapter the aim is to show how a set of data can be summarized using tables, charts and graphs. Very often, it is very difficult to see the 'wood for the trees' when confronted with a large mass of disorganized data. To overcome this problem, there are various ways of organizing data and presenting them in tabular and graphical forms. The purposes of these are threefold:

■ to summarize the data

■ to detect underlying trends and relationships

■ to provide tools of analysis

There are very many tabular and graphical methods used for these purposes. In this chapter we deal with the most commonly used ones, namely:

1. frequency distribution tables
2. line graphs
3. pie charts
4. bar charts
5. histograms
6. cross-tabulations – contingency tables
7. frequency polygons
8. cumulative frequency curves (ogives)
9. Lorenz curves
10. rate of change graphs (logarithmic scales)

The following diagram provides an overview of the whole of this chapter.

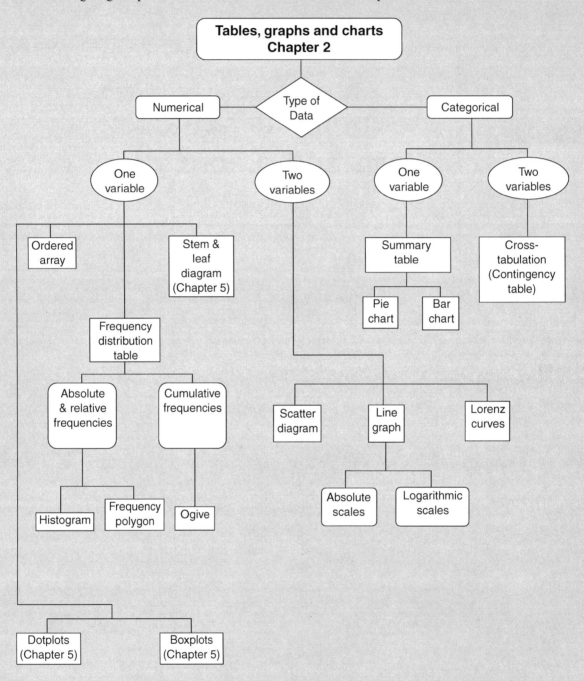

The choice of method depends on the nature of the data and on the purpose of the demonstration. But remember the danger of creating a misleading – or deliberately deceptive – impression. In this chapter we illustrate the 'right' and 'wrong' ways of describing data using tables, charts and graphs and illustrate the application of some specialized graphs.

Learning Outcomes   After working through this chapter you will be able to:

- employ tabular methods for organizing and classifying sets of raw data
- understand the terminology used in tabular and graphical techniques
- select and use appropriate tabular and graphical methods for descriptive purposes
- use tabular and graphical summaries as tools of analysis
- appreciate the danger of concealing important elements and characteristics of the underlying information when summarizing large volumes of data
- appreciate the uses and abuses associated with diagrammatic methods of summarizing data.

Principles

Frequency distribution tables

A *frequency distribution table* is a convenient way of summarizing a mass of raw data. This is achieved by organizing or arranging the data into groups. Depending on the nature of the data these groups may be categories (e.g. airline passengers grouped according to first class, club class and economy class) or may represent value ranges of the variable in question (e.g. three groups of luggage weights: under 20 kg, 20–40 kg, over 40 kg). Having established the groups into which we wish to allocate the raw data, it is then a simple matter to count the number of observations which fall in each group. The presentation of groups alongside the corresponding group frequencies is called a *frequency distribution table*, or simply a *frequency table*.

Table 2.1 provides a set of raw data representing the weights of luggage checked in by 65 passengers on an aircraft. A helpful first step is to arrange the data from smallest to largest into an *ordered array*. Table 2.2 shows the ordered array for the 65 luggage weights. It will be seen that this enables us to see at a glance the overall range of the data (minimum of 7 kg to a maximum of 55 kg) but it does not provide a useful summary of the data – the data remain in a raw, though ordered, form.

The next step is to arrange the data into groups and to count the number (i.e. frequency) of observations in each group. This produces a *frequency distribution* as shown in Table 2.3 (to be more precise, this should be referred to as a *grouped* frequency distribution). The table consists of the following elements:

▪ a number of groups (*classes*), each of which covers a range of the luggage weights (*a class interval*), shown in the first column

▪ *class frequencies*, giving the number of luggage weights falling within each class interval, shown in the second column

▪ *relative frequencies*, in which each absolute frequency is expressed as a percentage of the total frequency, shown in the third column. This is, of course, an optional feature.

Table 2.1 Raw data: luggage weights (rounded to nearest kilogram)

8	51	20	37	19	40	11	43
30	23	25	7	20	9	21	29
55	55	40	34	49	33	7	7
17	45	50	23	49	20	26	47
27	20	46	42	39	12	24	34
50	42	26	10	51	49	48	14
21	25	27	32	20	12	35	27
20	47	18	55	15	12	17	31
28							

Table 2.2 Ordered array of luggage weights

7	7	7	8	9	10	11	12
12	12	14	15	17	17	18	19
20	20	20	20	20	20	21	21
23	23	24	25	25	26	26	27
27	27	28	29	30	31	32	33
34	34	35	37	39	40	40	42
42	43	45	46	47	47	48	49
49	49	50	50	51	51	55	55
55							

Table 2.3 Frequency distribution of luggage weights

Weight classes (kilograms)	Class frequencies (number)	Relative frequencies (%)
9 or less	5	7.7
10–19	11	16.9
20–29	20	30.8
30–39	9	13.8
40–49	13	20.0
50 or more	7	10.8
Total	65	100.0

It will be seen that this table reduces the 65 observations down to six classes and is clearly much more informative than the raw data. The table shows that the luggage weights vary from 9 kg or less to 50 kg or more with nearly two-thirds falling within the three classes 20–29 kg, 30–39 kg and 40–49 kg. The relative frequencies make it easy to see the relative importance of the different weight classes.

The purpose of a tabular summary is to reveal an overall pattern and it is therefore important to choose a number of classes that is not too big or too small. There is no hard and fast rule about the number or classes but it is conventional to have around 5–15 (depending on the range and volume of the raw data). A certain amount of trial and error is often necessary in order to arrive at an acceptable summary distribution. It is often useful to make the class intervals uniform, i.e. of equal width (this is helpful for interpretation and if the data are used for calculation), but it may sometimes be necessary to use intervals unequal in size and/or to leave the top and/or bottom classes open-ended as illustrated in Table 2.3.

Naturally, it is important to ensure that the upper and lower limits of each class are defined unambiguously so that it is clear into which class any observation is meant to fall and that there is no overlap (i.e. classes should be *mutually exclusive*). In this connection, it is important to consider whether the variable is *discrete* or *continuous* and whether or not its values have been rounded. Discrete variables denote counts (e.g. number of children in a family, number of firms in an industry) while continuous variables denote measurement (e.g. weights, heights, lengths). In the example above, the luggage weights (a continuous variable) have been rounded to the nearest kilogram to produce a discrete variable. Thus the second class in Table 2.3 (10–19 kg) covers all luggage weights from 9.5 kg up to but less than 19.5 kg. The values 9.5 kg and 19.5 kg here are referred to as the *true class limits* or *class boundaries* of this class. True class limits for the other classes are similarly defined. Alternatively, the classes could have been defined, for example, as '10 and under 20 kg', '20 and under 30 kg', etc. The true class limits would then have been different. Thus, for example, the values from 9.5 but less than 10 would now be included in the *first* rather than the *second* class, with the second class itself ('10 and under 20 kg') including all weights from 10.00 kg up to and including 19.99 kg. Needless to say, given the raw data in Table 2.1, which have been rounded to the nearest kilogram, this alternative arrangement of classes is not possible here.

Frequency distributions – summary of terminology

Classes	reduction of the data into a smaller number of groups;
Class intervals	range of values which define the width for each class;
Class limits	the boundaries of each class, referred to as *lower* and *upper* class limits respectively;
Open-ended classes	the first and last classes may be left without the lower or upper limit specified, giving open-ended classes;
Class frequencies	the number of observations which fall in each class;
Relative class frequencies	each class frequency expressed as a percentage of the total frequency.

A frequency distribution is usually only the first step in summarizing a collection of data. Very often, the next step is to present the data in some sort of graphical form. This is particularly useful for conveying information very quickly and succinctly and for comparing one distribution with another. In the remainder of this chapter, we therefore turn to demonstrating the most common graphical methods, starting with *line graphs*.

Line graphs

Line graphs are, perhaps, the most familiar form of graphical presentation of data and are used for showing movements in a variable over time. They are particularly useful, therefore, in helping to highlight trends. To illustrate the principles of constructing line graphs, we use the data in Table 2.4 below which report the average price of new houses, year by year, from 1994 to 2000.

Four line graphs, based on these data, are shown in Figure 2.1 overleaf. It should be noted that, although these use the same set of data, they give very different impressions of the movement in house prices over the period. This arises simply as a result of the stretching or condensing of the axes or the suppression of the origin (zero point). Thus, comparing parts (a) and (b) in Figure 2.1, the vertical axis in (b) is stretched relative to (a) while the horizontal axis in (a) is stretched relative to (b). The effect, therefore, is to give the impression of a moderate rate of increase in part (a) and a steeper increase in part (b). In part (c), the origin has been suppressed and this again creates the impression of a relatively steep rate of increase in house prices. Sometimes, of course, the range of data values is such that they can only be suitably graphed by suppressing the origin. Where this is done, however, attention should be drawn to it by showing a break in the axis, as illustrated in part (d). Part (d) represents what may be regarded as 'best practice'. It is important, therefore, to exercise great care in both the presentation and interpretation of such graphs.

Pie charts

A *pie chart* is a commonly used graphical presentation, in the form of a circle, for showing relative frequencies in the case of categorical data. The relative frequencies are used to divide the circle into segments that correspond to the relative frequency for each category of the variable in question.

Figure 2.2 (on page 20) shows two pie charts illustrating housing tenure in 1961 and 1991 respectively. Each shows a breakdown of housing tenure into three categories: owner-occupied, public-sector rented and private-sector rented. Comparison of the two charts shows considerable growth in owner occupation between 1961 and 1991 at the expense of renting, but, in particular, private renting. The two charts also demonstrate that the total number of dwellings increased over this period from roughly 16.3 million in 1961 to around 23.0 million in 1991 – an increase of 41 per cent. Thus, the total area of pie chart (b) must be shown as 41 per cent bigger than the area of pie chart (a). The area of the pie is, of course, measured by πr^2 where π is the mathematical constant ($=3.142$) and r is the radius. Hence, r needs to be adjusted accordingly. This is an important point to note when

Table 2.4 Average price of new houses

Year	1994	1995	1996	1997	1998	1999	2000
Average price (£)	37,308	43,646	51,290	64,615	74,976	78,917	76,442

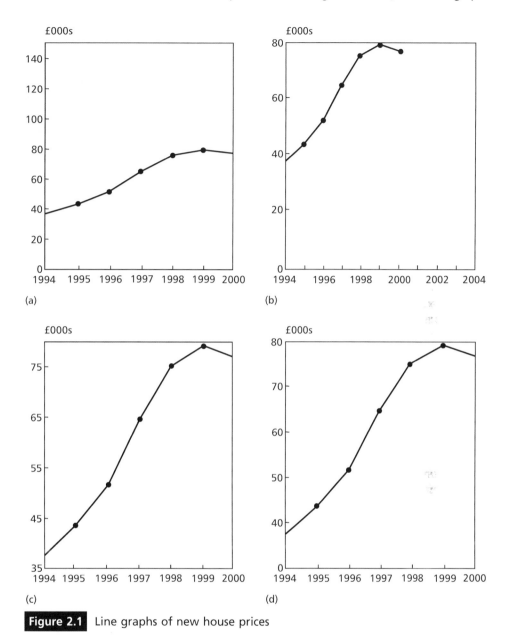

Figure 2.1 Line graphs of new house prices

making comparisons using pie charts – it is a point that is often forgotten and the result is that a misleading impression is given.

Bar charts

Bar charts are another popular form of graphical illustration and are an alternative to pie charts when the variable of interest can be sub-divided, but they are not confined to the illustration of categorical data. They can also be used to illustrate quantitative data when the variable is discrete. Bar charts can be drawn horizontally or vertically.

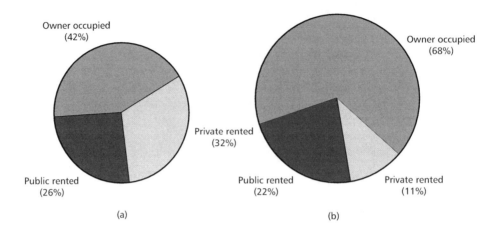

Owner occupied
(42%)

Owner occupied
(68%)

Private rented
(32%)

Public rented
(26%)

Public rented
(22%)

Private rented
(11%)

(a)   (b)

Figure 2.2   Pie charts of housing tenure: (a) 1961, total number of dwellings 16.273 million; (b) 1991, total number of dwellings 22.986 million

Figure 2.3 shows the same data as in the pie chart in Figure 2.2(b) in bar chart form. The height of each bar represents the number of dwellings (frequency) in each type of housing tenure. Alternatively, of course, the percentages (relative frequencies) could have been shown, instead of the absolute numbers.

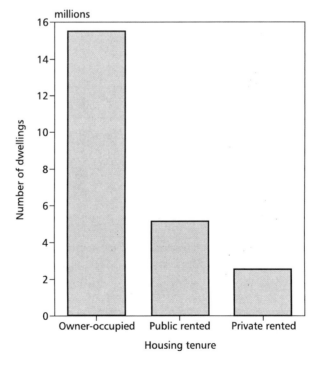

Figure 2.3   Bar chart of housing tenure

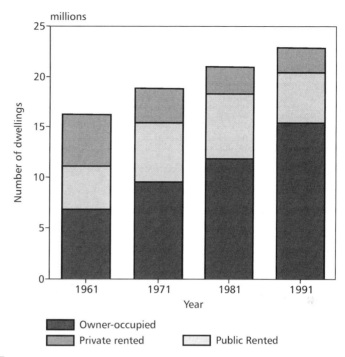

Figure 2.4 Component bar chart of growth of housing stock and composition

Figure 2.4 demonstrates the use of bar charts to illustrate non-categorical data – in this case growth in the total number of dwellings in a country over time. This figure also illustrates another extension of bar charts to show a breakdown of the data into component sub-categories (in this case, owner-occupied, public-sector rented and private-sector rented). Such bar charts are, therefore, referred to as *component* or *stacked bar charts*.

Cross-tabulations – contingency tables

Data are often classified according to more than one characteristic and it may be desired often to cross-classify the data according to two characteristics simultaneously. For example, the data on housing tenure in Figure 2.4 may be cross-classified according to type of dwelling and displayed in a cross-tabulation as in Table 2.5 below. Such a table is also called a *contingency table*.

Table 2.5 Cross-tabulation of housing tenure by type of dwelling (percentages)

| Tenure | Type of dwelling | | | | |
	Detached	Semi-detached	Terraced	Flats & maisonettes	Total
Owner-occupied	30	25	6	6	67
Private rented	3	3	2	3	11
Public rented	1	5	6	10	22
Total	34	33	14	19	100

Histograms

Histograms are similar to bar charts except that no gaps are left between the bars. They are only used for quantitative variables which are continuous in nature. To demonstrate their construction, consider the data in Table 2.6 which show the distribution of the number of loans by banks, grouped by size of loan.

The histogram for these data is shown in Figure 2.5. Note that there is no gap between the bars and that the upper and lower boundaries of each bar are located on the horizontal axis at the *true class limits* of each class. Note also that the height of each bar corresponds to its respective class frequency. This, however, is only appropriate when the class intervals are equal.

When the intervals are not equal, it is necessary to adjust the heights by computing *frequency densities* – this is to ensure that the area enclosed by each bar is proportional to

Table 2.6 Distribution of bank loans by size of loan

Size of loan £	Number of loans
10,000–19,999	983
20,000–29,999	1950
30,000–39,999	2546
40,000–49,999	1350
50,000–59,999	580

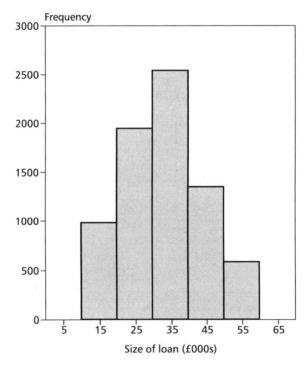

Figure 2.5 Histogram of equal intervals: distribution of bank loans by size

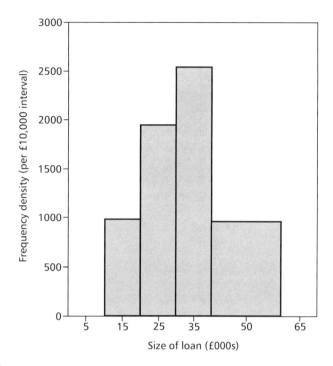

Figure 2.6 Histogram of unequal intervals: distribution of bank loans by size

its frequency. A frequency density is simply the class frequency per unit class interval. For example, if one class is assigned a class interval which is twice the width of the other classes, then the frequency of the class is halved. This point is illustrated in Figure 2.6, using the same data as before (see Table 2.6) but with the last two classes amalgamated. The total frequency of this amalgamated class is 1930 but, because the class width has been doubled, the frequency plotted for it in Figure 2.6 is half of this number (i.e. 965).

A final point to note about histograms is that if there are open-ended class intervals then it is not possible to complete the histogram by closing it at the relevant end (or ends); all that can be done in such cases is to indicate on the histogram itself the number of observations (frequency) that fall within the open-ended class(es).

Frequency polygons

Frequency polygons are useful alternatives to histograms, especially when two distributions need to be compared on the same diagram. This is because when two histograms are superimposed some bars will overlap and make it difficult to distinguish one histogram from the other. A frequency polygon is simply formed by plotting the frequencies of each class against their *mid-points* and joining the plots by straight lines. In addition, these lines are extended down to the horizontal axis at both ends as illustrated in Figure 2.7, which shows a frequency polygon for the same data as the histogram in Figure 2.5. Note that the polygon lines extend down to join the horizontal axis at the mid-points of standard-width classes at each end of the distribution (zero frequency, of course). This ensures that the area enclosed by the frequency polygon is equal to that enclosed by the histogram. This area represents the total frequency. It should be appreciated that frequency polygons cannot be drawn if there are open-ended class intervals.

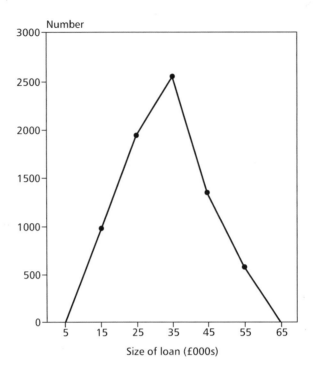

Figure 2.7 Frequency polygon: distribution of bank loans by size

To illustrate the use of frequency polygons to compare *two* distributions, consider the information given in Table 2.7 which shows the distribution of mortgage loans granted to first-time house buyers and existing home-owners during a particular year. This information is illustrated in Figure 2.8(a) which shows the absolute frequencies (number of loans in each category) while Figure 2.8(b) shows the corresponding relative frequencies (percentages). It will be seen from part (a) that in the case of existing home-owners the loans are roughly concentrated in the range £40,000–£80,000, and that there are more existing home-owner borrowers than first-time buyers. In addition, first-time buyers tend to borrow smaller amounts than existing home-owners. Part (b) focuses on the relative

Table 2.7 Distribution of mortgage loans

Size of mortgage (£)	First-time buyers		Existing home-owners	
	Number	%	Number	%
10,000–19,999	5600	8	1800	2
20,000–29,999	3500	5	3600	4
30,000–39,999	3500	5	3600	4
40,000–49,999	6300	9	4500	5
50,000–59,999	21 000	30	13 500	15
60,000–69,999	14 000	20	31 500	35
70,000–79,999	9800	14	18 000	20
80,000–89,999	6300	9	13 500	15
Total	70 000	100	90 000	100

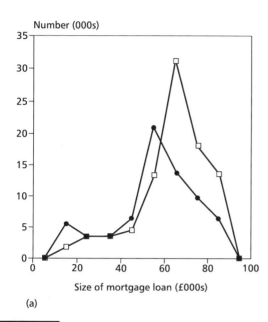
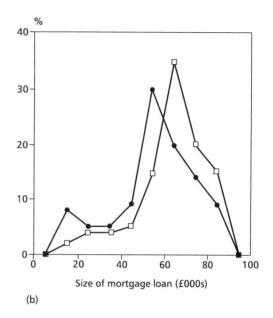

(a)

(b)

Figure 2.8 Comparative frequency polygons: distribution of loans by borrower type; ● first-time buyers; □ existing home-owners: (a) absolute frequencies; (b) relative frequencies

distribution within each borrower group and shows, again, that more existing home-owners tend to take larger loans.

Cumulative frequency curves (ogives)

Cumulative frequency curves (often called *ogives*) are a particularly useful device for answering questions relating to a frequency distribution such as that shown in Table 2.6. The following types of question can be addressed:

(i) how many loans of £25,000 or more were made?

(ii) how many loans of £21,000 or less were made?

As the name implies, a cumulative frequency curve is obtained by plotting *cumulative*, rather than individual, class frequencies. Naturally, the frequencies may be cumulated from the top of the table to the bottom, or vice versa. In the former case, we obtain a curve showing the number of observations *equal to* or *less than* the upper class limit of each corresponding class – referred to as the 'or less' curve. Similarly, the latter case gives the number of observations *equal to* or *greater than* the lower class limit of each corresponding class – referred to as the 'or more' curve. In each case, these limits refer to the *true class limits* as defined earlier in this chapter.

Table 2.8 shows the cumulative frequency distributions, based on the data on bank loans in Table 2.6, for both the 'or less' and 'or more' cases. The second column in the table shows, for example, that 983 loans were for £19,999 *or less*, 2933 loans were for £29,999 *or less* and so on. The third column shows that 7409 loans were for £10,000 *or more*, 6426 loans were for £20,000 *or more* and so on.

Cumulative frequency curves are shown in Figure 2.9 for the data given in Table 2.8. In practice, only one of these curves need be drawn because each is the complement of the other. We show both, however, for the sake of completeness.

Table 2.8 Cumulative frequency distributions for bank loans (data as in Table 2.6)

Size of loan (£)	Cumulative frequencies	
	or less	or more
10,000–19,999	983	7409
20,000–29,999	2933	6426
30,000–39,999	5479	4476
40,000–49,999	6829	1930
50,000–59,999	7409	580

Note that, as indicated earlier, the cumulative frequencies have been plotted against the upper and lower class limits, respectively, and *not* against the class mid-points. Using this diagram we are now able to answer directly the questions posed above.

(i) There were approximately 5300 loans made of £25,000 or more, indicated by point A on the 'or more' curve.

(ii) There were approximately 1300 loans made of £21,000 or less, indicated by point B on the 'or less' curve.

It is worth noting that the two graphs in Figure 2.9 must necessarily cross at the mid-point of the distribution. The value indicated by the intersection is known as the *median* – a summary measure of the distributions which is discussed in the next chapter. The median in this example is a loan of roughly £33,000. Finally, it should be appreciated that use of

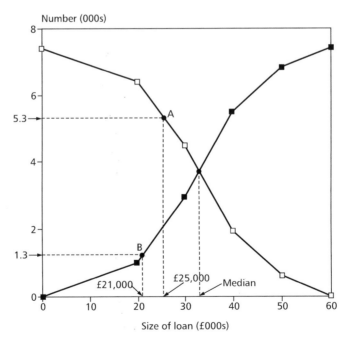

Figure 2.9 Cumulative frequency curves: distribution of bank loans; ■ 'or less' curve; □ 'or more' curve

cumulative frequency curves to obtain answers to particular questions provides a quick way of arriving at the answers, but these are, necessarily, only estimates. Accurate answers require the use of the original, raw, data.

Lorenz curves

Lorenz curves are an extension of cumulative frequency curves which allow further questions to be addressed, namely how equal or unequal is the distribution? Such questions are particularly relevant in studies of income and wealth distribution, the influence of taxation and so on. For example, consider the data given in Table 2.9 which show the distribution of income, before and after tax, together with the total amount of income earned by the people in each income range for a particular year.

A Lorenz curve is constructed by plotting the cumulative *percentage* frequency (in this case, number of people) against the corresponding cumulative *percentage* value of the variable in question (in this case, income). Cumulation is always carried out in ascending order of the variable (i.e. from the lowest to the highest values of income in this case). The relevant cumulations are shown in Table 2.10. The information given in this table now allows us to draw two Lorenz curves for the distribution of income before and after tax, respectively. These are given in Figure 2.10.

Before considering what the figure shows, note the following features.

▪ The cumulative percentage frequencies (columns (a) and (b) in Table 2.10) are plotted along the horizontal axis (by convention), unlike the way they are plotted in the cumulative frequency curves discussed earlier.

▪ The diagonal line shown in the diagram is referred to as the 'line of equal distribution'; it represents a Lorenz curve in a situation of a perfectly equal distribution – that is to say, it represents a situation in which 10 per cent of the people earn 10 per cent of the total amount of income, 20 per cent earn 20 per cent, 30 per cent earn 30 per cent and so on.

▪ The two broken lines represent the actual Lorenz curves. One is obtained by plotting column (a) against column (c) from Table 2.10, giving the Lorenz curve for the distribution *before* tax. The other is obtained by plotting column (b) against column (d), giving the Lorenz curve for the distribution *after* tax.

The interpretation of a Lorenz curve is that the more the curve diverges from the line of equal distribution, the more unequal is the actual distribution of the variable in question.

Table 2.9 Distribution of income

Income range (£000s)	Numbers (millions) in income ranges		Total income (£ billions)	
	Before tax	After tax	Before tax	After tax
Less than 1.5	11.2	13.1	10.5	12.3
1.5–<2.0	3.5	4.4	6.1	7.7
2.0–<2.5	3.4	3.9	7.7	8.8
2.5–<4.0	7.1	5.7	23.1	18.5
4.0 and over	3.1	1.2	18.6	6.7
Total	28.3	28.3	66.0	54.0

Table 2.10 Cumulative distribution of income

	Cumulative percentages[a]			
	Numbers in income ranges		Total income	
Income range (£000s)	Before tax (a)	After tax (b)	Before tax (c)	After tax (d)
Less than 1.5	39.6	46.3	16.0	22.8
1.5–<2.0	52.0	61.8	25.3	37.0
2.0–<2.5	64.0	75.6	36.9	53.2
2.5–<4.0	89.1	95.7	71.9	87.7
4.0 and over	100.0	100.0	100.0	100.0

[a] The cumulative percentages are derived from the data in Table 2.9 by first expressing each entry as a percentage of its corresponding column total and then cumulating these percentages. Thus for column (a) above, we have

$$39.6\% = (11.2/28.3) \times 100$$
$$52.0\% = 39.6\% + (3.5/28.3) \times 100$$
$$64.0\% = 52.0\% + (3.4/28.3) \times 100 \text{ etc.}$$

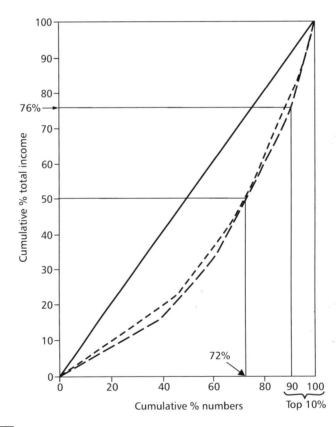

Figure 2.10 Lorenz curves: —— line of equal distribution; — — before-tax distribution; –– after-tax distribution

Lorenz curves are particularly useful for making comparisons of relative degrees of inequality. Thus, in this case it will be seen that the distribution of income after tax is more equal than the distribution before tax. The tax system in operation here may, therefore, be said to be progressive.

Finally, Lorenz curves may also be used to answer questions such as the following.

(i) What is the average income of the top 10 per cent of income earners, before tax?

(ii) What proportion of people receive half of the total income earned, after tax?

The answers to these questions are illustrated in Figure 2.10. In the case of question (i), the top 10 per cent of earners before tax receive $(100 - 76)$ per cent $= 24$ per cent of the total income. In money terms, the *average* income of these people is given by:

$$\frac{24 \text{ per cent of £66 billion}}{10 \text{ per cent of 28.3 million}} = \frac{£15.84 \text{ billion}}{2.83 \text{ million}} = £5597$$

In the case of question (ii), it will be seen that half of the total income after tax is earned by approximately 72 per cent of the lower income earners and, by symmetry, roughly 28 per cent of the higher income earners.

Rate of change graphs (logarithmic scales)

The final graphical form of presentation considered in this chapter is one designed to show the rate of change of a variable over time. Naturally, one can plot percentage rates of change rather than the absolute values of a variable but the same objective can be readily achieved by plotting the logarithms ('log-values') of the observations against time or, alternatively, plotting the actual values on special *semi-logarithmic* graph paper. Both of these methods are illustrated below. The reason why this is helpful is that such graphs show *proportionate* changes directly. Consider the data in Table 2.11 which show the absolute values of a variable growing by 100 per cent in each time period and the corresponding logarithms of the values.

The absolute values in column (a) are growing at a constant rate – doubling each time. A line graph of these values is shown in Figure 2.11(a). It will be seen that the graph gets steeper and steeper, appearing to suggest, therefore, that the rate of growth is increasing over time. Figure 2.11(b) shows a line graph of the logarithms of the values (given in column (b) of the table), while Figure 2.11(c) plots the absolute values using a logarithmic

Table 2.11 Values of a variable growing by 100 per cent in each time period

Time period	Absolute value (a)	Log value to base 10 (b)
1	10	1.0
2	20	1.3
3	40	1.6
4	80	1.9
5	160	2.2

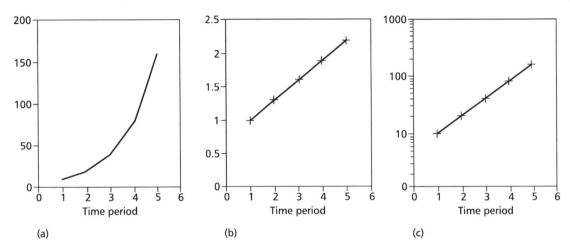

Figure 2.11 Rate of change graphs: (a) absolute values; (b) log values to base 10; (c) logarithmic scale

scale (referred to as semi-logarithmic graph paper). Note that many computer packages compute logarithms and generate logarithmic scales automatically. It will be seen that, by contrast, the graphs in Figures 2.11(b) and 2.11(c) appear as straight lines, indicating thereby that the rate of change over time is in fact constant.

The advantage of logarithmic graphs is that rates of change for one or more variables can be compared directly simply by observing the slopes of these graphs – equal slopes denote equal rates of change, regardless of the absolute magnitudes of the variables involved. One limitation of the use of logarithmic scales, however, is that it is impossible to show zero and negative values of the variable because such values have no logarithm.

| **Applications and Worked Examples** | **Worked example 2.1   Line graphs** |

Many illustrations of the principles and applications of tabular and graphical methods of summarizing data have already been given earlier and hence there is little to be gained from producing similar illustrations again.

General descriptions of the MINITAB and Excel computer programs and their application are given in Appendices B and C respectively.

In this section, therefore, we focus solely on the use of MINITAB and Excel to generate the various charts discussed and illustrated in this chapter.

Computer solution using MINITAB

The data used are the same as in Table 2.4 (p. 18) showing average house prices with years and prices entered in C1 and C2 of the worksheet. The procedure is as follows:

Choose **Graph → Plot. . .**
Under **Graph variables**: in **Y** enter **prices**, in **X** enter **year**

Under **Data display:** in **Display** enter **connect** (to connect up the points) and in **For each** enter **graph**
Choose **Edit Attributes**
In **Line type** enter **Solid**
Click **OK**
Choose **Annotation → Title. . .** and enter title (if desired), Click **OK**

Worked example 2.1   Line graphs *(continued)*

Choose **Annotation → Marker** (this is optional) here we have entered the following: in **Points – C1 C2** (i.e. year, prices); in **Type – Circle** and in **Colour – Black**
Click **OK**

Authors' comment: Note that, by default, the origin is suppressed. The graph is similar to that in Fig. 2.1(c).

This gives the following output:

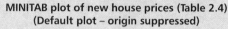

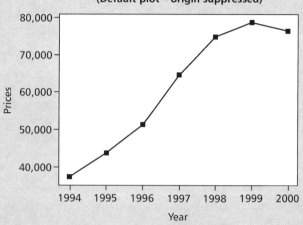

To control scaling of the **Y** axis (the **X** axis is controlled similarly):
Choose **Frame → Min and Max**
In **Minimum for Y** enter **0**
In **Maximum for Y** enter **80,000**
Click **OK**

Authors' comment: Note that this graph is identical to that in Figure 2.1(d) except for the break in the **Y** axis shown there.

This gives the following graph:

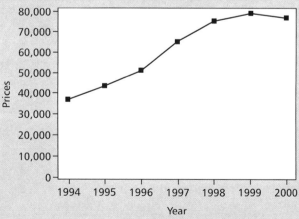

Computer solution using Excel

With data on house prices from Table 2.4 (p. 18) entered in two columns of the Excel worksheet and highlighted, click on **Chart Wizard**.

1. Select **XY scatter** and sub-type as required (here we have chosen **scatter with data points connected by lines**) and click **NEXT**.

2. Check to see that chart source data are pre-entered correctly and click **NEXT**.
3. Select/Enter chart options (titles, axes, gridlines etc.) as desired, click **NEXT**.
4. Position the chart as desired (as new sheet or as object in the same worksheet) and click **FINISH**.

The output is as follows:

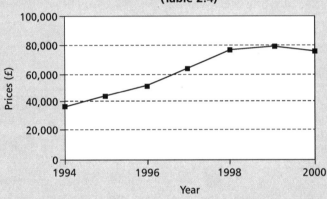

Excel chart of new house prices (Table 2.4)

Worked example 2.2   Pie charts

Computer solution using MINITAB

The data are the same as used in Figure 2.2 and are shown in the MINITAB Worksheet below. The procedure is as follows:

Choose **Graph → Pie Charts…**
Check **Chart table** and in **Categories in** select columns '**1961**' and '**1991**' (i.e. C1-T and C3-T) and in **Frequencies in** select C2 and C4

In **Order of categories** select **Worksheet**
In **Title** enter title (if desired)
Click **OK**
Choose **Options**
In **Label slices with category and** select **Percent**
Check **Add lines connecting labels to slices**
Click **OK**

Worked example 2.2   Pie charts *(continued)*

The output is as follows:

Worksheet

C1-T	C2	C3-T	C4
1961		1991	
Own-occupied	6834660	Own-occupied	15630480
Pte-rented	5207360	Pte-rented	2528460
Pub-rented	4230980	Pub-rented	5056920

Housing tenure 1961 and 1991

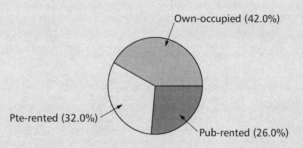

1961

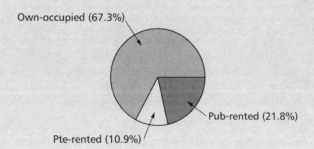

1991

Authors' comment: Compare the pie charts with those in Figure 2.2. Note that the MINITAB pies are not scaled according to the size of the databases.

Computer solution using Excel

To produce two pie charts on the same page one must construct the pie charts separately and then, using the multiple graph capability of Excel, merge them.

With data, as in Figure 2.2 (p. 20), entered in columns C1–C4 of the Excel worksheet, highlight data in C1 and C2 (i.e. categories and values for 1961), and click on **Chart Wizard**.

1. Select **Pie** and the sub-type desired and click **NEXT**.
2. Check to see that **Chart source data** are pre-entered correctly and click **NEXT**.
3. Select/Enter **Chart options** (we have entered **title**, unchecked **show legend**

button and, under **Data labels**, chosen **show label and percent**). Click **NEXT**.
4. Position the chart as desired (as new sheet or as object in the same worksheet) and click **FINISH**.

The output is that for 1961 in the chart below.

Repeat this procedure for 1991 (data in C3 and C4) of worksheet then drag the two charts to position them together on the same page, as shown below.

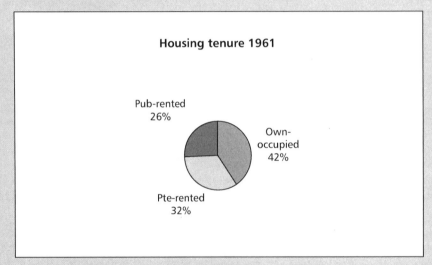

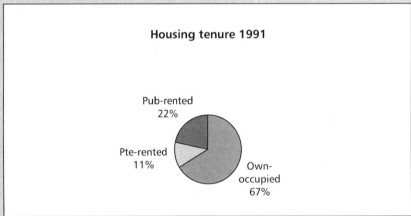

Authors' comment: Note that as in MINITAB, the two pie charts are not scaled proportionately according to the relative sizes of the two data sets.

Worked example 2.3 Bar charts

Computer solution using MINITAB

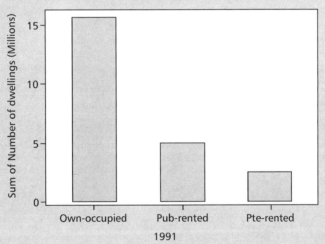

Data: As in Worked Example 2.2 for 1991
Choose **Graph** → **Chart**...
In **Function** enter **Sum**, in **Y** enter **C4** and in **X** enter **1991** (i.e. C3-T)
In **Display** enter **Bar** and in **For each** enter **Graph**
Choose **Edit Attributes**
In **Fill type** select **Solid** and in **Backcolour** select **Light grey**

Click **OK**
Choose **Annotation** → **Title** and enter title (if desired)
Click **OK** twice

Authors' comment: Compare Figure 2.3.

The output is as follows:

MINITAB bar chart of housing tenure in 1991
(Data as in Worked example 2.2 for 1991)

Stacked bar chart

Data: As in Worked Example 2.2 entered in the MINITAB worksheet as shown below. The procedure is as follows:

Choose **Graph** → **Chart**...
In **Function** enter **Sum**, in **Y** enter **Number** (i.e. C3) and in **X** enter **Type** (i.e. C2)
In **Display** enter **Bar**, in **For each** enter **Group** and in **Group variables** enter **Type** (C2)

Choose **Options**
Check **Stack** and enter **Type** (C2)
Choose **Edit Attributes** and select **fill type** and **colours** as desired for each of the three group types
Click **OK**
Choose **Annotation** → **Title** and enter title (if desired)
Click **OK** twice

The output is as follows:

Worksheet

C1 Year	C2 Type	C3 Number	C4	C5 Typecode	C6 Category
1961	1	6834660		1	Owner-occupied
1961	2	5207360		2	Private-rented
1961	3	4230980		3	Public-rented
1991	1	15630480			
1991	2	5056920			
1991	3	2528460			

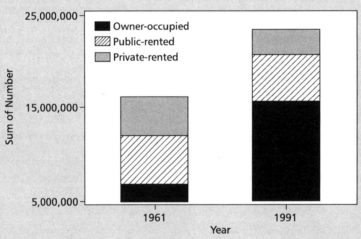

MINITAB Stacked Bar Chart of Housing Tenure in 1961 and 1991 (Data as in Worked example 2.2)

Authors' comment: Compare bar charts for 1961 and 1991 in Figure 2.4.

Computer solution using Excel

With data, as in Worked example 2.2 (p. 32), entered in columns C1–C4 of the Excel worksheet, highlight data in C3 and C4 (i.e. categories and values for 1991), and click on **Chart Wizard**.

1. Select **Column** (this produces a set of vertical bars, unlike the **bar** command which produces horizontal bars) and then the sub-type desired and click **NEXT**.

Worked example 2.3 Bar charts *(continued)*

2. Check to see that **Chart source data** are pre-entered correctly and click **NEXT**.

3. Select/Enter **Chart options** (we have entered **titles** and unchecked **show legend button**). Click **NEXT**.

4. Position the chart as desired (as new sheet or as object in the same worksheet) and click **FINISH**.

The output is as follows:

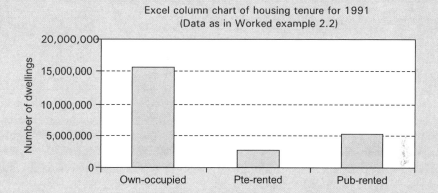

Excel column chart of housing tenure for 1991
(Data as in Worked example 2.2)

Excel stacked column chart

With data entered in the Excel worksheet as shown below, the commands to produce a stacked bar (column) chart in Excel are the same as above except for the choice of appropriate **Chart sub-type** and leaving **legend** selected at Step 3.

	Own-occupied	*Pte-rented*	*Pub-rented*
1961	6834660	5207360	4230980
1991	15630480	2528460	5056920

The chart is shown below:

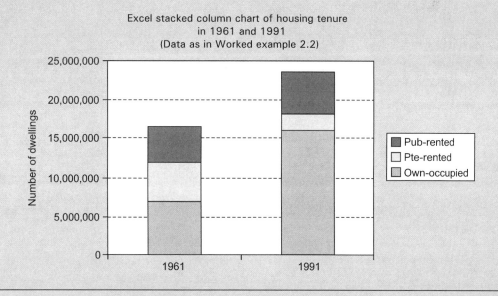

Excel stacked column chart of housing tenure
in 1961 and 1991
(Data as in Worked example 2.2)

Worked example 2.4 Histograms and frequency distributions

Data: the data on luggage weights in Table 2.1 (p. 16). This is used here, rather than the data used in Figures 2.5 and 2.6 to illustrate histograms, because data were already pre-grouped as a frequency distribution whereas raw data are needed to demonstrate the use of MINITAB both in constructing histograms and in facilitating the construction of frequency distributions themselves.

Computer solution using MINITAB

(a) MINITAB default solution

With the data in C1 of MINITAB worksheet
Choose **Graph** → **Histogram**
In **Graph variables** enter C1
Under **Data Display**: in **Display** select **Bar** and in **For each** select **Graph**
Choose **Edit Attributes** [if desired, this is optional] for example here under **Fill type** we select **Solid** and under **Backcolour** we select **Light grey**
Click **OK**

Choose **Annotation** → **Title...** and insert title (as desired)
Click **OK** twice

Authors' comment: Note that the number of bars in the histogram are chosen by MINITAB by default and the intervals are of equal size but both of these may be controlled – as shown in parts (b) and (c) below.

The output is as follows:

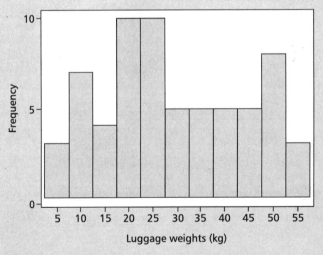

(a) MINITAB default histogram of luggage weights (Table 2.1)

(b) Histogram constrained to six equal intervals

To control the number of bars, choose, in the procedure shown above, **Options**
Under **Definition of Intervals**: in **Number of** **intervals** enter say '6' (alternatively midpoint equally-spaced positions may be specified)
Click **OK** twice
The output is as follows:

Worked example 2.4   Histograms and frequency distributions *(continued)*

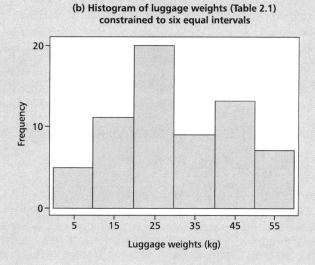

(b) Histogram of luggage weights (Table 2.1)
constrained to six equal intervals

Authors' comment: For the frequency distribution defined by this histogram, see under 'Frequency distribution' in part (d) below.

(c) Histogram with unequal intervals

Mid-points require equally spaced intervals. For unequal intervals choose 'cutpoints'.

It is then necessary to change the y axis to density. Using the same data as before, the procedure is as follows:

The procedure is the same as that above except for that part under **Options**:

Under **Type of Histogram** check **Density**

Under **Type of Intervals** check **Cutpoint**

Under **Definition of Intervals** check **Midpoint/Cutpoint positions** and enter the positions required (e.g. 0 5 10 20 30 45 60)

Click **OK** twice

The output is as follows:

(c) Histogram of luggage weights (Table 2.1)
with unequal intervals

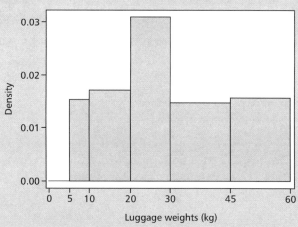

Authors' comment: With a *density* scale on the *y axis*, the total area under the histogram is one. The area of one bar is the proportion of the observations in that bar.

Note that a range of other options is available in MINITAB for producing different types of presentation, e.g. percentage scale, cumulative frequencies etc.

(d) Frequency distribution

A summary of the frequency distribution in histograms with equal intervals may be obtained as follows:

Choose **Graph → Character Graphs → Histogram...**
In **Variable** enter C1
In **First mid-point** enter '5' – the first mid-point in histogram (b) above – and in **Interval width** enter '10' – again the value in histogram (b) above
Click **OK**

The output is as follows:

Histogram of luggage N = 65

Mid-point	Count
5.0	5 *****
15.0	11 ***********
25.0	20 ********************
35.0	9 *********
45.0	13 *************
55.0	7 *******

Authors' comment: Note that the output is labelled 'Histogram' but provides a count of the number of observations in each interval and is therefore useful in compiling frequency distribution tables. In this respect the use of MINITAB to construct *dotplots* and *stem and leaf displays* is also useful in revealing the distribution of observations within a data set and is thus helpful in deciding on appropriate class boundaries.

Computer solution using Excel

Histograms are not constructed in Excel using the Chart Wizard. A two-stage procedure is required using a data analysis tool. At the first stage a frequency distribution is created and charted. At the second stage the chart itself is adapted.

The procedure is as follows:

Stage 1
Choose **Tools → Data Analysis → Histogram**
In **Input Range** enter the range of cells that you wish to analyze by highlighting the appropriate range of cells on the worksheet (i.e. the data on luggage weights from Table 2.1 entered in cells A2:A66 here).

In **Bin Range** enter a cell range with an optional set of boundary values that define classes in the frequency distribution. In this example we have entered the values 9, 19, 29, 39, 49, 59 in cells B2–B7; this defines a distribution with the class boundaries: 0–9, 10–19, 20–29, 30–39, 40–49, 50–59. Excel counts the number of observations in each of these classes and puts the result in the specified output range and adds an open-ended class labelled 'more' at the end (here this class has zero frequency).

Worked example 2.4   Histograms and frequency distributions *(continued)*

Authors' comment: If you omit the 'bin range' Excel creates, by default, a set of evenly distributed bins (classes) between the data's minimum and maximum values. The number of bins is the square root of the number of input values.

In **Output Range** enter the reference for the upper left cell of the range where you want the output range to appear (say, C2)

Check the **Chart Output** box and click **OK**
The output is as follows:

Bin	Frequency
9	5
19	11
29	20
39	9
49	13
59	7
More	0

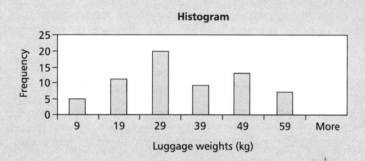

Stage 2

Click on one of the bars in the chart above and then:

Choose **Format Data Series** → **Options**

In **Gap width** enter a zero ('0') and click **OK**
The output is as follows:

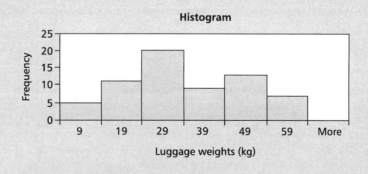

Worked example 2.5   Frequency polygon

Computer solution using MINITAB

Data: Worked Example 2.4 histogram (b). The relevant data are entered in C2 (mid-points of luggage weights) and C3 (frequencies) of the worksheet as shown below.

Choose **Graph → Plot**...
In **Y** enter **C3** and in **X** enter **C2**
In **Display** select **Connect**
Choose **Annotation → Polygon**...
In **Points** enter C2 C3
Click **OK**
Choose **Annotation** again and enter title and other annotations as desired
Click **OK** twice
The worksheet and output are as follows:

Worksheet

C2 Luggage weights (kg) (Mid-points)	C3 Frequency
2.5	0
7.5	5
15	11
25	20
35	9
45	13
55	7
65	0

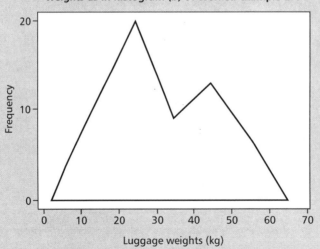

Frequency polygon of distribution of luggage weights as in histogram (b) of Worked example 2.4

Worked example 2.5   Frequency polygon *(continued)*

Computer solution using Excel

With data entered in columns A and B of the worksheet, click on **Chart Wizard**

Select **XY(Scatter)** and desired **sub-type** and click **Next**

Insert **Data range** by highlighting data in columns A and B and click **Next**

Insert **Titles**, uncheck **Show legend**, format gridlines if desired etc.

Click **Finish**

The output chart is shown below along with the worksheet.

Worksheet

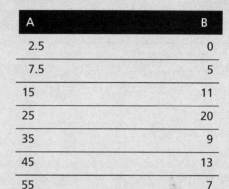

A	B
2.5	0
7.5	5
15	11
25	20
35	9
45	13
55	7
65	0

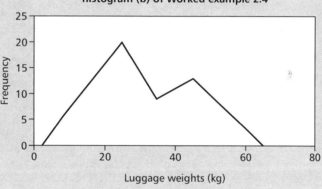

Excel frequency polygon of luggage weights as in histogram (b) of Worked example 2.4

Worked example 2.6   Cumulative frequency curve

Data: Luggage weights in Table 2.1 (p. 16)

Computer solution using MINITAB

Choose **Graph** → **Histogram…**

Under **Graph variables** in **X** enter **C1** (luggage weights)

In **Display** select **Connect**

Choose **Options** and check **Cumulative Frequency**

Click **OK**

Choose **Annotation**→**Title**... and enter title
    as desired

Click **OK** twice

The output is as follows:

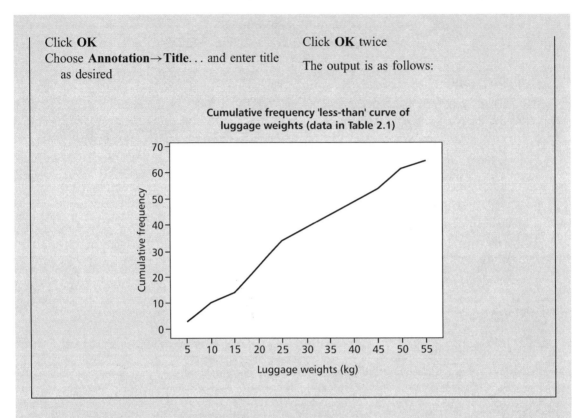

Computer solution using Excel

The procedure is the same as in Excel Worked Example 2.4 except that, in addition, at step 2 the **Cumulative Percentage** check box should be checked.

The output is as follows (table plus chart):

Luggage weights	Bin range	Output table		
		Bin	Frequency	Cumulative %
65 values as in Table 2.1 in this column	9	9	5	7.69%
	19	19	11	24.62%
	29	29	20	55.38%
	39	39	9	69.23%
	49	49	13	89.23%
	59	59	7	100.00%
		More	0	100.00%

Worked example 2.6 Cumulative frequency curve *(continued)*

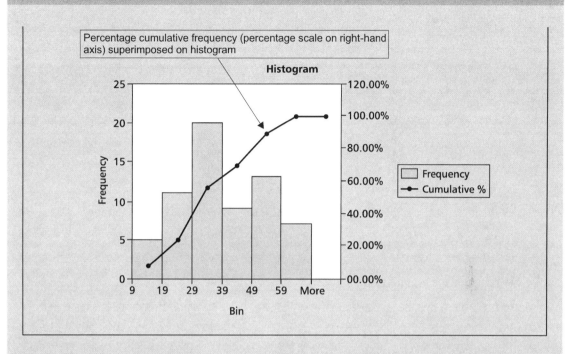

Percentage cumulative frequency (percentage scale on right-hand axis) superimposed on histogram

Worked example 2.7 Logarithmic scale

Computer solution using MINITAB

A graph on a logarithmic scale – as in Figure 2.11(c) – is constructed here using logarithmic transformation within MINITAB. With the absolute values in column (a) of Table 2.11 (p. 29) entered in column C1 of a MINITAB worksheet, the procedure is as follows:

Choose **Graph** → **Time Series Plot**. . .
In **Graph 1** enter **C1** and in **Display** for item 1 enter **Connect**
In **Graph 2** enter **C1** and in **Display** for item 2 enter **Symbol** [this allows the construc-

tion of a line graph with symbols]
Choose **Edit Attributes** and select line type and symbol type, in turn, as desired
Click **OK**
Choose **Options** and **Under Transformation for Y axis** check **Logarithmic**
Click **OK**
Choose **Annotations** → **Title**. . . and enter title as desired
Click **OK** twice

The output is as follows:

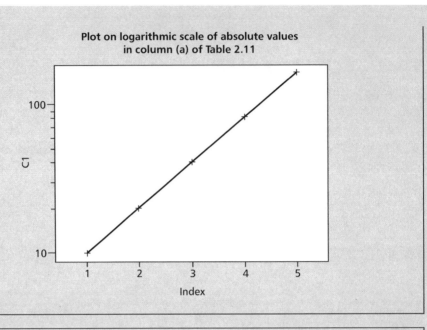

Plot on logarithmic scale of absolute values
in column (a) of Table 2.11

Computer solution using Excel

With the absolute values (10, 20, 40, 80, 160) in column A of worksheet, as shown, Click on **Chart Wizard**

Choose **Line**, and **Chart sub-type**, as desired, and click **NEXT**

Insert titles etc., uncheck **Show legend** and click **NEXT**

Click on **FINISH** then, to convert Y axis to a logarithmic scale: right click on the Y axis, left click on **Format** axis, left click on **Scale** and check the **Logarithmic scale** box

Click **OK**

The output is as follows:

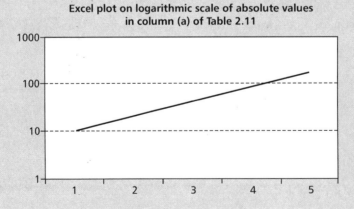

Excel plot on logarithmic scale of absolute values
in column (a) of Table 2.11

Key terms and concepts

Array A set of data in which the individual values are arranged in either ascending or descending order.

Bar chart Similar to a line chart but used when discrete data are grouped into classes with each class corresponding to a rectangle, the base of which is the class interval and the height of which is the class frequency.

Class frequencies The number of observations contained in each class.

Class intervals The range of values which define the width of each class (group).

Class limits (class boundaries) The upper and lower limits for the classes of a frequency distribution.

Classes The groups into which a set of data may be classified.

Component (stacked) bar chart A bar chart in which the component sub-categories of a set of data are represented by bars which are stacked one on top of the other.

Contingency table A table in which data are cross-classified according to two characteristics simultaneously.

Continuous variable A quantitative variable which may be measured on a continuous scale and can have any value within an interval.

Cumulative frequency curve (ogive) A graph of the cumulative frequency distribution.

Cumulative frequency distribution A tabular summary of a set of data that shows the total number of data items with values *either* less than or equal to the upper limit of the class *or* greater than or equal to the lower limit of each class.

Discrete variable A quantitative variable in which the scale of measurement varies in discrete steps.

Frequency density The class frequency per unit class interval.

Frequency distribution table A tabular summary of a set of data showing the frequency (or number) of items in each of several non-overlapping classes.

Frequency polygon A graph formed by plotting class frequencies against the mid-points of the corresponding classes, connecting the points to form a graph and extending it to each end of the distribution to meet the horizontal axis at the mid-point at what would have been the next class below or above respectively.

Histogram A chart similar to a bar chart and may be used in the cases of continuous or discrete grouped data but with no gaps left between the bars.

Line chart A graphical presentation for showing the information contained in a frequency distribution where the variable is discrete and the data are ungrouped.

Line graph A graphical presentation of data often used for showing movements in a variable over time.

Logarithmic scale A scale in which the actual values of a variable are measured as the logarithms of the values.

Lorenz curve A curve showing the degree of inequality in a frequency distribution.

Open-ended classes The first and/or the last classes in a frequency distribution in which the lower or upper limits respectively are not specified.

Pie chart A pictorial device for presenting categorical data in which a circle is divided into wedges (like slices of a pie), each wedge corresponding to the relative frequency of each class.

Rate of change (logarithmic graph) A graphical presentation in which absolute values of a variable are plotted against a logarithmic scale or the logarithms of the variable (log values) are plotted on a conventional scale.

Relative frequencies Each class frequency expressed as a percentage of the total number of observations.

Self-study exercises

2.1. A shop sells the following numbers of gallons of milk weekly over a 24-week period:

28, 36, 41, 23, 45, 23, 24, 45, 20, 26, 53, 54,
28, 39, 45, 49, 52, 57, 43, 21, 29, 60, 42, 33.

(a) Construct a frequency distribution using class intervals of 20–24, 25–29, and so on.

(b) Construct a histogram.

(c) Draw a cumulative frequency curve.

(d) Estimate the minimum weekly sales in the top 10 per cent of weeks.

2.2. The table below shows a frequency distribution of the gross weekly earnings for a sample of 1100 workers. Construct a histogram for these data (hint: note the presence of unequal class intervals).

Gross weekly earnings (£)	Number in sample
Less than 50	80
50–<100	110
100–<150	240
150–<200	350
200–<300	120
300–<400	92
400–<550	90
550–<1000	18
Total	1100

2.3. Compare the two sets of data given in the following table, which show the composition of the labour force in two factories, using (a) pie charts and (b) stacked bar charts (hint: note that the workforces differ in size).

	Number of workers	
Category of worker	Factory A	Factory B
Skilled manual	270	200
Unskilled manual	600	100
Managerial	50	40
Clerical	80	60
Total	1000	400

2.4. The table below shows the number of taxpayers receiving mortgage interest relief and the average value of tax relief by range of income. Use cumulative frequency and Lorenz curves to answer the following questions:

(a) Is the distribution of tax relief more or less evenly distributed than income?

(b) (i) What is the level of income such that there are as many taxpayers earning more than this amount as there are earning less?

(ii) What is the name given to this measure? (see Chapter 3).

(c) How many taxpayers are in the bottom 60 per cent according to income?

(d) For taxpayers in the bottom 60 per cent according to income, how much tax relief do they obtain:

(i) in total?

(ii) on average?

Income range (£000s)	Number of taxpayers (000s)	Average value of tax relief (£)
3–<5	660	122
5–<7	850	263
7–<9	1105	327
9–<12	1950	382
12–<15	760	465
15–<85	495	834

2.5. The following data show imports and gross domestic product (GDP) measured in constant (1995) prices.

(a) Draw line graphs of these two series on the same chart to illustrate their relative rates of growth over time (hint: consider what type of scale is appropriate here).

(b) Comment briefly on what the graphs show.

Year	Imports (£ billion)	GDP (£ billion)
1993	8.8	39.4
1994	9.4	40.4
1995	10.1	41.9
1996	10.2	42.6
1997	10.9	43.6
1998	11.4	44.5
1999	12.7	45.2
2000	14.4	47.9

2.6. What is wrong with each of the following sets of class intervals?

(a)   0 to 10
10 to 20
20 to 30
30 to 40

(b) 40–<50
51–<60
61–<70

(c) 10–19
20–29
30–39

(d) <60
61–80
>81

| Chapter 3 | # Describing data: central tendency |

In the last chapter we examined ways of visually describing a set of data using tables and graphs. We now introduce techniques for numerically describing a set of data by using summary measures. There are two main features of a set of data to be measured, namely its *central tendency* (i.e. average value) and its *dispersion* (i.e. variability of the data). In this chapter we deal with the first of these while the next chapter deals with the second feature.

There are four measures of central tendency which are commonly used, the choice depending on the nature of the data and the purpose of measurement. These are:

- arithmetic mean
- median
- mode
- geometric mean

Principles

Arithmetic mean

The arithmetic mean is the most popular measure of central tendency and is merely the average of the data. It is often simply referred to as the *mean*. Given a sample of data such as 2, 6, 6, 6, 4, 1, 3, the mean of this sample, denoted $\bar{x}$ (pronounced 'x bar'), is given by

$$\bar{x} = \frac{2 + 6 + 6 + 6 + 4 + 1 + 3}{7} = \frac{28}{7} = 4$$

In general, observations of x are denoted x_i where the subscript, i, denotes the 1st, 2nd, 3rd etc. observations. Given a data set $x_1, x_2, \ldots, x_n$, where n is the number of data values, then

Arithmetic mean

$$\bar{x} = \frac{x_1 + x_2 + \cdots + x_n}{n} = \frac{\Sigma x_i}{n}$$

where 'Σ' (pronounced 'sigma') is a shorthand notation for summation. The attraction of using the mean is that it is a single summary measure which is easy to calculate and readily understood. However, since it uses all the values being examined, it can be distorted by extreme values (or so-called *outliers*). For this reason it is often appropriate to calculate a *trimmed mean* which disregards a proportion of the highest and lowest values in the data set (say, the top and bottom 5 per cent). The trimmed mean gives a robust measure of central location which is therefore not affected by a few extreme values.

Median

The median (denoted Md) of a set of data is the middle value when the data are arranged from smallest to largest in an array. For the data above, the ordered array is:

1, 2, 3, 4, 6, 6, 6

Thus, the median value in this case equals 4 (i.e. the $[(n + 1)/2]$th value). When there is an odd number of observations, the median is easily identified, especially with a small data set. Naturally, for a large data set, computers do all the work. For an even number of observations, the median is taken as the mean of the *two* central values.

The median is particularly attractive when the data set is skewed, i.e. when the observations tend to be concentrated at one end of the range of values. In this case the arithmetic mean would be distorted. For example, consider the following numbers:

2, 7, 13, 28, 150

The mean equals $200/5 = 40$ which is greater than the lowest four values but much lower than the highest value and is thus not representative. The median value is 13 which is more closely aligned with the first four values and is hence a better measure of central tendency in this case.

Mode

The mode (denoted Mo) is simply the most frequently occurring observation. Thus for the data 1, 2, 3, 4, 6, 6, 6, the mode is 6. Examples of situations where the mode is particularly relevant are:

- in the clothing industry where the key concern is not, say, the mean or median waist size but the most common waist size;
- in examining qualitative data where we are only interested in the frequency of occurrence of events. For example, we can note that the modal medal of all the medals won by the USA at the Olympic Games was silver. The mean and median are, of course, not relevant in this situation.

It should be noted that in some data sets there may be no mode, whereas in other sets there may be more than one mode. If a mode clearly exists, then the distribution of values is said to be *unimodal*. A distribution is referred to as *bimodal* if there is a two-way tie for the most frequently occurring value. If a set of data is not exactly bimodal but contains two values that are more dominant than the others, some researchers take the liberty of referring to the data set as bimodal even though there is not an exact tie for the mode. Data sets with more than two modes are referred to as *multimodal*.

Mean, median and mode for grouped data

So far in this chapter we have only dealt with raw or ungrouped data. As we saw in Chapter 2, a large set of data may be presented in grouped (or class) form as a frequency distribution. Commonly, this is the only form of data available to the user. For grouped data, the mean, median and mode are calculated as follows.

Mean for grouped data

If n data values have been grouped into classes, and the mid-points of the class intervals are denoted by x_i and the frequencies for each class are denoted by f_i, then

$$\bar{x} = \frac{\Sigma f_i x_i}{n}$$

This produces a *weighted arithmetic mean* (or 'weighted average') in which the mid-points of the classes are weighted by their respective frequencies.

Median for grouped data

The median for grouped data must be found by interpolation. It is first necessary to identify the class in which the median value falls and then to interpolate within this class as follows:

$$\text{Md} = L + \frac{f_c}{f_m} W$$

where L is the lower limit of the median class, f_m is the frequency of the median class, W is the width of the median class and f_c is the number of observations that must be covered in the median class in order to reach the middle of the distribution. Note that the mid-point of the distribution is located by $n/2$ and *not* by $(n + 1)/2$. The latter is not appropriate here because, once the data have been grouped, it is no longer possible to identify a discrete observation (or pair of observations) at the centre of the distribution.

Mode for grouped data

For grouped data with equal class intervals, the modal class is the class with the largest frequency. The mode is then defined as follows:

$$\text{Mo} = L + \frac{\Delta_1}{\Delta_1 + \Delta_2} W$$

where L is the lower limit of the modal class and Δ (pronounced delta) denote differences as follows: Δ_1 is the difference between the frequency of the modal class and the frequency of the previous class, Δ_2 is the difference between the frequency of the modal class and the frequency of the next class and W is the width of the modal class. The rationale of this formula is illustrated in Figure 3.1.

It should be appreciated that when calculating the mean, median and mode for grouped data we can only approximate the values. The 'true' values can only be derived

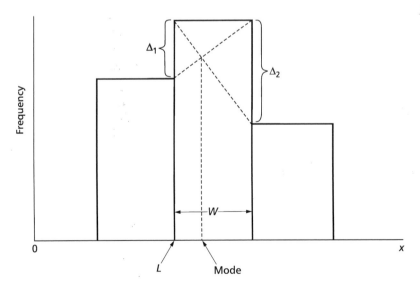

Figure 3.1 Mode for grouped data

from the raw, ungrouped data. Examples of calculations are given in the Applications and Worked Examples section below.

Relationship between mean, median and mode

Figure 3.2 shows the relationship between the mean, median and mode for sets of data that are (a) symmeterical, (b) positively skewed and (c) negatively skewed. It will be seen that the measures coincide in a symmetrical distribution but diverge in a non-symmetrical distribution. The mode is the same in all three cases but the mean is pulled in the direction of the skewness and the median falls in between the other two measures.

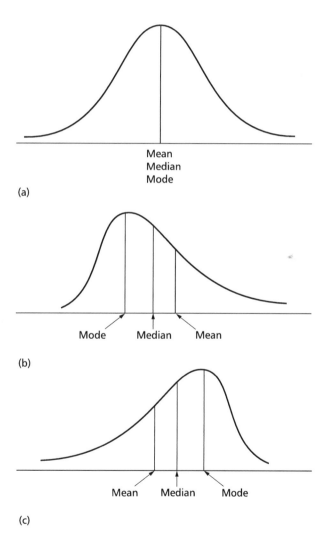

Figure 3.2 Relationship between mean, median and mode: (a) a symmetrical distribution; (b) a positively skewed distribution; (c) a negatively skewed distribution

Geometric mean

The geometric mean (denoted g) is used when averaging rates of change. It is defined as

Geometric mean

$$g = \sqrt[n]{(x_1 \times x_2 \times x_3 \times \cdots \times x_n)}$$

where x_1, x_2, etc. represent rates of change between successive observations and $\sqrt[n]{}$ is the nth root (readily calculated using a calculator or computer).

Compound formula

A compound formula provides an alternative to the geometric mean for calculating average rates of change and is the only method of calculation available when only the original and final values of a series are known. The formula, which may be better known as the formula for calculating compound interest, is as follows:

$$A(1 + r)^n = F$$

where A is the original value, r is the rate of increase per time period, n is the number of time periods and F is the final value. It will be appreciated that this formula can be used to find the value of any one of the four elements (A, r, n or F) when we known the values of the other three.

Applications and Worked Examples   **Worked example 3.1   Ungrouped data**

General descriptions of the MINITAB and Excel computer programs and their application are given in Appendices B and C respectively.

For the following data set, find the values of the mean, median and mode:

8, 13, 22, 1, 22, 85, 15, 7, 14, 3, 22.

Worked solution

Mean: $\bar{x} = \dfrac{\Sigma x_i}{n} = \dfrac{212}{11} = 19.273$

Median: $Md = [(n + 1)/2]$th value

$= $ 6th value $= 14$

Mode: Mo (value with highest frequency)

$= 22.$

Computer solution using MINITAB

Ⓜ

The mean and median may be obtained using separate commands and also using a general 'descriptive statistics' command; the mode must be obtained indirectly. With the data

(8, 13, 22, 1, 22, 85, 15, 7, 14, 3, 22) entered in column C1 of the MINITAB worksheet (the worksheet is not shown here), the procedures are as follows:

Mean

Choose **Calc → Column stats → Mean**
Select input variable (C1)
Click **OK**
The resulting output is shown below:

Column Mean

Mean of C1 = 19.273

Median

Choose **Calc → Column stats → Median**
Select input variable (C1)
Click **OK**
The resulting output is shown below:

Column Median

Median of C1 = 14.000

Mode

Choose **Stat → Tables → Tally**

Select variable (C1) and check the 'Counts' button
Click **OK**
The resulting output is shown below:

Authors' comment: This choice gives frequencies of values and thus allows the mode to be determined by inspection.

Summary Statistics for Discrete Variables

[C1]	Count	
1	1	
3	1	
7	1	
8	1	
13	1	
14	1	
15	1	
22	3	← Mode = 22
85	1	
N =	11	

Descriptive statistics

Choose **Stat → Basic Statistics → Display Descriptive Statistics**
Select variable (C1)
Click **OK**

Authors' comment: The resulting output is shown below (note that it gives both the mean and the median along with other output considered in Chapter 4).

Descriptive Statistics

Trimmed mean

Variable	N	Mean	Median	TrMean	StDev	SE Mean
C1	11	19.27	14.00	14.00	23.03	6.94

See Chapter 4

Variable	Minimum	Maximum	Q1	Q3
C1	1.00	85.00	7.00	22.00

Worked example 3.1: Ungrouped data *(continued)*

Computer solution using Excel

Mean

Use the function: **AVERAGE**

With the data (8, 13, 22, 1, 22, 85, 15, 7, 14, 3, 22) entered in Column A of the worksheet, position cursor on cell where you wish the answer to appear

Click on = sign in the formula bar, choose **AVERAGE** and enter data range in box labelled **Number 1** (either by typing A1:A11 or by highlighting data in the worksheet)

The answer (19.273) appears immediately on the **Function Palette** and, after clicking **OK**, in the chosen cell on the worksheet.

Median

Use the function: **MEDIAN**

With the data (8, 13, 22, 1, 22, 85, 15, 7, 14, 3, 22) entered in Column A of the worksheet, position cursor on cell where you wish the answer to appear

Click on = sign in the formula bar, choose **MEDIAN** and enter data range in box labelled **Number 1** (either by typing A1:A11 or by highlighting data in the worksheet)

The answer (14.0) appears immediately on the **Function Palette** and, after clicking **OK**, in the chosen cell on the worksheet.

Mode

Use the function: **MODE**

With the data (8, 13, 22, 1, 22, 85, 15, 7, 14, 3, 22) entered in Column A of the worksheet, position cursor on cell where you wish the answer to appear

Click on = sign in the formula bar, choose **MODE** and enter data range in box labelled **Number 1** (either by typing A1:A11 or by highlighting data in the worksheet)

The answer (22) appears immediately on the

Function Palette and, after clicking **OK**, in the chosen cell on the worksheet.

Descriptive statistics

Use the **Data Analysis Tools**

With the data (8, 13, 22, 1, 22, 85, 15, 7, 14, 3, 22) entered in Column A of the worksheet:

Choose **Tools → Data Analysis → Descriptive Statistics**

In **Input Range** enter the range of data in column A (either by typing A1:A11 or by highlighting the data in the worksheet)

In **Output Options** check the **Summary statistics** box and Click **OK**

The output is as follows:

Column 1

Mean	19.27273
Standard Error	6.94405
Median	14
Mode	22
Standard Deviation	23.03081
Sample Variance	530.4182
Kurtosis	8.140315
Skewness	2.699833
Range	84
Minimum	1
Maximum	85
Sum	212
Count	11

Authors' comment: Note that, as in MINITAB, the Excel output gives a variety of measures, in addition to the mean, median and mode. The meaning of some of these is self-evident; some of the others are considered elsewhere in this book (Chapter 4 in particular).

It will be seen that the command **Display Descriptive Statistics** in MINITAB produces a variety of output including a value for the trimmed mean (as well as other measures discussed later in Chapter 4). In this example, the trimmed mean equals 14 and is obtained by excluding the lowest and highest observations (corresponding to the bottom and top 5 per cent of the number of observations).

Interpretation

The results show that the distribution of the data is positively skewed as the mean is greater than the median. Trimming the mean removes the extreme values and in this case gives a value equal to the median.

Worked example 3.2 Grouped data

The table below shows a frequency distribution for grouped data. Calculate the mean, median and mode.

Frequency distribution

Classes	Class frequencies (f_i)
0 −<5	2
5 −<15	6
15 −<25	20
25 −<35	15
35 −<45	8
45 −<100	4
Total	$n = 55$

Worksheet

Class mid-points (x_i)	$f_i x_i$	Cumulative frequency
2.5	5	2
10	60	8
20	400	28
30	450	43
40	320	51
72.5	290	55
	1525	

Worked solution

Mean: $\bar{x} = \dfrac{\Sigma f_i x_i}{n} = \dfrac{1525}{55} = 27.7273$

Median: $Md = L + \dfrac{f_c}{f_m} W$

The median class is the class which contains the middle value $n/2 = 55/2 = 27.5$th. From the worksheet above, the cumulative frequency column shows that this is in the class '15 − <25'. Thus:

$$Md = 15 + \left(\frac{27.5 - 8}{20}\right)10 = 15 + \left(\frac{19.5}{20}\right)10$$

$$= 24.75.$$

Worked example 3.2   Grouped data *(continued)*

(Recall that since we are dealing with grouped data now we cannot identify the actual median – the $[(n+1)/2]$th value – and so the median, obtained by interpolation, must be taken at the exact centre of the distribution of values, defined as the $(n/2)$th position. Note that $f_c = 27.5 - 8 = 19.5$, where 8 is the sum of the class frequencies below the median class.)

Mode

The modal class is the class with the greatest frequency. This is the '15–<25' class. Hence:

$$\mathrm{Mo} = L + \left(\frac{\Delta_1}{\Delta_1 + \Delta_2}\right) W$$

$$= 15 + \left\{\frac{(20 - 6)}{(20 - 6) + (20 - 15)}\right\} 10$$

$$= 15 + \left(\frac{14}{14 + 5}\right) 10 = 22.37.$$

Interpretation

The data are positively skewed as the mean is greater than the median. However, the numerical results are only approximations since they are based on grouped data and, in particular, they rely on an implicit assumption that the underlying raw observations are evenly distributed within each class.

Computer solution using MINITAB

MINITAB does not automatically calculate the mean, median and mode for grouped data. However, MINITAB can be used to do the arithmetic involved for calculating the mean and we show this below.

Worksheet

Mid-points	Frequency
2.5	2
10.0	6
20.0	20
30.0	15
40.0	8
72.5	4

Choose **Calc → Calculator**

In the dialogue box select the **Function** required ('Sum' in this case)

Complete the **Expression**: Sum (c1*c2)/Sum (c2)

Complete **Store result in variable** box: e.g. K1

Click **OK**

The results can be displayed in two ways:

1. Choose **Window** (on Menu bar) → **Info**. This gives results as follows:

Column	Count	Missing	Name
C1	6	0	
C2	6	0	
Constant	Value	Name	
K1	27.7273		

This is the mean

2. Alternatively, choose **Editor** → **Enable Command Language**, then type in the Session Window:

MTB > PRINT k1

This produces:

Data Display

K1	27.7273
MTB >	

E

Computer solution using Excel

Mean from grouped data

With mid-points in column A and frequencies in column B of the Excel worksheet use a **Formula** to carry out the arithmetic involved as follows:

In cell C1 enter $= \mathbf{A1*B1}$ then, using the Autofill facility, drag the fill handle on C1 down the column to cell C6 (this multiplies all the corresponding cells in columns A and B down to A6 and B6.

Then in a cell where you want the result to appear (say D1) type the formula:

$= \mathbf{Sum(c1:c6)/Sum(b1:b6)}$ and press **RETURN**

The input and output appears as follows:

Input columns		Output	
A	B	C	D
2.5	2	5	27.72727273 ← Mean
10	6	60	
20	20	400	
30	15	450	
40	8	320	
72.5	4	290	

Mid-points   Frequencies   Products of A1 × B1, A2 × B2 etc.

Worked Example 3.3 Averaging proportions

A firm's labour costs as a proportion (percentage) of annual turnover for the last four years are shown below together with the value of turnover in each year.

Year	1997	1998	1999	2000
Turnover (£ million)	2.2	2.1	2.5	3.4
Labour costs (per cent)	35	31	30	28

What is the average proportion of labour costs over the four years?

Solution

A simple average of the four percentages is inappropriate because the base of the percentage (turnover) is different each year and thus a weighted average is needed to allow for this fact as follows:

$$\frac{35(2.2) + 31(2.1) + 30(2.5) + 28(3.4)}{2.2 + 2.1 + 2.5 + 3.4}$$
$$= 30.6 \text{ per cent.}$$

Interpretation

On average, labour costs represent 30.6 per cent of turnover over the four years. A simple average of the four proportions would give 31 per cent.

Worked example 3.4 Geometric mean: averaging growth rates

Retail prices increase as follows:

Year 1	Year 2	Year 3	Year 4
10%	7.6%	5.2%	4.1%

Calculate the average annual rate of increase.

Worked solution

Expressing each of the annual percentages as an index relative to the previous year we have

$$g = \sqrt[4]{(1.10 \times 1.076 \times 1.052 \times 1.041)}$$
$$= 1.06701.$$

Interpretation

On average, the price level each year is 1.06701 times the level of the previous year, giving an average annual rate of increase equal to 6.701 per cent.

Computer solution using MINITAB

Choose **Calc → Calculator**
Choose **Editor → Enable Command Language**

Enter commands in Session window as below:

```
MTB > LET K1 = (1.10*1.076*1.052*
      1.041)**(1/4)
MTB > PRINT K1
```

The output is as follows:
Data Display

K1	1.06701
MTB >	

Authors' comment: Geometric mean of average price level compared with previous year.

Computer solution using Excel

(E)

The solution requires the calculation of the geometric mean.

With the values (1.1, 1.076, 1.052, 1.041) entered in cells A1–A4, select another worksheet cell where you wish the answer to appear, then click on the equals sign (=)

on the formula bar, locate the Function name **GEOMEAN** and click on it.

Enter the **Input range** by typing A1:A4 or highlighting these cells on the worksheet. This returns the answer on both the **Formula Palette** and the worksheet as: 1.067008 – interpreted as above.

Worked Example 3.5   Compound formula

If the population has grown from 26 million in 1908 to 56 million in 2000, what has been the annual average percentage rate of growth?

$$A(1 + r)^n = F$$
$$26(1 + r)^{92} = 56$$

$$\text{Thus } r = \sqrt[92]{\left(\frac{56}{26}\right)} - 1 \approx 0.0084.$$

Worked solution

Using the compound formula:

Computer solution using MINITAB

(M)

Choose **Calc → Calculator**
Choose **Editor → Enable Command Language**

Enter commands in Session window as below:

```
MTB >
MTB > LET K1 = ((56/26)**(1/92))-1
MTB > PRINT K1
```

The output is as follows:

Data Display

K1	0.00837460
MTB >	

Authors' comment: i.e. approx 0.84 per cent.

Worked Example 3.5   Compound formula *(continued)*

Computer solution using Excel

E

The following formula (see the manual Worked solution for details – p. 63) is used as follows: $=((56/26)^{\wedge}(1/92)) - 1$, giving the answer: 0.008375 (i.e. approx 0.84%).

Authors' comment: Note that Excel uses the caret ($^{\wedge}$) for exponentiation, unlike MINITAB which uses a double asterisk ($**$).

Interpretation

The population has grown over the period at a compound rate of roughly 0.84 per cent per annum.

Key terms and concepts

Arithmetic mean A measure of the central location of a data set. It is computed by summing all the values in the data set and dividing by the number of items:

$$\bar{x} = \frac{\Sigma x_i}{n}$$

Bi-modal distribution A distribution which has two modes.

Compound formula A formula for calculating compound rates of change or the outcome of compound rates of change:

$$A(1 + r)^n = F$$

Geometric mean A measure of central location of a data set used when it is desired to produce an average of rates of change:

$$g = \sqrt[n]{(x_1 \times x_2 \times x_3 \times \cdots \times x_n)}$$

Median A measure of central location of a data set. It is the value which splits the data set into two equal groups – one with values greater than or equal to the median, and one with values less than or equal to the median.

Mode A measure of central location of a data set, defined as the most frequently occurring data value.

Multi-modal distribution A distribution which has more than two modes.

Outliers Extreme values in a set of data.

Skewed distribution A distribution in which one tail is skewed either to the right (positively skewed) or to the left (negatively skewed).

Trimmed mean A measure of central location which disregards a proportion of the highest and lowest values in a data set.

Uni-modal distribution A distribution which has only one mode.

Weighted arithmetic mean A measure of the central location of a data set used when observations on particular values may occur more than once or when we need to reflect relative frequencies or the relative importance of certain values:

$$\bar{x} = \frac{\Sigma f_i x_i}{n}$$

Self-study exercises

3.1. Using the data in Self-study exercise 2.1 in Chapter 2, calculate the mean, median and mode.

3.2. A quality inspector found the following number of defective parts on 15 different days on an assembly line production:

3, 10, 8, 4, 6, 10, 12, 6, 10, 7, 11, 9, 1, 13, 4.

(a) Calculate the arithmetic mean number of defective parts.

(b) What is the median number of defective parts?

(c) What is the value of the mode?

3.3. Managers in the engineering sector of the economy have a lower mean annual income than managers in the retailing sector but a higher median income than those in the retailing sector. Explain how this can occur.

3.4. The frequency distribution given in the following table refers to the number of hours worked per week by employees in a local factory.

(a) Calculate the mean, median and mode for the number of hours worked by employees in this factory.

(b) Explain why the three values are different and hence explain which is the most appropriate choice of average in this case.

Hours per week	Number of employees
0 – <20	30
20 – <30	70
30 – <40	100
40 – <50	200
50 – <70	100
70 and over	0

3.5. The table below shows various data relating to all firms in three industries. Compute the following:

(a) average weekly earnings across all industries combined.

(b) average return on capital across all industries combined.

Industry	Number of firms	Total number of employees (000s)	Total capital (£m)	Average weekly earnings (£)	Average return on capital (%)
Food	68	200	110	105.8	5.4
Drink	27	46	70	98.0	6.7
Tobacco	5	54	20	101.1	4.3

3.6. The number of cans of various soft drinks sold from a dispensing machine during a particular week are as follows:

Cola   324

Pepsi   162

Lucozade 85

Fanta   323

Sprite   314

What is the mode for these data?

3.7. A firm's annual sales (in volume terms) grew as follows over the period 1996–2000.

1996–7	1997–8	1998–9	1999–2000
3.7%	1.3%	3.2%	1.5%

Calculate the average annual rate of increase in sales volume.

3.8. If prices increase at 5 per cent per year how long will it take for prices to double?

3.9. If a person is charged interest at 2 per cent per month what is the APR (i.e. the annual percentage rate)?

3.10. **(a)** A country's population grew from 38.2 million in 1911 to 55.5 million in 2001. What was the average annual percentage rate of increase?

(b) If this increase were maintained after 2001 what would the population be in the year 2010?

Chapter 4 | Describing data: variability

Aims

In the previous chapter we dealt with various measures of central tendency i.e. averages. While a single measure such as the mean, median or mode is useful it does not tell us anything about the extent of the dispersion or variability of the data set around this measure. Of course, if all the data values are identical, then the mean, median and mode would coincide and the variability would be zero. This is rarely the case, however, and hence we need a measure of variability that will increase as the data become more dispersed.

Knowledge of variability is important for a number of reasons. For example, in industrial processes it may be possible to control the extent of variability of, say, the quality of final products by better inspection control procedures. The most important measures of variability (dispersion) are as follows:

- range and modified ranges
- mean deviation
- standard deviation
- variance
- coefficient of variation

Of these, the most commonly used measures are the standard deviation and the variance.

Learning Outcomes

After working through this chapter you will be able to:

- understand what is meant by variation (dispersion) in a set of data
- appreciate the meaning and purpose of different measures of variability
- calculate the different measures of variability using grouped and ungrouped data
- compare the variability of different data sets
- understand which measure of variation it is appropriate to use in particular circumstances
- understand and use basic symbols and formulae for expressing variability measures.

Principles

Range and modified ranges

The *range* of a set of data is simply the difference between the largest and smallest values. While the range is easy to calculate, its use is limited because it relies on only two values at the extremes of the distribution of data values. Because of this limitation, *modified ranges* are sometimes employed instead. There are several of these, all of which focus on a central proportion of the data set, thus excluding the influence of extreme values. In this respect they are similar in nature to the trimmed mean considered in Chapter 3.

Modified ranges are based on various measures which divide a data set into equal parts such as quarters (*quartiles*), tenths (*deciles*) and hundredths (*percentiles*). Collectively, these are referred to as *quantiles*.

Quartile measures are defined, therefore, as follows:

- first (or lower) quartile, Q_1: located at the $[(n + 1)/4]$th observation
- third (or upper) quartile, Q_3: located at the $[3(n + 1)/4]$th observation

where n is the number of observations. The second quartile (Q_2) is, of course, the same as the median which lies at the mid-point of the range of values (see Chapter 3).

Decile and percentile measures are similarly defined: deciles divide a data set into ten equal parts and percentiles into 100 equal parts. Thus, for example, the first decile value (D_1) is located at the $[(n + 1)/10]$th observation etc. and the 99th percentile value (P_{99}) is located at the $[99(n + 1)/100]$th observation.

The types of modified ranges commonly used are *interquartile*, *interdecile* and *interpercentile* ranges. Like the ordinary range, these measures are simply the differences between the upper and lower quartiles, deciles and percentiles respectively. In addition, a *semi-interquartile range*, defined as $(Q_3 - Q_1)/2$, is often used, in association with the median, as a measure of variation.

Mean deviation

The mean deviation of a data set focuses on the average of the deviations of every observation x_i from the mean $\bar{x}$. As the sum of such deviations always equals zero by virtue of the definition of the mean, the mean deviation is calculated using the *absolute* values of the deviations. Thus:

Mean deviation

$$\frac{\Sigma|x_i - \bar{x}|}{n}$$

where $|x_i - \bar{x}|$ denotes the absolute value of each deviation (i.e. ignoring negative signs). This measure of variability is not widely used, preference being given to two related measures in which the offsetting effects of positive and negative deviations are eliminated by squaring. These measures are *standard deviation* and *variance*.

Standard deviation

After summing the squared deviations it might seem obvious that the next step is to divide by n as with the mean deviation above. However, this only applies when we are dealing with a statistical 'population' (as in a census) as opposed to a 'sample' (as in a survey). A sample standard deviation (denoted s) is often used as an estimator of the population standard deviation (denoted by the Greek lower case letter sigma, σ). These are obtained by dividing the sum of the squared deviations by $n - 1$ and N respectively. Thus:

Sample standard deviation

$$s = \sqrt{\frac{\Sigma(x_i - \bar{x})^2}{n - 1}}$$

where $\bar{x}$ is the sample mean and n is sample size.

Population standard deviation

$$\sigma = \sqrt{\frac{\Sigma(x_i - \mu)^2}{N}}$$

where μ (pronounced mu) is the population mean and N is the population size.

Variance

The variance is simply the standard deviation squared. Thus the sample and population variances are given by:

Sample variance

$$s^2 = \frac{\Sigma(x_i - \bar{x})^2}{n - 1}$$

and the population variance is given by

Population variance

$$\sigma^2 = \frac{\Sigma(x_i - \mu)^2}{N}$$

Typically, the standard deviation is used in preference to the variance as a measure of dispersion because the units of measurement in the variance are squared units and hence are difficult to interpret.

It is worth noting that many electronic calculators and computer software packages include pre-programmed functions to calculate the standard deviation and variance automatically with divisors of both N (for populations) and $n - 1$ (for samples) as required. In the absence of computer facilities, the above formulae are laborious to use when the data sets are large and the values themselves are large numbers. Fortunately, it is possible to simplify the formulae into equivalent expressions which are quicker and easier to use as follows:

Simplified computational formulae

Sample standard deviation

$$s = \sqrt{\frac{\Sigma(x_i^2) - (\Sigma x_i)^2/n}{n - 1}}$$

Population standard deviation

$$\sigma = \sqrt{\frac{\Sigma(x_i^2) - (\Sigma x_i)^2/N}{N}}$$

The corresponding variance formulae are, naturally, simply the square of each of these expressions as before. These formulae are much quicker to use because they eliminate the need to calculate deviations, requiring only the computation of x_i^2 values and the summations of x_i^2 and x_i.

Coefficient of variation

The coefficient of variation (denoted CV) is a measure of *relative* dispersion. It is appropriate for comparing the variability of different data sets where the units of measurement may be the same or different. It is defined as the standard deviation of the data set as a percentage of its mean and, like the other measures of variability, it can be calculated for samples and populations.

Sample coefficient of variation

$$CV = \frac{s}{\bar{x}} \times 100$$

Population coefficient of variation

$$CV = \frac{\sigma}{\mu} \times 100$$

Applications and Worked Examples   **Worked example 4.1   Ungrouped data**

General descriptions of the MINITAB and Excel computer programs and their application are given in Appendices B and C respectively.

For the following set of ten sample data values calculate the range, interquartile range, semi-interquartile range, mean deviation, standard deviation, variance and coefficient of variation:

4 7 8 9 22 26 29 37 40 48 ($n = 10$).

Worked solution

Range: $48 - 4 = 44$
Interquartile range: $Q_3 - Q_1$

where
$Q_3 = [3(n + 1)/4]$th $= 8.25$th observation $= 37 + 0.25(40 - 37) = 37.75$ and
$Q_1 = [(n + 1)/4]$th $= 2.75$th observation $= 7 + 0.75(8 - 7) = 7.75$.
Thus $Q_3 - Q_1 = 37.75 - 7.75 = 30$.

Semi-interquartile range (or quartile deviation):
$$\frac{Q_3 - Q_1}{2} = 15$$

Mean deviation:
$$\frac{\Sigma |x_i - \bar{x}|}{n}$$
where $\bar{x} = 23$ and the absolute deviations $|x_i - \bar{x}|$ are

19, 16, 15, 14, 1, 3, 6, 14, 17, 25.

Worked example 4.1   Ungrouped data *(continued)*

Thus

$$\frac{\Sigma|x_i - \bar{x}|}{n} = \frac{130}{10} = 13$$

Standard deviation (*s*):

$$\sqrt{\left[\frac{\Sigma(x_i - \bar{x})^2}{n-1}\right]}$$

where

$$\Sigma(x_i - \bar{x})^2 = 19^2 + 16^2 + 15^2 + 14^2 + 1^2$$
$$+ 3^2 + 6^2 + 14^2 + 17^2 + 25^2$$
$$= 2194.$$

Thus

$$s = \sqrt{\left(\frac{2194}{9}\right)} = 15.61$$

Using the short method of computation:

$$s = \sqrt{\left[\frac{\Sigma(x_i^2) - (\Sigma x_i)^2/n}{n-1}\right]}$$

where

$$\Sigma(x_i^2) = 16 + 49 + 64 + 81 + 484 + 676$$
$$+ 841 + 1369 + 1600 + 2304$$
$$= 7484$$
$$(\Sigma x_i)^2 = (230)^2 = 52\,900.$$

Thus

$$s = \sqrt{\frac{7484 - (230)^2/10}{10 - 1}} = 15.61$$

(as before).

Variance (*s²*):

$$\frac{\Sigma(x_i - \bar{x})^2}{n-1} = \frac{2194}{9} = 243.78$$

Coefficient of variation (CV):

$$\frac{s}{\bar{x}} \times 100 = \frac{15.61}{23} \times 100 = 67.87 \text{ per cent.}$$

Computer solution using MINITAB Ⓜ

With the data entered in column C1 of MINI-TAB worksheet use the **Describe** procedure as follows:

Choose **Stat → Basic Statistics → Display Descriptive Statistics**

Enter variable name or column number in dialogue box

Click **OK**

The output is as follows:

Descriptive Statistics

Variable	N	Mean	Median	TrMean	StDev	SE Mean
C1	10	23.00	24.00	22.25	15.61	4.94

Variable	Minimum	Maximum	Q1	Q3	
C1	4.00	48.00	7.75	37.75	Explained in Chapter 11

Authors' comment: Note that, in addition to measures of central tendency, MINITAB gives the sample standard deviation, the minimum and maximum values and the first and third quartiles. These allow the other measures of variability – range, interquartile range, variance and coefficient of variation – to be calculated, but not the mean deviation.

Computer solution using Excel

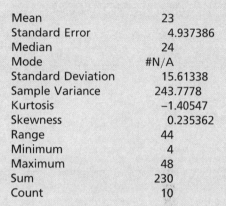

The command **Descriptive statistics**, available as one of the **Data Analysis Tools**, produces most of the measures required.

With the data (4, 7, 8, 9, 22, 26, 29, 37, 40, 48) entered in Column A of the worksheet, choose **Tools → Data Analysis → Descriptive Statistics**

Enter **Input Range** (either by typing A1:A10 or by highlighting the data in the worksheet)

In **Output Options** check the **Summary statistics** box and Click **OK**

The output is as follows:

Descriptive statistics

Column 1

Mean	23
Standard Error	4.937386
Median	24
Mode	#N/A
Standard Deviation	15.61338
Sample Variance	243.7778
Kurtosis	−1.40547
Skewness	0.235362
Range	44
Minimum	4
Maximum	48
Sum	230
Count	10

Authors' comment: Note that all the measures are produced except quartiles, mean deviation and coefficient of variation. However, the mean deviation may be obtained using the function **AVEDEV** and the quartiles may be obtained using the function **QUARTILE**. Most of the other measures may also be produced using the function facility.

Interpretation

The measure of variability appropriate to a problem depends on which measure of central location is used (see Chapter 3). Thus the standard deviation, mean deviation and variance would all be used in association with the mean, while the interquartile range or semi-interquartile range would be used in association with the median. The range is sometimes quoted with both the mean and the median. There is no measure of variability which is allied to the mode.

It will be seen from the solutions above that the data set could be described as having either a mean of 23 and a standard deviation of 15.61 (or variance of 243.78) or a median of 24 and an interquartile range of 30 (or semi-interquartile range of 15). This means that approximately 50 per cent of the observations lie within the range 24 ± 15 (this range is exact only for symmetrical distributions). The coefficient of variation (64.39 per cent) shows that the standard deviation is quite high relative to the mean and thus the data show considerable variation around the mean.

Worked example 4.2   Grouped data

For the sample of grouped data given below, calculate the interquartile range, semi-interquartile range, mean deviation, standard deviation, variance and coefficient of variation.

Classes	Class frequency
0–<5	2
5–<10	5
10–<15	8
15–<20	6
20–<25	4
25–<45	1

Solution

The table below provides details of the appropriate worksheet used to calculate the various measures.

Interquartile range: $Q_3 - Q_1$

As in the case of the median (Q_2), these values must be found by interpolation. Thus Q_3 is the $(3n/4)$th observation, i.e. the 19.5th observation; this lies in the '15 – < 20' class. By interpolation

$$Q_3 = 15 + \left(\frac{19.5 - 15}{6}\right)5 = 18.75$$

Similarly, Q_1 is the $(n/4)$th observation, i.e. the (6.5)th observation which lies in the '5 and under

10' class. By interpolation

$$Q_1 = 5 + \left(\frac{6.5 - 2}{5}\right)5 = 9.5$$

Thus, $Q_3 - Q_1 = 18.75 - 9.5 = 9.25$

Semi-interquartile range:

$$\frac{Q_3 - Q_1}{2} = 4.625$$

Mean deviation:

$$\frac{\Sigma f_i |x_i - \bar{x}|}{n}$$

where $\bar{x} = 372.5/26 = 14.33$ and the absolute deviations $f_i |x_i - \bar{x}|$ are 23.66, 34.15, 14.64, 19.02, 32.68, and 20.67. Thus

$$\frac{\Sigma f_i |x_i - \bar{x}|}{n} = \frac{144.82}{26} = 5.57.$$

Standard deviation (s):

$$\sqrt{\left[\frac{\Sigma f_i(x_i)^2 - (\Sigma f_i x_i)^2/n}{n-1}\right]}$$

$$= \left[\frac{6631.25 - (372.5)^2/26}{26 - 1}\right]$$

$$= 7.2 \text{ (approximately).}$$

Variance (s^2) $= (7.2)^2 = 51.8$ (approximately).

Coefficient of variation (CV) $= \dfrac{s}{\bar{x}} \times 100$

$$= \frac{7.2}{14.33} \times 100 = 50.2 \text{ per cent.}$$

Frequency distribution			Worksheet		
Classes	Class frequencies (f_i)	Class mid-point (x_i)	$f_i x_i$	$f_i(x_i^2)$	Cumulative frequency
0–<5	2	2.5	5.0	12.50	2
5–<10	5	7.5	37.5	281.25	7
10–<15	8	12.5	100.0	1250.00	15
15–<20	6	17.5	105.0	1837.50	21
20–<25	4	22.5	90.0	2025.00	25
25–<45	1	35.0	35.0	1225.00	26
Total	$n = 26$		372.5	6631.25	

Interpretation

The grouped data have a mean of 14.33 with a standard deviation of 7.2. The coefficient of variation (50.2 per cent) shows that the standard deviation is quite high relative to the mean.

Worked example 4.3 Comparing variability of different data sets

Two data sets of heights and weights for a group of people have the following means and standard deviations. Do heights vary more than weights?

	Mean ($\bar{x}$)	Standard deviation (s)
Height	176 cm	14 cm
Weight	80 kg	10 kg

Solution

To make comparisons of variability in two different data sets we can calculate the coeffi-cient of variation for each as follows:

$$CV = \frac{s}{\bar{x}}$$

Height: $\dfrac{14}{176} \times 100 = 7.95$ per cent

Weight: $\dfrac{10}{80} \times 100 = 12.50$ per cent.

Interpretation

As 7.95 per cent is less than 12.50 per cent, we can conclude that the heights for this particular group of people are not as variable as their weights.

Key terms and concepts

Coefficient of variation A measure of relative dispersion for a data set, found by dividing the standard deviation by the mean and multiplying by 100 to express the coefficient as a percentage:

Population coefficient of variation

$$CV = \frac{\sigma}{\mu} \times 100$$

Sample coefficient of variation

$$CV = \frac{s}{\bar{x}} \times 100$$

Interdecile range A measure showing the difference between the upper and lower decile values.

Interpercentile range A measure showing the difference between the upper and lower percentile values.

Interquartile range A measure of dispersion which covers the central 50 per cent of the observations in a data set i.e. the difference between the lower quartile (the value below which the lowest 25 per cent of the observations lie), denoted Q_1, and the upper quartile (the value above which the highest 25 per cent of the observations lie), denoted Q_3.

Mean deviation A measure of dispersion of a data set expressed as the average of the absolute differences between each observation (x_i) and the arithmetic mean of all observations ($\bar{x}$), expressed as:

$$\frac{\Sigma|x_i - \bar{x}|}{n}$$

Modified range A range which focuses only on a central portion of a data set (i.e. excluding extreme values).

Quantiles Measures which divide a data set into a number of equal parts.

Quartiles, deciles and percentiles Measures which divide a data set into 4, 10 and 100 parts respectively, each part containing the same number of observations.

Range A measure of dispersion for a data set, defined to be the difference between the highest and lowest values.

Semi-interquartile range Simply defined as the interquartile range divided by two (also referred to as *quartile deviation*).

Standard deviation A measure of dispersion for a data set, found by taking the positive square root of the population or sample variance.

Sample standard deviation

$$s = \sqrt{\frac{\Sigma(x_i - \bar{x})^2}{n - 1}} = \sqrt{\frac{\Sigma(x_i)^2 - (\Sigma x_i)^2/n}{n - 1}}$$

Population standard deviation

$$\sigma = \sqrt{\frac{\Sigma(x_i - \mu)^2}{N}} = \sqrt{\frac{\Sigma(x_i^2) - (\Sigma x_i)^2/N}{N}}$$

Variance A measure of dispersion for a data set, found by summing the squared deviations of the data values about the mean and then dividing the total by N if the data set is a population or by $n - 1$ if the data set is for a sample.

Sample variance: $\quad s^2 = \dfrac{\Sigma(x_i - \bar{x})^2}{n - 1}$

Population variance: $\quad \sigma^2 = \dfrac{\Sigma(x_i - \mu)^2}{N}$

Self-study exercises

4.1. A rugby team scored the following number of points in each of their last ten matches respectively:

18, 3, 21, 15, 9, 84, 27, 10, 42, 6.

Compute the following descriptive statistics of variability for the set of points:

(a) range

(b) upper and lower quartiles

(c) interquartile range

(d) semi-interquartile range

(e) upper decile

(f) standard deviation

(g) variance

(h) coefficient of variation

4.2. In a survey undertaken by a transport company the data in the table below were the recorded weights and values of goods transported by days of the week.

(a) For both values and weights, calculate

    (i) mean deviation

    (ii) standard deviation

    (iii) variance

    (iv) coefficient of variation

(b) Is the variation between days of the week larger in value or in weight?

	Monday	Tuesday	Wednesday	Thursday	Friday
Value (£)	4900	6400	6400	7200	3100
Weight (tonnes)	16	21	18	23	12

4.3. The data given in the following table show weekly expenditure on milk by 50 households.

(a) Arrange the data as a grouped frequency distribution using seven uniform class intervals as follows: £2 but under £3; £3 but under £4; etc. up to £8 but under £9.

(b) Calculate from the frequency distribution:

    (i) the upper and lower quartiles

    (ii) the bottom decile

(c) Compare these values with the *true* values derived from the raw (ungrouped) data and explain why they differ.

(d) On the basis of the quartiles derived from the frequency distribution construct a measure of variability and explain its meaning.

Weekly household expenditure on milk (£)									
6.0	5.9	3.5	2.9	8.7	7.9	7.1	5.0	5.2	3.9
3.7	6.1	5.8	4.1	5.8	6.4	3.8	4.9	5.7	5.5
6.9	4.0	4.8	5.1	4.3	5.4	6.8	5.9	6.9	5.4
2.4	4.9	7.2	4.2	6.2	5.8	3.8	6.2	5.7	6.8
3.4	5.0	5.2	5.3	3.0	3.6	3.8	5.8	4.9	3.7

4.4. The mean and variance of heights and weights for a sample of males and females are given below. Based on this information, answer the following questions:

(a) Do men vary more in height than in weight?

(b) Do women vary more in height than in weight?

(c) Do men vary more in height than women?

(d) Do women vary more in weight than men?

	Height (metres)		Weight (kilograms)	
	Mean	Variance	Mean	Variance
Males	1.80	0.16	76	36
Females	1.65	0.20	57	36

Chapter 5 | Exploratory data analysis

Aims
Exploratory data analysis (EDA) is a fairly new technique which provides a graphical means of exploring the underlying structure of a data set in more detail than the numerical summary measures of central tendency and dispersion described in Chapters 3 and 4. Within EDA there are two popular diagrams referred to as *stem and leaf* diagrams and *boxplots*. Stem and leaf diagrams are similar to histograms in that they show the shape of the data at a glance, though in a more complex way as we shall illustrate. Boxplots are graphical displays of the location of the quartiles and the overall spread of the data.

These techniques are extremely useful in summarizing reasonably sized data sets. They are mainly used with a computer-based statistical package such as MINITAB and we focus attention on the techniques here by illustrating their application using this package. In addition, we also take the opportunity to illustrate another MINITAB facility, known as *dotplot*, which simply plots the data along a horizontal scale to give a visual indication of the spread of the data set.

- appreciate the purpose and use of exploratory data analysis (EDA) in examining the underlying structure of a data set
- understand the terminology used in EDA
- understand the meaning of the most common EDA techniques
- apply the various techniques to sets of data
- interpret the output of the EDA techniques.

Principles and applications

Stem and leaf diagrams

To illustrate the use of stem and leaf analysis we take the test scores obtained by 50 students shown in Table 5.1 overleaf. These data are set in column C1 of a MINITAB worksheet. The tool is not available in Excel.

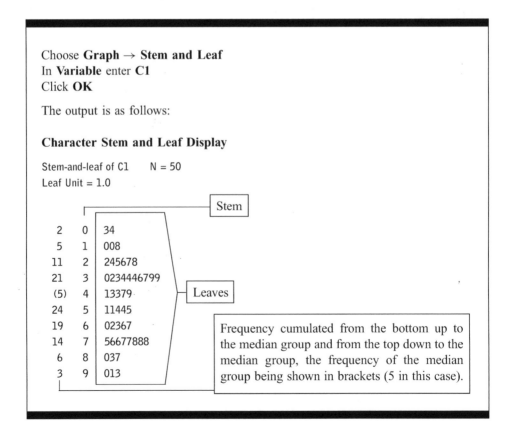

Choose **Graph → Stem and Leaf**
In **Variable** enter **C1**
Click **OK**

The output is as follows:

Character Stem and Leaf Display

Stem-and-leaf of C1   N = 50
Leaf Unit = 1.0

```
               ┌──────────────────────────┤ Stem
    2     0 │ 34
    5     1 │ 008
   11     2 │ 245678
   21     3 │ 0234446799
  (5)     4 │ 13379        ┤ Leaves
   24     5 │ 11445
   19     6 │ 02367
   14     7 │ 56677888
    6     8 │ 037
    3     9 │ 013
```

Frequency cumulated from the bottom up to the median group and from the top down to the median group, the frequency of the median group being shown in brackets (5 in this case).

Table 5.1 Test scores obtained by 50 students

54	25	43	3	28	39	78	32	54	93
27	33	22	78	75	83	62	76	77	67
77	80	4	26	18	10	34	30	36	43
78	41	24	91	90	63	87	55	60	49
37	39	51	66	10	76	34	47	51	34

The MINITAB command **Stem and Leaf** produces the required results in three parts.

■ *Stem*, shown in the second column, gives the range of 10s over which the full data set is distributed – in this example, from under 10 (i.e. zero 10s) up to the 90s.

■ *Leaf*, shown in the variable-length rows to the right of the stem, gives the last digit of each data value – thus the stem consists of the remaining digit(s) after the stem.

■ *Cumulative frequencies*, shown in the first column of the output along with the identification of the median class (shown in brackets).

Hence, it will be seen that for the test scores achieved by the 50 students, there were two below 10 (i.e. 3 and 4) and at the other end of the range there were three in the 90s (i.e. 90, 91 and 93). The median lies in the 40s range and there were five scores in the median class (i.e. 41, 43, 43, 47 and 49).

A stem and leaf diagram is a form of frequency histogram but there is now no loss of information since the printout not only shows the shape of the distribution but it also allows us to read off every value directly. It is particularly useful as a preliminary exercise to drawing up a frequency distribution table in terms of deciding how many and how large the classes should be (see Chapter 2).

Boxplots

A boxplot, also known as a *box and whisker plot*, produces another kind of visual summary of the data structure. The procedure and output, using MINITAB and the data shown in Table 5.2 (entered in column C1 of the MINITAB worksheet) are given below. This tool is not available in Excel.

Table 5.2 Recorded rainfall (mm) in 24 cities

100	86	90	90	120	40	60	92
50	140	90	98	86	42	80	85
215	160	64	78	114	21	72	84

Choose **Graph** → **Boxplot**
In **Graph variables** under **Y** enter **C1**
Click **OK**

The output is as follows:

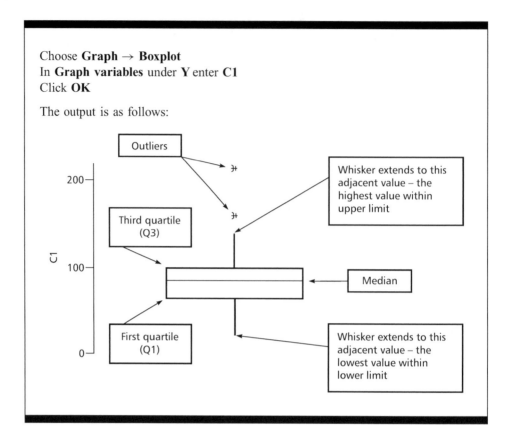

The boxplot consists of a box, whiskers and outliers. A line is drawn across the box at the median. By default, the bottom of the box is at the first quartile (Q1) and the top is at the third quartile (Q3) value. The whiskers are the lines that extend from the top and bottom of the box to the adjacent values. The adjacent values are the lowest and highest observations that are still inside the region defined by the following limits:

Lower Limit: $Q1 - 1.5\,(Q3 - Q1)$
Upper Limit: $Q3 + 1.5\,(Q3 - Q1)$

Note that $(Q3 - Q1)$ is, of course, the interquartile range.

The symbols (⋇) denote points which are beyond the lower and upper limits and are considered as outliers. An outlier may be an observation for which the value has been incorrectly recorded; if so, it can be corrected before proceeding with further analysis. Alternatively, it may be an item that was incorrectly included in the data set and so should be removed. On the other hand, it may simply be an unusual item that has been correctly recorded and does belong in the data set and therefore should not necessarily be excluded.

For the above data, it will be seen from the boxplot that the median value is roughly 86 (by visual inspection only), and the interquartile range is roughly 34 (i.e. from 66 to 100 approximately). Two outliers are identified, these are the two largest values: 215 and 160.

Dotplots

A dotplot displays a dot for each observation along a number line. If there are multiple occurrences of an observation, or if observations are too close together, then dots will be stacked vertically. If there are too many points to fit vertically in the graph then each dot may represent more than one point. In this case, a message will be displayed on the graph denoting the maximum number of observations that the dots represent. Using the same data as for the boxplot above, the procedure and output are as follows:

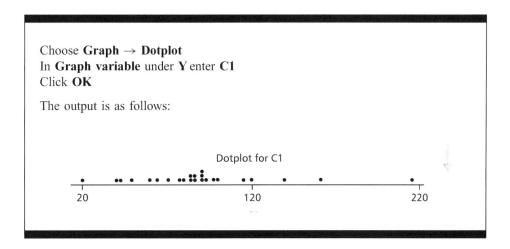

Choose **Graph** → **Dotplot**
In **Graph variable** under **Y** enter **C1**
Click **OK**

The output is as follows:

Dotplot for C1

The **Dotplot** command also allows you to generate dotplots for several variables at once and to create dotplots for each group within variables, but this is not illustrated here.

Clearly, the dotplot facility gives a good visual impression of the structure of the data set but is less complex and gives less information than the other two exploratory data analysis techniques.

Key terms and concepts

Boxplot A graphical display of the location of the quartiles of a data set and the overall spread of the data (also referred to as *box and whisker plot*).

Dotplot A horizontal display of the individual values of a data set.

Exploratory data analysis A technique which provides a graphical means of exploring the underlying structure of a data set.

Stem and leaf diagram A form of frequency histogram which not only shows the shape of a distribution but allows one to read off every value directly.

Self-study exercises

5.1. Examination marks for a class of 25 students studying statistics were as follows:

43	57	92	24	88
94	76	54	42	57
25	50	86	60	45
53	17	57	67	78
61	56	46	70	82

Provide a stem and leaf diagram for these data.

5.2. The performances of 12 secretaries are assessed on the basis of the average number of typing errors per page recorded during a particular day. The performances are as follows:

2.4	3.6	4.8	6.3	5.3	2.1
18.5	0.0	4.4	1.0	14.0	2.8

(a) Provide a boxplot of these data and comment on the results.

(b) How many observations extend beyond the whiskers and which ones should be regarded as possible outliers?

5.3. Produce a dotplot for the data given below:

25	24	40	25	4
15	25	6	6	25
6	10	7	74	10
10	89	78	25	9
15	25	78	10	80

Part II

Probability and probability distributions

I have set my life upon a cast, And I will stand the hazard of the die.

WILLIAM SHAKESPEARE

The following diagram provides an overview of the whole of this part of the book.

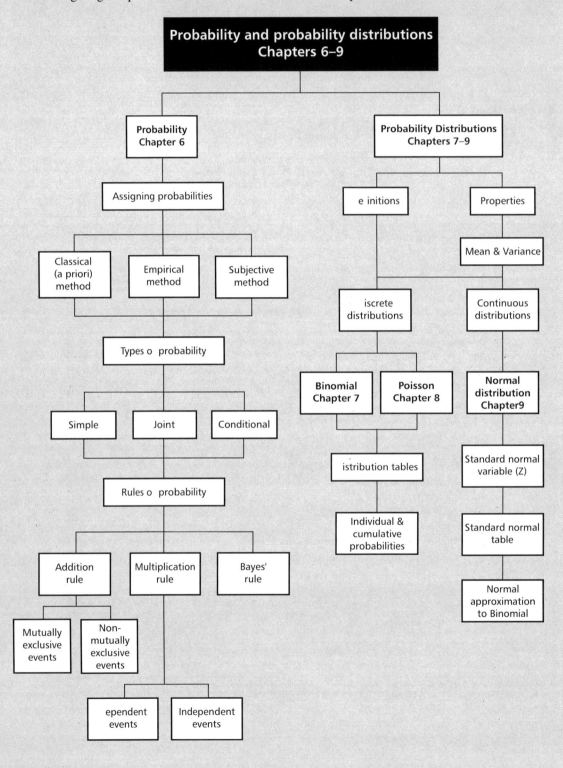

Chapter 6 | Probability concepts

Aims

The previous chapters have been concerned with the description, in tabular, graphical and numerical form, of a data set usually representing a sample of observations drawn from a population (i.e. a survey as opposed to a census). We have been dealing with actual recorded values. However, it is often the case that we wish to *infer* something about the underlying nature of the population using samples of data. In these situations we are dealing with *inference* or *uncertainty* – i.e. *probabilities*. It is for this reason that the study of probability concepts is a vital topic and one which underlies much of the analysis in many of the remaining chapters of this book.

Situations in which probability concepts are relevant are very varied. For example, it is obviously useful for management to have some idea of the likelihood (i.e. probability) of a marketing campaign being successful in terms of increased product sales; manufacturers are unlikely to know in advance the frequency of defective components in a production line, but by taking random samples it is possible to estimate the probability of possible outcomes occurring, such as one, two or three defects being discovered in a whole batch; people's behaviour is likely to be influenced by knowledge of the probability of being killed while at home, compared with the probability of being killed while crossing the road or flying in an aeroplane, etc.

In this chapter we set out the fundamental principles of probability and show how these can be applied in decision-making situations. Careful and systematic study of these principles will provide a basis for future chapters in this book which deal with statistical inference and, ultimately, the basis for sound decisions. The key concepts covered in turn below are:

- assigning probabilities
- rules of probability
- permutations and combinations

Probability is a topic which at first sight may seem daunting and confusing. But it is not complex mathematically – it simply requires careful thought and logical thinking.

After working through this chapter you will be able to:

- appreciate the important role that probability plays in drawing inferences in the face of uncertainty
- understand the key concepts and terminology used in probability analysis
- assign probabilities using the classical, empirical and subjective approaches
- apply the addition rule of probability for mutually and non-mutually exclusive events
- understand the meaning of conditional probability
- apply the multiplication rule of probability for dependent and independent events
- apply Bayes' rule for the calculation of probabilities in its general and extended forms
- use tree diagrams and decision trees in the context of decision-making
- understand the meaning of permutations and combinations and their calculation.

Principles

Assigning probabilities

Probability is a numerical measure of the likelihood of an event occurring. It is measured on a scale from 0 to 1, with zero indicating the impossibility of an event taking place and unity representing certainty. In addition, the sum of the probabilities of all possible outcomes must equal 1. In notation form, where P stands for probability,

$\Sigma P(\text{outcomes}) = 1$

Probability, of course, increases as one moves up the scale from 0 to 1, so that the expression 'there is a 50:50 chance' represents a probability of 0.5.

There are three approaches to assigning probability values to events (two objective and one subjective) depending on the nature of the situation. These are:

- classical approach
- empirical approach
- subjective approach.

Classical approach

This approach relies solely on abstract reasoning because logic by itself is sufficient to determine the probability of the event being considered. Therefore, the approach only applies to situations in which a particular outcome can be assigned a probability on *a priori* grounds. For example, in tossing a coin only two outcomes are possible, head or tail, and if these outcomes are *equally likely* (i.e. the coin is 'fair') then $P(\text{Head}) = P(\text{Tail}) = 0.5$.

In many situations, however, the assumption of 'equally likely outcomes' is not a reasonable one. In these situations one of the other two approaches must be used.

Empirical approach

In this approach empirical evidence obtained from records or surveys (so-called 'repeated trials') is used to assign a probability to a particular outcome as follows:

$$P(\text{outcome}) = \frac{\text{Number of times the outcome occurs}}{\text{Number of trials}} = \frac{f_i}{N}$$

i.e. $P(\text{outcome})$ is calculated as the relative frequency of the outcome occurring. For example, if records of the number of hours for which a particular type of electric light bulb lasts show that out of every 1000 bulbs tested, 100 last for less than 500 hours, then the probability of any one bulb, selected at random, having a life of less than 500 hours is

$$P(\text{life} < 500) = \frac{100}{1000} = 0.1$$

Naturally, it is not always possible to obtain empirical data to calculate probabilities. This is especially true of future events. In this case the subjective approach has to be used.

Subjective approach

It is possible to assign a probability to an outcome according to what we *subjectively* feel is likely to happen based on our degree of belief about the outcome itself. This may be a 'gut feeling' but may also be based upon informed judgement. Of course, using this approach different people are likely to assign different probabilities to the same outcome (e.g. not everyone assigns the same probability to a particular person winning a particular race).

Rules of probability

We are frequently confronted with situations in which we wish to determine the probability not of single events (as above) but of *related* events. For example, for two events A and B, we may be interested in knowing whether *both* A *and* B will occur or, alternatively, whether *at least one* of them will occur. To answer such questions requires the introduction of two fundamental rules, or laws, of probability. These are:

- the addition rule of probability
- the multiplication rule of probability.

The addition rule

The addition rule applies when we are considering two (or more) possible outcomes (events) and wish to determine the probability that *at least one* of them will occur. There are two variations of the addition rule, however, depending on whether the events are *mutually exclusive* or not. Events are said to be mutually exclusive if they cannot occur at the same time (e.g. a head and a tail on a single toss of a coin). Otherwise events are said to be non-mutually exclusive. Thus we have the following two rules.

Addition rule for mutually exclusive events

If A and B are two mutually exclusive events, then the probability of obtaining *either* A *or* B is equal to the probability of obtaining A *plus* the probability of obtaining B, i.e.

$P(\text{A or B}) = P(\text{A}) + P(\text{B})$

Addition rule for non-mutually exclusive events

If A and B are non-mutually exclusive events, then 'A or B' means that A occurs *or* B occurs or both A *and* B occur simultaneously. In this case, we must subtract the *joint occurrence* of A and B from the sum of their probabilities to avoid double-counting, i.e.

$P(\text{A or B}) = P(\text{A}) + P(\text{B}) - P(\text{A and B})$

Note that if A and B are mutually exclusive, then $P(\text{A and B}) = 0$

These two addition rules of probability can be illustrated clearly using *Venn diagrams* as shown in Figures 6.1 and 6.2. In these diagrams an event is represented by a circle. Mutually exclusive events can then be shown as *non-overlapping* circles (Figure 6.1), while non-mutually exclusive events are shown as overlapping circles (Figure 6.2).

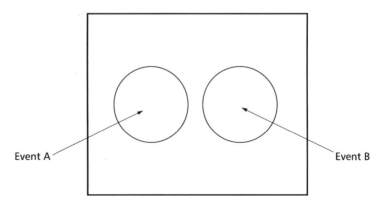

Figure 6.1 Mutually exclusive events

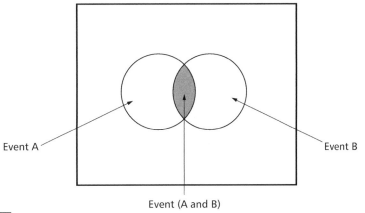

Figure 6.2 Non-mutually exclusive events

In Figure 6.2 the shaded area (the intersection of the circles) represents the *joint occurrence* of events A *and* B, i.e. *P*(A and B).

Generalization of addition rules

The two variations of the addition rule can be generalized to more than two events as follows.

> ## Generalization of addition rules
>
> *Mutually exclusive events:*
>
> $$P(A \text{ or } B \text{ or } C \text{ or} \ldots) = P(A) + P(B) + P(C) + \ldots \text{etc.}$$
>
> *Non-mutually exclusive events:*
>
> $$P(A \text{ or } B \text{ or } C) = P(A) + P(B) + P(C) - P(A \text{ and } B)$$
> $$- P(A \text{ and } C) - P(B \text{ and } C) + P(A \text{ and } B \text{ and } C)$$
>
> and so on for more than three events.

These generalizations are also illustrated using Venn diagrams in Figure 6.3.

In Figure 6.3(b) it should be noted that *P*(A and B and C) – i.e. the shaded area – is in effect *included* three times over through the summation of *P*(A), *P*(B) and *P*(C) and then *subtracted* three times through the subtraction of *P*(A and B), *P*(A and C) and *P*(B and C). Hence it must be included back in again at the end of the expression by adding *P*(A and B and C).

The multiplication rule

The multiplication rule is used to find the *joint occurrence* of two or more events, e.g. *P*(A *and* B) as in Figure 6.2 or *P*(A *and* B *and* C) as in Figure 6.3(b). The answers to such problems depend on whether or not the events in question are *dependent* or *independent*.

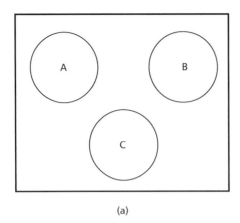

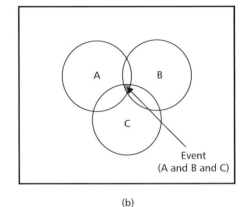

(a)   (b)

NOTE
(a) mutually exclusive events, $P(A \text{ or } B \text{ or } C) = P(A) + P(B) + P(C)$;
(b) non-mutually exclusive events, $P(A \text{ or } B \text{ or } C) = P(A) + P(B) + P(C) - P(A \text{ and } B) - P(A \text{ and } C) - P(B \text{ and } C) + P(A \text{ and } B \text{ and } C)$.

Figure 6.3 Generalization of addition rules

Two events are said to be dependent when the occurrence of one affects the probability of occurrence of the other. In contrast, they are said to be independent when the occurrence of one does not affect the probability of occurrence of the other. Hence, the multiplication rule in each case is as follows.

Multiplication rule for dependent events

$P(A \text{ and } B) = P(A) \times P(B|A)$;

where $P(B|A)$ denotes the probability of B occurring, given that A has already occurred (note that the vertical line in this expression is read as 'given').

Multiplication rule for independent events

$P(A \text{ and } B) = P(A) \times P(B)$

Conditional probability and Bayes' rule

Note that the expression $P(B|A)$ above denotes *conditional probability*, i.e. the probability of B is affected by the occurrence of A. Naturally, if A and B are independent then $P(B|A) = P(B)$.

A useful extension of this idea in the context of several sequential events is covered by a rule known as Bayes' rule. This provides a general method for revising *prior* probabilities (such as the probability of event A occurring, i.e. $P(A)$) in the light of new information becoming available (such as the fact that certain other events B and C etc., to which A is related, have already occurred) to provide what is referred to as *posterior* probability. In a *two-event* case, Bayes' rule is, in general form, as follows:

Bayes' rule – general form

$$P(B|A) = \frac{P(A \text{ and } B)}{P(A)}$$

This is, of course, simply a rearrangement of the multiplication rule for dependent events shown above.

Bayes' rule can be readily extended to the case of more than two (say n) mutually exclusive events, $B_1, B_2, B_3, \ldots, B_n$, and where one of the n events must occur. Thus the probability of any particular event B_i occurring given that A has occurred, expressed as $P(B_i|A)$, is given by the extended form of Bayes' rule as follows:

Bayes' rule – extended form

$$P(B_i|A) = \frac{P(B_i)P(A|B_i)}{P(B_1)P(A|B_1) + P(B_2)P(A|B_2) + \cdots + P(B_n)P(A|B_n)}$$

$$= \frac{P(B_i)P(A|B_i)}{\sum_{i=1}^{n} P(B_i)P(A|B_i)} \quad \text{where } i = 1, 2, \ldots, n$$

where the expression $\sum_{i=1}^{n}$ denotes summation over all values from $i = 1$ to $i = n$.

Tree diagrams and decision trees

The application of Bayes' rule in both its general and extended forms is given in the applications section. The rule may look complicated at first sight but it may be readily understood using what are known as *tree diagrams*. These are a useful way of clarifying a problem and their use is also illustrated below in the applications section. Tree diagrams themselves may also be extended to many decision-making problems, especially in business in the context of decisions about investment and expected returns. In this context they are generally referred to as *decision trees*. Again, an example is given below in the applications section.

Permutations and combinations

By definition, the probability of an event A occurring is equal to the number of outcomes relating to A divided by the total number of possible outcomes, i.e. $P(A) = f_i/N$, as in the empirical (relative frequency) formula for assigning probabilities given earlier. Finding values of f_i and N is usually relatively easy. When the problems are simple, the number of outcomes can easily be counted. However, for other problems which may involve a very large number of possible outcomes, it may not be so easy to count them. We conclude this section, therefore, by noting the mathematical techniques involving *permutations* and *combinations* which are at our disposal to determine the total number of possible outcomes. Most counting problems fall into one or other of these types. The application of the mathematical techniques is very easy – the testing aspect of any problem is in deciding *which* method is appropriate (i.e. permutations or combinations).

Permutations refer to the number of ways in which a set of objects can be arranged *in order* (the order being crucial), while combinations refer to the number of ways in which a set of objects can be arranged *regardless of order.*

The mathematical expressions for permutations and combinations involve some new notation, namely $n!$ (pronounced 'n factorial') which denotes the product of $n(n-1)(n-2)(n-3)\ldots(2)(1)$. Thus $3! = 3 \times 2 \times 1 = 6$. Note that the factorial sequence stops at 1. Zero factorial is defined as equal to 1, i.e. $0! = 1$.

The corresponding mathematical expressions for the number of possible permutations and combinations are as follows:

Formula for the number of permutations

$$^nP_r = \frac{n!}{(n-r)!}$$

where ^nP_r denotes the number of permutations possible in selecting r objects from a total of n objects;

Formula for the number of combinations

$$^nC_r = \frac{n!}{r!(n-r)!}$$

where ^nC_r denotes the number of combinations possible in selecting r objects from a total of n different objects.

The calculation of factorial expressions for large numbers is not as daunting as it might appear at first sight since, in the cases of both permutations and combinations, a large part of the numerators and denominators can be cancelled out, as will be seen below in the relevant worked examples. In any case many electronic calculators and computer software programs have pre-programmed factorial functions to make life even easier.

Applications and Worked Examples	Worked example 6.1 Application of addition rule – non-mutually exclusive events
General descriptions of the MINITAB and Excel computer programs and their application are given in Appendices B and C respectively.	A company employs 50 people, ten of whom are female. There are ten male executives and only five female executives employed. If a member of staff is selected at random to be

made redundant what is the probability that the person selected will be a female (F) *or* an executive (E)?

Solution

The required probability may be expressed as *P*(F *or* E). The events are non-mutually exclusive because it is possible to be both a female and an executive. It thus involves the addition rule of probability:

$$P(F \text{ or } E) = P(F) + P(E) - P(F \text{ and } E)$$
$$= 10/50 + 15/50 - 5/50$$
$$= 20/50 = 0.4$$

This is also shown in the Venn diagram:

Interpretation

This tells us that there is a 40 per cent probability of an employee, selected at random for redundancy, being a female or an executive. The person could even be a female executive but not necessarily.

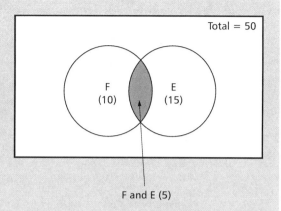

F and E (5)

Worked example 6.2   Application of addition rule – mutually exclusive events

What would the answer for Worked Example 6.1 be if there were no female executives in the workforce?

Solution

In this case, *P*(F or E) would involve mutually exclusive events, as shown in the following Venn diagram. Thus $P(F \text{ or } E) = P(F) + P(E) = 15/50 + 10/50 = 0.5$.

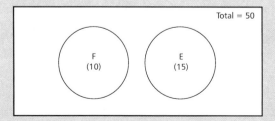

Interpretation

This tells us that there is a 50 per cent probability of a randomly selected employee being a female or an executive when there are no female executives in the workforce.

Worked example 6.3 Generalization of addition rule – non-mutually exclusive events

A potential second-hand car buyer estimates that 20 per cent of the cars currently advertised for sale are in the price range he can afford, 70 per cent are of the desired size and 30 per cent are of the desired colour. He also estimates that 25 per cent of the cars are suitable with respect to size and colour, 15 per cent are suitable with respect to size and price and 5 per cent are suitable with respect to colour and price. Only 1 per cent of the advertised cars meet all the three criteria he desires. What is the probability that he will find a car which meets *at least one* of his desired criteria?

Solution

Let us use the following notation to denote the relevant events: A_1, size suitable; A_2, colour suitable; A_3 price suitable. From the information given above we can write:

$P(A_1) = 0.70$ $P(A_1 \text{ and } A_2) = 0.25$
$P(A_2) = 0.30$ $P(A_1 \text{ and } A_3) = 0.15$
$P(A_3) = 0.20$ $P(A_2 \text{ and } A_3) = 0.05$
$P(A_1 \text{ and } A_2 \text{ and } A_3) = 0.01$

The statement 'at least one' must be interpreted as meaning that exactly one or exactly two or exactly three of the buyer's requirements with respect to size, price and colour are met. The required probability is $P(A_1 \text{ or } A_2 \text{ or } A_3)$. Using the generalized addition rule, this is given as follows:

$$
\begin{aligned}
P(A_1 \text{ or } A_2 \text{ or } A_3) = {} & P(A_1) + P(A_2) + P(A_3) \\
& - P(A_1 \text{ and } A_2) \\
& - P(A_1 \text{ and } A_3) \\
& - P(A_2 \text{ and } A_3) \\
& + P(A_1 \text{ and } A_2 \text{ and } A_3) \\
= {} & 0.70 + 0.30 + 0.20 \\
& - 0.25 - 0.15 - 0.05 \\
& + 0.01 \\
= {} & 0.76
\end{aligned}
$$

Interpretation

There is thus a 76 per cent chance of this buyer finding a car with at least one of the characteristics he desires.

Worked example 6.4 Multiplication rule – independent events

A woman has applied for a job with two firms A and B. She estimates that her chance of getting a job offer from A is 0.4 and from B is 0.3. Assuming that the offer of a job from one firm is independent of an offer from the other firm, what is the probability that:

(a) she gets an offer from both firms?
(b) she will get at least one offer?
(c) neither firm offers her a job?
(d) firm A does not offer her a job but firm B does?

Solutions

Let $P(A)$ and $P(B)$ denote probabilities of a job offer from A and B respectively. Thus:

(a) $P(A \text{ and } B) = P(A) \times P(B) = 0.4 \times 0.3$
$= 0.12$
(b) $P(A \text{ or } B) = P(A) + P(B) - P(A \text{ and } B)$
$= 0.4 + 0.3 - 0.12 = 0.58$ (by the addition rule for non-mutually exclusive events since a job offer from A does not preclude an offer from B)
(c) $1 - P(A \text{ or } B) = 1 - 0.58 = 0.42$
(d) $[1 - P(A)] \times P(B) = (1 - 0.4) \times 0.3$
$= 0.18$

Worked example 6.5   Multiplication rule – dependent events

The workforce of a firm employing 200 people is comprised of four groups, namely male manual workers, male supervisors, female manual workers and female supervisors, as follows:

	Male (M)	Female (F)
Supervisors (S)	20	50
Manual workers (W)	100	30

If a person is selected at random, find the conditional probabilities: (a) $P(M|S)$; (b) $P(F|S)$; (c) $P(S|F)$.

Solutions

(a) $P(M|S) = \dfrac{P(S \text{ and } M)}{P(S)} = \dfrac{(20/200)}{(70/200)} = \dfrac{2}{7}$

(b) $P(F|S) = \dfrac{P(S \text{ and } F)}{P(S)} = \dfrac{(50/200)}{(70/200)} = \dfrac{5}{7}$

(c) $P(S|F) = \dfrac{P(F \text{ and } S)}{P(F)} = \dfrac{(50/200)}{(80/200)} = \dfrac{5}{8}$

Worked example 6.6   Multiplication rule – sampling without replacement

Given the data in Worked Example 6.5 above, if two employees are selected at random find the joint probability that the first person is male and the second person is female.

Solution

Selection of the first person reduces the number of people from which the second person is selected. This process is one of sampling *without* replacement (for further details see Chapter 11). The problem may be denoted as:

$$P(M_1) \times P(F_2|M_1) = \frac{120}{200} \times \frac{80}{199} = \frac{48}{199}$$

where the subscripts 1 and 2 denote the order of selection. Note that since we are sampling without replacement, having selected one person already then: $P(F_2|M_1) = 80/(200-1) = 80/199$.

Worked example 6.7   Bayes' rule

A hardware store purchases light bulbs in bulk from three different suppliers, denoted A, B and C. These supply 60 per cent, 30 per cent and 10 per cent of the store's requirements respectively. On average, the proportion of faulty bulbs supplied by each of the three suppliers is 2 per cent, 5 per cent and 8 per cent respectively. If the manager of the store chooses a bulb at random and finds it to be faulty, what is the probability that it came from supplier C?

Solution

The problem, in notation form, is equivalent to

$$P(C|F) = \frac{P(C)P(F|C)}{P(F)}$$

$$= \frac{P(C)P(F|C)}{P(A)P(F|A) + P(B)P(F|B) + P(C)P(F|C)}$$

where F denotes a faulty light bulb. The denominator represents the overall probability

Worked example 6.7   Bayes' rule *(continued)*

of selecting a faulty light bulb, regardless of which supplier is responsible for it. Hence,

$P(C|F)$

$$= \frac{(0.1)(0.08)}{(0.6)(0.02) + (0.3)(0.05) + (0.1)(0.08)}$$

$$= \frac{0.008}{0.035}$$

$$= 0.23 \text{ (approximately)}$$

Interpretation

Thus, given that a faulty light bulb has been found by the manager, the probability that it has originated from supplier C is 0.23 (i.e. a 23 per cent chance). Note that in applying Bayes' rule to problems such as this we must know not only the conditional probabilities $P(F|A)$, $P(F|B)$ and $P(F|C)$ but also the *prior* probabilities $P(A)$, $P(B)$ and $P(C)$.

Worked example 6.8   Tree diagrams

The logic of the solution to Worked Example 6.7, using Bayes' rule, is easily seen using a *tree diagram* as shown below. (G: good light bulb.)

Thus, from the tree diagram, we can now read off all the required values. Therefore,

$$P(C|F) = \frac{0.008}{0.012 + 0.015 + 0.008}$$

$$= 0.23 \text{ (approximately)}.$$

This is equivalent to the result derived using Bayes' rule in Worked Example 6.7.

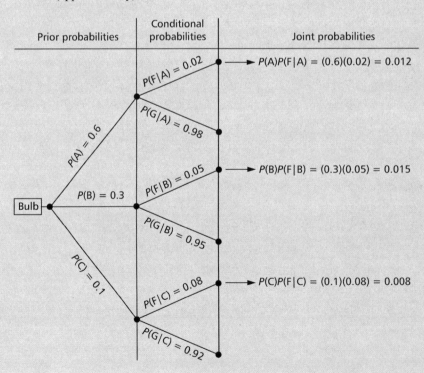

Worked example 6.9   Decision trees

The use of tree diagrams shown above can be extended to the analysis of business decisions involving alternative courses of action to which are attached expected pay-offs with associated probabilities of achievement. The objective often involves the maximization of, say, profits or the minimization of loss. To show the application of decision trees in these contexts, consider the following example of three alternative investment strategies A, B and C in research and development on a new drug. The drug manufacturer estimates that the potential profit pay-offs in £ (with associated probability) from each strategy are as shown in the following table. If the drug company aims to maximize profits, which investment strategy should be pursued?

Strategy	High (H) pay-off	Low (L) pay-off
A	200 (0.3)	−20 (0.7)
B	150 (0.5)	20 (0.5)
C	100 (0.4)	60 (0.6)

Solution

The solution to this problem is easily obtained from the figure below which is referred to as a *decision tree*. It will be seen that the expected total pay-off from investment strategy A is + £46 million, from B it is + £85 million and from C it is + £76 million. The company, if it wishes to maximize profits, should be in favour of strategy B.

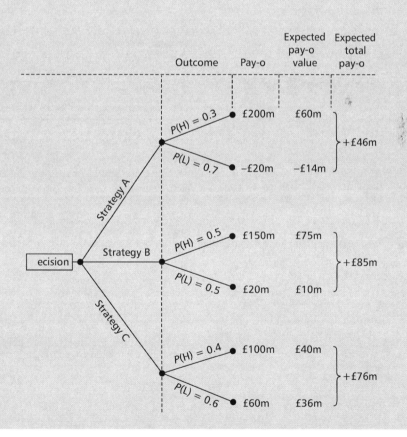

	Outcome	Pay-o	Expected pay-o value	Expected total pay-o
Strategy A	$P(H) = 0.3$	£200m	£60m	
	$P(L) = 0.7$	−£20m	−£14m	+£46m
Strategy B	$P(H) = 0.5$	£150m	£75m	
	$P(L) = 0.5$	£20m	£10m	+£85m
Strategy C	$P(H) = 0.4$	£100m	£40m	
	$P(L) = 0.6$	£60m	£36m	+£76m

Worked example 6.10   Permutations

Suppose that there are four jobs to be allocated among seven workers. If only one worker is to be assigned to each job, how many different possible assignments can be made?

Solution

In this case there are seven ways of assigning the first job after which there are six ways of assigning the next job and so on. Hence, the total number of possible assignments (i.e. permutations) is:

$$^nP_r = \frac{n!}{(n-r)!}$$

where n is number of workers ($= 7$) and r is number of jobs ($= 4$). Thus,

$$^7P_4 = \frac{7!}{(7-4)!} = \frac{7 \times 6 \times 5 \times 4 \times 3 \times 2 \times 1}{3 \times 2 \times 1}$$
$$= 840$$

Hence, there are 840 different ways of making the job assignments.

Computer solution using Excel

(There is no equivalent tool in MINITAB)

Use **PERMUT** function:
Click on $=$ on formula bar and choose
  PERMUT
In **Number** enter the total number of objects
  (7 in this case)

In **Number_chosen** enter the number of
  objects in each permutation (7 in this case)
This returns the answer: 840.

Worked example 6.11   Combinations

Five people have applied for the posts of either manager or trainer of a football team. All five applicants are suitably qualified for either post. How many different ways can these two posts be filled if the order of selection does not matter (i.e. we are only interested in the number of ways of selecting any two people, regardless of which post they fill)?

Solution

The number of combinations of two people to be selected from a total of five will be less than the

number of possible permutations. Permutations would be given by

$$^5P_2 = \frac{5!}{(5-2)!} = \frac{5 \times 4 \times 3 \times 2 \times 1}{3 \times 2 \times 1} = 20$$

where the appointment of two people, A and B, to the managerial and trainer's post in that order respectively would be regarded as different to the situation where their jobs are reversed, i.e. B and A. Hence, since combinations disregard the order (i.e. A, B is the same as B, A), then the number of combinations in this case is half the number of permutations, i.e.

$$^nC_r = \frac{n!}{r!(n-r)!} = \frac{5!}{2!(5-2)!} = \frac{5 \times 4 \times 3 \times 2 \times 1}{(2 \times 1)(3 \times 2 \times 1)} = 10$$

Computer solution using Excel

(There is no equivalent tool in MINITAB)

Use **COMBIN** function:
Click on = on formula bar and choose
   COMBIN

In **Number** enter the total number of objects (5 in this case)
In **Number_chosen** enter the number of objects in each combination (2 in this case)
This returns the answer: 10.

Key terms and concepts

Addition rule for mutually exclusive events If A and B are two mutually exclusive events, then the probability of obtaining *either* A *or* B is equal to the probability of obtaining A *plus* the probability of obtaining B:

$$P(A \text{ or } B) = P(A) + P(B)$$

Addition rule for non-mutually exclusive events If A and B are not mutually exclusive events, then we must subtract the probability of the joint occurrence of A and B from the sum of their probabilities:

$$P(A \text{ or } B) = P(A) + P(B) - P(A \text{ and } B)$$

Bayes' rule for conditional probabilities A general method for revising prior probabilities in the light of new information to provide posterior probabilities:

Two events: $P(B|A) = \dfrac{P(A \text{ and } B)}{P(A)}$

General form: $P(B_i|A) = \dfrac{P(B_i)P(A|B_i)}{\displaystyle\sum_{i=1}^{n} P(B_i)P(A|B_i)}$ where $i = 1, 2, \ldots, n$.

Classical approach A method of assigning probabilities which assumes equally likely outcomes.

Combinations These refer to the number of ways in which a set of objects can be arranged without regard to the order of their selection:

$$^nC_r = \frac{n!}{r!(n-r)!}$$

Conditional probability The probability of one event (say B) occurring given that another event (say A) has already occurred, denoted $P(B|A)$.

Decision tree Tree diagram used in the context of decision-making.

Dependent events Two events are dependent when the occurrence (or non-occurrence) of one event affects the probability of occurrence of the other event.

Empirical (relative frequency) approach A method of assigning probabilities based on experimental or historical data, where:

$$P(\text{outcome}) = \frac{f_i}{N}$$

Generalization of addition rules

Mutually exclusive events:

$$P(A \text{ or } B \text{ or } C \text{ or} \ldots) = P(A) + P(B) + P(C) + \cdots \text{etc.}$$

Non-mutually exclusive events:

$$P(A \text{ or } B \text{ or } C) = P(A) + P(B) + P(C)$$
$$- P(A \text{ and } B) - P(A \text{ and } C) - P(B \text{ and } C)$$
$$+ P(A \text{ and } B \text{ and } C) \text{ and so on for more than three events.}$$

Independent events Two events are independent when the occurrence (or non-occurrence) of one event has no effect on the probability of occurrence of the other event.

Multiplication rule for independent events

$$P(A \text{ and } B) = P(A) \times P(B)$$

Multiplication rule for dependent events

$$P(A \text{ and } B) = P(A) \times P(B|A)$$

Mutually exclusive events Events which cannot occur simultaneously.

Permutations These refer to the number of ways in which a set of objects can be arranged in order (the order being crucial):

$$^nP_r = \frac{n!}{(n-r)!}$$

Probability A numerical measure of the likelihood that an event will occur.

Subjective approach A method of assigning probabilities based upon judgement.

Tree diagram A graphical device helpful in defining sample points of an experiment involving multiple steps.

Venn diagram A pictorial device to illustrate the concepts of exclusive and non-mutually exclusive events.

Self-study exercises

6.1. Which one of the three methods of assigning probabilities would be used in each of the following situations:

(a) when the assumption of equally likely outcomes is used to assign probability values?

(b) when the results of experimentation, or historical data, are used to assign probability values?

(c) when probabilities are assigned based on judgement?

6.2. An experiment consists of three stages. There are two possible outcomes at the first stage, three possible outcomes at the second stage and three possible outcomes at the third stage. What is the total number of experimental outcomes?

6.3. Let A and B denote mutually exclusive events with $P(A) = 0.3$ and $P(B) = 0.5$. Find:

(a) $P(A \text{ and } B)$

(b) $P(A \text{ or } B)$.

6.4. If A and B are independent events with $P(A) = 0.05$ and $P(B) = 0.65$, find:

(a) the probability of A occurring given that B has already occurred, (i.e. $P(A|B)$;

(b) the probability of A and B occurring (i.e. $P(A \text{ and } B)$);

(c) the probability of either A or B occurring (i.e. $P(A \text{ or } B)$).

6.5. The employment records of a company give the details shown in the table below about the make-up of the workforce in terms of job category and the proportion holding a professional qualification.

(a) Given that E_1, E_2 and E_3 denote a skilled worker, a manager and an administrator respectively, and that Q denotes a worker having a professional qualification, state the following probabilities:

(i) $P(E_1)$

(ii) $P(E_2)$

(iii) $P(E_3)$

(iv) $P(Q|E_1)$

(v) $P(Q|E_2)$

(vi) $P(Q|E_3)$

(b) Construct a tree diagram for this situation.

(c) Using the tree diagram, calculate the probability that a randomly selected worker will be identified as having a professional qualification.

(d) Use Bayes' rule to calculate the probability that a worker identified as having a professional qualification is a manager.

Job of worker	Proportion of workforce (%)	Qualified (Q) (%)	Not qualified (NQ) (%)
Skilled	60	6	94
Managerial	4	12	88
Administrative	36	8	92
Total	100		

6.6. How many different arrangements can be made of the letters of the words:

(a) STUDY?

(b) EXERCISES?

(*Hint.* Note that in situations where some objects are alike – as in (b) – the number of arrangements is reduced.)

6.7. All the employees of PLC company are assigned identity codes. The code consists of the first letter of the employee's last name followed by four digits.

(a) How many different identity codes are possible?

(b) How many different identity codes are there for employees whose last name starts with an A?

6.8. A committee of five people is to be selected from a group of six workers. How many different ways are there of selecting the committee?

6.9. A group of ten workers is to be divided into two teams of five each and allocated to two assembly lines. How many different ways are there of dividing the workers?

6.10. A combination lock has four dials each with ten digits. How many different four-figure codes are possible? Why is the term 'combination' lock a misnomer in this case?

6.11. The British Lions rugby squad consists of 25 players, eight from England, seven from Wales, six from Scotland and four from Ireland. A team of 15 players is to be selected for the next match.

(a) How many possible selections of 15 players can be made?

(b) How many selections of 15 players are possible if there is to be at least one player in the team from each of the four countries?

| Chapter 7 | Binomial probability distribution |

Aims

Determining probabilities associated with complex as well as simple or arbitrary events can be greatly simplified if we can construct a general, mathematical, model which accurately describes the situation associated with a particular event in which we are interested. Such a model, used to obtain the probabilities of certain events occurring, is called a *probability distribution* (or probability function). Probability distributions have defined mathematical properties which allow us to draw statistical inferences in a wide range of problems.

In this and the next two chapters we introduce three common probability distributions. These are:

- the binomial distribution
- the Poisson distribution
- the normal distribution

The first two of these are examples of *discrete* probability distributions while the last one is a *continuous* distribution. For a discrete variable (e.g. x) we can compute the probability of it taking on particular discrete values, e.g. $P(x = 5)$. Recall that discrete variables are usually associated with the counting of events while continuous variables can take on any value within a range of values along a continuous scale. Continuous variables are usually associated with measurements (such as heights, weights, etc.).

This chapter sets out the principles relating to the first of the three distributions, namely the binomial probability distribution, as follows:

- the binomial distribution formula
- tables of the binomial distribution
- the cumulative binomial distribution
- properties of the binomial distribution

Principles

The binomial distribution formula

The binomial distribution is applicable to situations in which only *two* outcomes are possible in any one *experiment* or *trial*. An experiment is any process which generates well-defined outcomes (*events*). Thus the binomial distribution applies to situations which generate 'yes' or 'no' type responses such as in quality inspections (defective or not defective), births (male or female), consumer purchase decisions (buy or not buy), etc. These situations are referred to as *binomial experiments* and the distributions of the results follow that of the *binomial distribution*. It is common practice to label the two possible outcomes in such situations as 'Success' (S) and 'Failure' (F), with the probability of obtaining the outcome 'Success' denoted by p and the outcome 'Failure' denoted by q. Thus, since the two outcomes are, by definition, mutually exclusive, we have

$$p + q = 1 \quad \text{or} \quad q = 1 - p$$

In a binomial experiment, we are interested in the *number of successes* (or failures) occurring in n independent trials (such as n items inspected on a production line etc.). If we let x represent the random variable denoting the number of successes occurring in n such trials, then x can take on any of the discrete values $0, 1, 2, \ldots, n$. The probabilities associated with each of the possible outcomes that x may take (i.e. $x = 1, 2, \ldots, n$) will have a frequency distribution, which is the binomial probability distribution.

In general, we can use the formula for the binomial distribution to calculate the probability of achieving x successes in n independent trials with the probabilities of success and failure at each trial equal to p and q, respectively. This is given by the following formula:

Binomial distribution formula

$P(x \text{ successes}) = {}^{n}C_{x}p^{x}q^{n-x}$

where:

$${}^{n}C_{x} = \frac{n!}{x!(n-x)!}$$

The relevance of this formula to binomial situations is readily explained in terms of a tree diagram. For example, consider a situation where a woman has three children. At each birth the probability of having a boy or a girl may be taken as equal. What is the probability of the mother ending up with two boys and a girl? The answer may be seen from the tree diagram given in Figure 7.1. It will be seen from this figure that there are three combinations of two boys and one girl (i.e. ${}^{n}C_{x} = {}^{3}C_{2} = 3!/2!1! = 3$). Also, the probability of any of the three combinations occurring is equal to $(\frac{1}{2})(\frac{1}{2})(\frac{1}{2})$ since the probability of B or G is $\frac{1}{2}$. This is equivalent to:

$$p^{x}q^{n-x} = \left(\frac{1}{2}\right)^{2}\left(\frac{1}{2}\right)^{3-2} = \left(\frac{1}{2}\right)\left(\frac{1}{2}\right)\left(\frac{1}{2}\right) = \frac{1}{8}$$

and as this can occur in one of three possible ways the total probability of two boys and a girl being born equals $3(\frac{1}{8}) = \frac{3}{8}$. This could have been found directly using the binomial

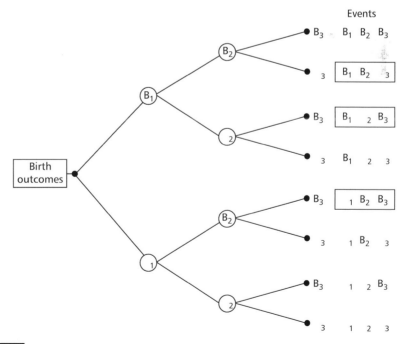

Figure 7.1 Tree diagram for boy and girl problem

distribution formula:

$$P(2B \text{ and } 1G) = {}^nC_x p^x q^{n-x} = 3\left(\frac{1}{2}\right)^2 \left(\frac{1}{2}\right)^{3-2} = \frac{3}{8}$$

where n denotes three births and x denotes 2 boys.

Tables of the binomial distribution

Given the formula for the binomial distribution, it is possible to calculate the probability of success or failure for any value of n, p and x and it is convenient to tabulate the results in order to avoid the need for repetitive calculation. One such table is given in Appendix A Table A1. As will be seen, this gives the individual binomial probabilities, $P(x \text{ successes})$, for various values of n from 1 to 20 and for various values of p from 0.05 to 0.5. For values of p exceeding 0.5, the value of $P(x \text{ successes})$ can be obtained by substituting $q = (1 - p)$ for p and finding $P(n - x)$. In addition, computer software packages, such as MINITAB and Excel, include routines for calculating the required probabilities (see the applications section below).

The cumulative binomial distribution

So far we have considered the use of the binomial formula to obtain *precisely* x successes (or, of course, $n - x$ failures). However, it often arises that we need to calculate the probability that the number of successes is *less than or equal to* (or greater than or equal to) some value, say k, i.e. we may wish to determine

$$P(x \leq k) \quad \text{or} \quad P(x \geq k)$$

The solutions to such problems merely involve summing the relevant binomial probabilities given in the binomial tables (Appendix A Table A1) over the appropriate range of the values for x. For example, with $n = 4$, $p = 0.3$, the probability of x being *less than or equal to* 2, i.e. $P(x \leq 2)$, is given by

$$P(x = 2) + P(x = 1) + P(x = 0) = 0.2646 + 0.4116 + 0.2401 = 0.9163$$

That is, a roughly 92 per cent chance. Alternatively, tables of *cumulative* binomial probabilities are available (but in the interests of brevity these are not included in this book). In addition, many statistical software packages, including MINITAB and Excel, also include appropriate routines for computing the probabilities (see below).

Properties of the binomial distribution

It is common to summarize a set of data using the mean and variance (or standard deviation) as shown in Chapters 3 and 4. A binomial distribution can be summarized in the same way, but in this case it can be shown that the two summary measures depend simply on the values of n and p as follows:

Mean $= np$

Variance $= npq$

Thus, by definition, the mean (np) of a binomial distribution represents the average number of 'successes' in n trials. For example, in all four-children families, the average (i.e. mean) number of boys would be $4 \times \frac{1}{2} = 2$ (assuming that the probability of a male birth is the same as a female birth, i.e. $\frac{1}{2}$).

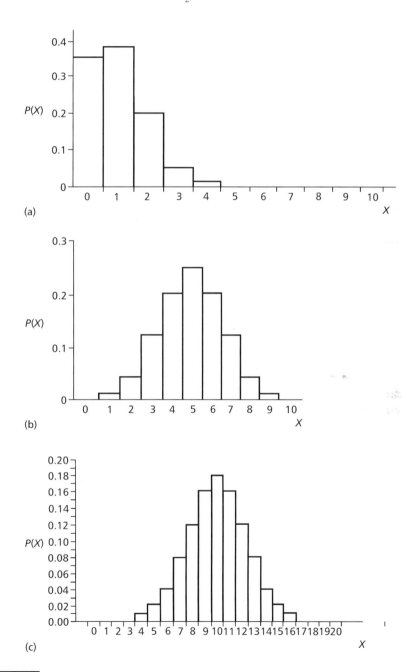

Figure 7.2 Binomial distributions for different values of n and p: (a) $n = 10$, $p = 0.1$; (b) $n = 10$, $p = 0.5$; (c) $n = 20$, $p = 0.5$

It should be appreciated that once we known the values of n and p, we can set up a frequency distribution table showing the values of $P(x)$ for all possible values of x. This table can be illustrated in the form of a histogram. Three histograms corresponding to three different binomial distributions for different combinations of n and p are shown in Figure 7.2. Note that since $p = 0.5$ in the case of parts (b) and (c) of the figure, the distributions are symmetrical, unlike that in part (a), where $p = 0.1$, which is skewed to the right.

Applications and Worked Examples   Worked example 7.1   Binomial distribution

General descriptions of the MINITAB and Excel computer programs and their application are given in Appendices B and C respectively.

If a large factory has a workforce of which 60 per cent are trade unionists, what is the probability that a random sample of ten workers includes three trade unionists?

Worked solution

The required probability is given by the binomial formula:

$$P(x = 3) = {}^nC_x(p^x)(q^{n-x}) = \frac{10!}{3!(10 - 3)!}\left(\frac{6}{10}\right)^3\left(\frac{4}{10}\right)^{10-3}$$

$$= 120(0.216)(0.00164)$$

$$= 0.0425$$

Thus there is a 4.25 per cent probability that the random sample includes three trade unionists. The reader should confirm this result from the binomial tables in Appendix A Table A1.

Computer solution using MINITAB

Choose **Calc → Probability Distributions → Binomial**

Select **Probability** (this calculates the pdf – probability density function)

In **Number of trials** enter $n (= 10$ in this case)

In **Probability of success** enter p ($= 0.6$ in this case)

In **Input column** enter the column to be evaluated. As in this case we wish to evaluate the probability of 3, the number 3 is input into a column of the worksheet, for example C1 and thus **C1** is entered

Click **OK**

The output is as follows:

Probability Density Function

Binomial with n = 10 and p = 0.600000

x	P(X = x)	
3.00	0.0425	$P(x = 3)$

Computer solution using Excel

Use the **BINOMDIST** function
Click = in the formula bar, select the **BINOMDIST** function and enter:
In **Number_s**, the number of successes in trials (3 in this case)
In **Trials**, the number of independent trials (10 in this case)
In **Probability_s**, the probability of success on each trial (0.6 in this case)
In **Cumulative**, enter **TRUE** for the cumulative distribution function or **FALSE** for the

probability mass function (False is entered in this case)
This returns the answer: 0.042467.

Authors' comment: The term *probability mass function* is the probability function describing the probability distribution of a *discrete* random variable, as opposed to the term *probability density function* which applies to a continuous random variable.

Worked example 7.2   Cumulative binomial distribution

Using the same information as in Worked example 7.1 above, calculate the probability that *no more than* two of the ten workers in the sample belongs to a trade union.

Worked solution

The probability of 'no more than 2' is equivalent to $P(x=0)+P(x=1)+P(x=2)$. From the binomial formula, this cumulative probability is given by

$P(x = \text{no more than } 2)$

$$= {}^{10}C_0\left(\frac{6}{10}\right)^0\left(\frac{4}{10}\right)^{10} + {}^{10}C_1\left(\frac{6}{10}\right)^1\left(\frac{4}{10}\right)^9$$

$$+ {}^{10}C_2\left(\frac{6}{10}\right)^2\left(\frac{4}{10}\right)^8$$

$$= 0.0001 + 0.0016 + 0.0106$$

$$= 0.0123$$

There is thus a 1.23 per cent chance of no more than two of the ten workers belonging to a union.

Computer solution using MINITAB

Choose **Calc → Probability Distributions → Binomial**
Select **Cumulative Probability** (this calculates the cdf – cumulative density function)
Enter n (=10 in this case) in the **Number of trials** box
Enter p (=0.6 in this case) in the **Probability of success** box
Complete **Input column** box, i.e. the column to be evaluated (in this case we wish to evaluate the probability of no more than 2, thus the number 2 is input to a column of

the worksheet, for example, C1 and thus **C1** is entered in the box)
Click **OK**

The output is as follows:

Cumulative Distribution Function

Binomial with n = 10 and p = 0.600000

x	P(X <= x)
2.00	0.123

← $P(x=\text{no more than } 2)$

Worked example 7.2   Cumulative binomial distribution *(continued)*

Computer solution using Excel

E

The solution here involves an identical procedure to that described in Excel Worked Example 7.1 above except that in **Cumulative**, enter **TRUE** (in order to obtain the cumulative distribution function).

This returns the answer: 0.012295.

Worked example 7.3   Properties of the binomial distribution

A government inspector responsible for checking cars for defective tyres finds a defective tyre in one out of every five cars examined. In a normal working week 80 cars are examined on average.

(a) What is the average number of cars per week found to have a defective tyre?

(b) What is the standard deviation of the number of cars per week with a defective tyre?

Solution

Using the properties of the binomial distribution:

$$\text{Mean} = np = 80 \times 0.2 = 16 \text{ cars}$$
$$\text{Standard deviation} = \sqrt{(\text{variance})} = \sqrt{(npq)}$$
$$= \sqrt{(80 \times 0.2 \times 0.8)}$$
$$= 3.58 \text{ cars (i.e. in round terms, 4 cars).}$$

Key terms and concepts

Binomial experiment A series of identical random trials in which each trial has only two possible mutually exclusive and complementary outcomes and in which all trial outcomes are statistically independent of one another.

Binomial probability distribution A probability distribution showing the probability of X successes in n trials of a binomial experiment:

$$P(X \text{ successes}) = {}^{n}C_{x}p^{x}q^{n-x}$$
$$\text{where } {}^{n}C_{x} = \frac{n!}{x!(n-x)!}$$

Continuous random variable A quantitative random variable that may be measured on a continuous scale and can take any value within an interval.

Discrete random variable A quantitative random variable in which the scale of measurement varies only in discrete steps – typically associated with counts.

Event The outcome of an experiment.

Experiment Any statistical process which generates well-defined outcomes (events).

Probability distribution An ordered listing of all possible values of a random variable and their associated probabilities.

Properties of the binomial distribution

Mean $= np$

Variance $= npq$

Random variable A numerical description of the outcome of an experiment (an *event*).

Trial A random experiment which results in a random outcome.

Self-study exercises

7.1. A fair coin is tossed ten times. What is the probability of obtaining:

(a) four heads?

(b) less than four heads?

7.2. Forty per cent of workers in a company belong to a trade union. If four workers are chosen to be members of a committee what is the probability that:

(a) none are members of a trade union?

(b) at least three are members of a trade union?

7.3. If the average number of rejects in the manufacture of a certain article is 5 per cent what are the probabilities of 0, 1, 2, 3, 4 rejects in a sample of ten articles taken at random? Confirm your answers from the binomial probability distribution table (Appendix A Table A1).

7.4. Assuming that one in 80 births is a case of twins, calculate the probability of two or more sets of twins in a certain town on a day when 30 births occur.

7.5. Suppose a garage examining cars for defective tyres finds a defective tyre in one out of every five cars examined. In a normal working week 80 cars are examined on average.

(a) What is the average number of cars per week found to have a defective tyre?

(b) What is the standard deviation of the number of cars per week with a defective tyre?

7.6. A random variable x is binomially distributed with a mean of 14.96 and a variance of 9.84. Find n and p.

7.7. A second-hand car dealer guarantees that he will repair, free-of-charge, any car that proves to be defective within one year of purchase. Past experience suggests that 10 per cent of cars have to be repaired under guarantee. What is the probability that from a random sample of 25 cars the number requiring repair under guarantee will be:

(a) none?

(b) one?

    (c) more than two?

    (d) If the average cost of a repair is £54, calculate the expected repair bill for this sample of 25 cars.

7.8. The probability of a person winning a game in an amusement arcade is always 0.2. How many people need to play the game to ensure that the probability that at least one person wins is 0.90?

Chapter 8 | Poisson probability distribution

Aims   The aim of this chapter is to introduce another type of discrete probability distribution called the *Poisson distribution*. This is concerned with the number of randomly occurring events per unit of time or space. It was originally derived as a *limiting form* of the binomial distribution in the sense that it acts as a good approximation for calculating binomial probabilities when the number of trials n is very large (tends towards infinity) and when the probability of success at each trial, p, is very small (tends towards zero). Calculation of probabilities in such situations using the binomial formula is extremely tedious. The use of the Poisson distribution, however, greatly facilitates computation as we shall see below.

The Poisson distribution is important in its own right, however, as a 'model' for random events and has a wide range of applications. Its main use is in problems dealing with rare events which occur independently and randomly in continuous situations such as space or time. A classic example, noted by the discoverer of the distribution in the last century, was the number of Prussian soldiers kicked to death by horses in the course of a year. These deaths were found to follow as a Poisson probability distribution. Nowadays, it is found that many applications arise in situations involving waiting time or queuing such as the number of customers passing through a supermarket checkout per hour, the number of cars arriving at a car park per day, the number of cash withdrawals from a cash dispenser per hour. These problems involve a variable generated over time. Other situations involve *space* (defined by length, area or volume) such as the number of leaks along an oil pipeline, the number of defects in the production of a particular area of plastic sheeting, the number of fish in a given volume of ocean, etc. The attention in all these situations is on the occurrence of relatively rare events because from a practical viewpoint it is easier to observe the (relatively small) number of times such events occur than it is to observe the number of times such events do not occur.

It is worth noting, therefore, that unlike the binomial distribution in which we are interested in observing *both* 'success' and 'failure', the application of the Poisson distribution is concerned only with occurrences rather than non-occurrences. For example, we can focus on the number of defects in the production of a plastic sheet but not on the number of non-defects.

In this chapter we set out the principles relating to the Poisson distribution as follows:

- the Poisson distribution formula
- tables of the Poisson distribution
- the cumulative Poisson distribution
- properties of the Poisson distribution

Learning Outcomes   After working through this chapter you will be able to:

- understand the meaning and derivation of the Poisson distribution as a limiting form of the binomial distribution
- use the Poisson distribution formula for deriving individual and cumulative probabilities
- employ the tables of the Poisson distribution to calculate individual and cumulative probabilities
- summarize the properties of a Poisson distribution in terms of its mean and variance.

Principles

The Poisson distribution formula

The Poisson distribution is appropriate only in certain circumstances determined by the following conditions:

- the random events must take place in a unit of time or space
- the number of events which might occur in any given unit of time or space must be theoretically unlimited
- the probability of occurrence in any single unit of time or space is independent of occurrence in any other unit of time or space

In general, if x is the random variable representing the number of rare events under investigation, then the probability of these x events occurring in some specified interval of time or space is given by the formula for the Poisson distribution, defined as:

Poisson distribution formula

$$P(x) = \frac{e^{-\mu}\mu^x}{x!}$$

where $P(x)$ is the probability of x occurrences in an interval of time or space, μ is the average number (or expected value) of occurrences of x in the specified interval and $e = 2.7183$ (a mathematical constant).

Before we consider the application of this formula to particular problems, it is worth noting that the value of x has no upper limit (i.e. it can occur an infinite number of times); thus, x can equal 0, 1, 2, etc. up to infinity.

Tables of the Poisson distribution

For relatively small values of μ and x, the above formula can be evaluated quickly using a pocket calculator or personal computer. For larger values, the problem of calculation becomes tedious. Consequently, sets of Poisson probability distribution tables are available which give the probability of x occurrences for selected values of μ. Appendix A Table A2 gives such a table for values of μ ranging from 0.005 to 8.0. Alternatively, computer software packages, such as MINITAB, include routines for calculating the probability values (see the applications section below).

The cumulative Poisson distribution

Just as we can derive a *cumulative* binomial probability distribution (as shown in Chapter 7), so too we can derive the *cumulative Poisson distribution* for calculating the probability that a random variable x is less than or equal to some value k. For example,

$$P(x \leq 2) = P(x = 0) + P(x = 1) + P(x = 2)$$

Each of the individual probabilities on the right-hand side can be computed separately using the Poisson formula or read directly from the Poisson probability tables in Appendix A Table A2. The individual probabilities can then be summed to give $P(x \leq 2)$. Tables are also readily available which give such cumulative Poisson probabilities (but for brevity these are not reproduced in this book). As with the cumulative binomial distribution, cumulative values for the Poisson distribution can be obtained directly using computer software packages such as MINITAB and Excel (see below).

Properties of the Poisson distribution

Just as with the binomial distribution, we can define the values of the summary measures – mean and variance – for the Poisson distribution. In this case, however, both these summary measures are equal to μ. Thus

Mean $= \mu$
Variance $= \mu$

Hence, once we know the value of μ, the Poisson distribution of x for values of x equal to or greater than zero can be illustrated in the form of a histogram (just as we did in Chapter 7 for the binomial distribution). Three examples are shown in Figure 8.1 corresponding to situations where μ equals 0.5, 1 and 4. Note that, as shown in this figure, the Poisson

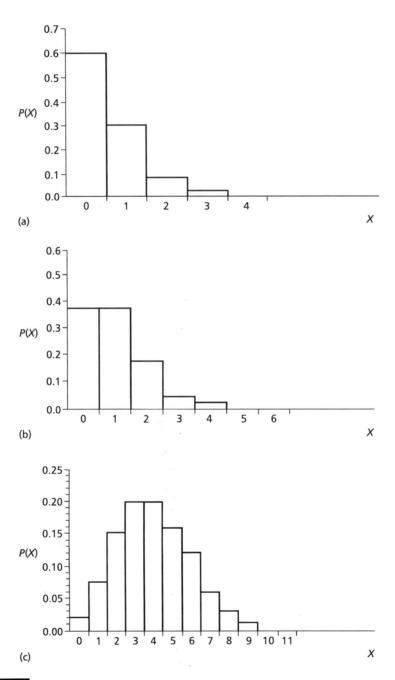

Figure 8.1 Poisson distributions for different values of μ:
(a) $\mu = 0.5$; (b) $\mu = 1$; (c) $\mu = 4$

distribution is always skewed to the right, but the degree of skewness decreases as μ increases.

The Poisson approximation to the binomial

Under certain conditions the Poisson distribution provides a close approximation to the binomial. This is the case when the number of trials, n, is large and the probability of success, p, is small. As a rule of thumb, the approximation is acceptable when $n \geq 20$ and $p \leq 0.05$. In these circumstances, the mean of the binomial distribution (np) and the variance (npq) are approximately equal. Because tables of the binomial distribution are often not available for values of n over 20, the Poisson approximation is especially useful in these circumstances.

| **Applications and Worked Examples** | **Worked example 8.1 Poisson distribution** |

General descriptions of the MINITAB and Excel computer programs and their application are given in Appendices B and C respectively.

In a study of activity at a car wash it is found that the average number of cars arriving between 8 a.m. and 9 a.m. on Monday morning is five. What is the probability that on any *particular* Monday morning five cars will arrive between 8 a.m. and 9 a.m.?

Worked solution

The Poisson distribution is appropriate here because:

(a) the problem involves independent events occurring in a unit of time and
(b) the total number of events is (theoretically) unlimited.

If we let x denote the number of cars arriving at the car wash between 8 a.m. and 9 a.m. on Monday morning, then

$$P(x = 5) = \frac{e^{-\mu}\mu^x}{x!} = \frac{(2.7183^{-5})5^5}{5!} = 0.1755$$

Thus there is just over a 17.5 per cent chance of five cars arriving between 8 a.m. and 9 a.m. on a particular Monday morning. This result may be confirmed using the Poisson probability tables in Appendix A Table A2.

Computer solution using MINITAB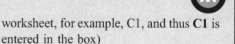

Choose **Calc → Probability Distributions → Poisson**

Select **Probability** (this calculates the pdf – probability density function)

In the **Mean** box enter value of the mean (= 5 in this case)

Complete **Input column** box, i.e. the column to be evaluated (in this case we wish to evaluate the probability of 5, thus the number 5 is input into a column of the worksheet, for example, C1, and thus **C1** is entered in the box)

Click **OK**

The output is as follows:

Probability Density Function

Poisson with mu = 5.00000

x	P(X = x)
5.00	0.1755	← $P(x = 5)$

Worked example 8.1   Poisson distribution *(continued)*

Computer solution using Excel

Use the **POISSON** function

Click = in the formula bar, select the **POISSON** function and enter:

In **X**, enter the number of events (5 in this case)

In **Mean**, enter the value of the mean (5 in this case)

In **Cumulative**, enter **TRUE** for the cumulative Poisson probability or **FALSE** for the Poisson probability mass function – **FALSE** is entered in this case.

This returns the answer: 0.1755.

Authors' comment: The term *probability mass function* is the probability function describing the probability distribution of a *discrete* random variable, as opposed to the term *probability density function* which applies to a continuous random variable.

Worked example 8.2   Cumulative Poisson distribution

Using the same data as in Worked example 8.1, what is the probability that fewer than three cars will arrive between 8 a.m. and 9 a.m. on a particular Monday?

Worked solution

The required probability may be expressed as $P(x < 3)$. This is, of course, the sum of the individual probabilities that 0 or 1 or 2 cars will arrive, i.e. $P(x \leq 2)$. This is calculated as

$$P(x < 3) = P(x = 0) + P(x = 1) + P(x = 2)$$

From Appendix A Table A2 or the computer printout in Worked example 8.1, the required probabilities are:

$$0.0067 + 0.0337 + 0.0842 = 0.1246$$

Thus there is a chance of just over 12 per cent that fewer than three cars will arrive between 8 a.m. and 9 a.m. on a particular Monday.

Computer solutions

Instead of using the probability density function (pdf) shown in the computer solution to Worked example 8.1 to find $P(x = 0)$, $P(x = 1)$ $P(x = 2)$, we can use instead a cumulative (probability) density function (cdf) to find $P(x < 3)$ directly. This is shown in the printouts below.

Computer solution using MINITAB

Choose **Calc → Probability Distributions → Poisson**

Select **Cumulative Probability** (this calculates the cdf – cumulative density function)

Enter value of mean (= 5 in this case) in the **Mean** box

Complete **Input column** box, i.e. the column to be evaluated (in this case we wish to evaluate the probability of fewer than 3 (i.e. 2 or less), thus the number 2 is input into a column of the worksheet, for example C1, and thus **C1** is entered in the box)

Click **OK**

The output is as follows:

Cumulative Distribution Function

Poisson with mu = 5.00000

x	P(X <= x)	
2.00	0.1247	← $P(x \le 2)$

Computer solution using Excel

Use the **POISSON** function
Click = in the formula bar, select the **POISSON** function and enter:
In **X** the number of events (5 in this case)
In **Mean** the value of the mean (5 in this case)
In **Cumulative** enter **TRUE** for the cumulative Poisson probability or **FALSE** for the

Poisson probability mass function – **FALSE** is entered in this case
This returns the answer: 0.124652

Note that the computer solution of 0.1247 differs slightly from the worked solution above (0.1246). This is due to rounding.

Worked example 8.3 Properties of the Poisson distribution

With respect to the examples above, what are the mean and standard deviation of the probability distribution?

mean $= \mu = 5$

standard deviation $= \sqrt{(\text{variance})} = \sqrt{\mu}$
$= \sqrt{5} = 2.236.$

Solution

From the properties of the Poisson distribution we get:

Key terms and concepts

Poisson probability distribution A probability distribution showing the probability of x occurrences of an event over a specified interval of time or space.

$$P(x) = \frac{e^{-\mu}\mu^x}{x!}$$

Properties of the Poisson distribution

Mean $= \mu$

Variance $= \mu$

Self-study exercises

8.1. The average number of errors on a page of a book is 1.4. What is the probability that a page will contain four errors?

8.2. A supplier of pharmaceutical products supplies chemists in a town with boxes of pills on demand. The number of boxes supplied each week follows a Poisson distribution with a mean of 0.6.

    (a) What is the probability that two or more boxes are requested in any one week?

    (b) What is the minimum level of stock that should be held by the supplier to ensure, with at least 90 per cent confidence, that all demands can be met?

8.3. During peak traffic periods, accidents occur on a certain stretch of road at the rate of two per hour. The morning peak lasts for one and a half hours and the evening peak lasts for two hours.

    (a) On any one day what is the probability that there will be no accidents during the morning peak period?

    (b) What is the probability of two accidents during the evening peak period?

    (c) What is the probability of four or more accidents during the morning peak period?

    (d) On any one day what is the probability that there will be no accidents at all?

8.4. A car hire firm has two cars which it hires out day by day. The number of demands for a car on each day is distributed as a Poisson distribution with a mean of 1.5.

    (a) Calculate the proportion of days on which neither car is used.

    (b) Calculate the proportion of days on which some demand is refused.

    (c) If each car is used an equal amount, on what proportion of days is only one of the cars not in use?

    (d) What proportion of demands has to be refused?

8.5. It is known that a certain drug sometimes causes serious side-effects but that they are rare. The probability that a patient will suffer from these side-effects is put at 0.002. If the drug is administered to 3000 patients, what is the probability that three patients will suffer from these side-effects?

8.6. A manufacturer makes a component and finds that on average 4 per cent of the components are defective. Components are packed in cartons of 40. What is the probability that the carton contains:

    (a) at least one defective?

    (b) exactly two defectives?

8.7. On average a certain event happens once in 40 trials. What is the probability of three successes in the next 20 trials?

8.8. Sheet steel is produced in 500 metre rolls, each of which includes two flaws on average. What is the probability that a particular 100 metre segment will include no flaws?

Chapter 9

Normal probability distribution

Aims

In this chapter we introduce a probability distribution which relates to continuous variables (i.e. measurements as opposed to counts) called the *normal distribution*. This distribution is important not only in its own right but because it also represents the foundation of statistical inference. It has been shown that many natural phenomena conform to the normal distribution, for example, the heights and weights of people, errors in the reading of instruments, the lives of products such as television tubes and deviations of measurements around established norms. Statisticians have also been able to demonstrate that several important *sample* statistics (such as the mean) tend towards a normal distribution as the sample size increases. This remains the case even if the *population* from which samples have been drawn is *not* normal. Even in the case of the discrete binomial distribution, as we allow the number of trials, n, to increase indefinitely, the distribution will tend towards a normal distribution. For this reason, the normal distribution is often used as an approximation to the binomial.

In this chapter we set out the principles relating to the normal distribution as follows:

- the normal distribution formula
- properties of the normal distribution
- the standard normal distribution table
- normal approximation to the binomial distribution

| Learning Outcomes | After working through this chapter you will be able to: |

- appreciate the conditions which give rise to a normal distribution
- understand the meaning of the normal distribution formula
- know what is meant by a probability density function and cumulative density function
- summarize the properties of a normal distribution in terms of its mean and variance
- understand the meaning, interpretation and use of the standard normal variable (Z)
- use the standard normal tables for deriving individual and cumulative probabilities
- appreciate the normal distribution as an approximation to the binomial distribution
- apply the continuity correction factor when using the normal distribution as an approximation to the binomial.

Principles

The normal distribution formula

The formula which defines the normal distribution is complex. We state the expression here but, fortunately, it is never necessary to use it in practice because it is possible to solve all relevant probability problems using a *single* table of probabilities for the so-called *standard normal distribution* (see below). The formula for the normal distribution is

The normal distribution formula

$$f(x) = \frac{1}{\sqrt{(2\pi\sigma^2)}} e^{-\frac{1}{2}[(x-\mu)/\sigma]^2}$$

where $f(x)$ is the probability density function (pdf) which determines the shape of the normally distributed variable x; μ is the mean of x; σ is the standard deviation of x; $\pi = 3.1416$ (a mathematical constant); and e = 2.7183 (a mathematical constant).

The normal distribution formula, calculated for all values of x, produces a smooth, 'bell-shaped' curve as shown in Figure 9.1. The precise location and shape of this curve depends only on the value of the two parameters μ and σ. It will be seen that the curve is unimodal and symmetrical around the mean μ. Note also that the tails of the curve are *asymptotic* to the horizontal axis; i.e. they approach but never touch the axis. Therefore, in theory, x can assume infinitely large positive or negative values. In practice, of course, we do not deal with infinity and hence the distribution is used to approximate the actual distribution. For all practical purposes, this makes little or no difference.

Since the shape of any particular normal curve is dependent on the values of the parameters μ and σ for a particular set of x values, there will be a whole family of normal distributions and curves, depending on the particular values which these two parameters take. Three examples are shown in Figure 9.2 where the distributions A and B have different means ($\mu = 5$ and 15 respectively) but the same standard deviation ($\sigma = 4$). Distribution C has the same mean as B but a smaller standard deviation ($\sigma = 2$). It will be seen that different values for μ and σ give squashed or stretched versions of the same bell shape.

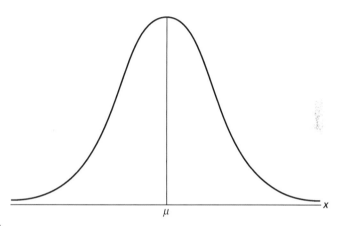

Figure 9.1 The normal distribution

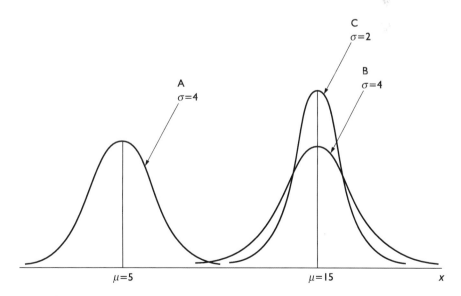

Figure 9.2 Normal distributions: different means and standard deviations

Properties of the normal distribution

The properties of the normal distribution are such that the proportion of all the observations of a normally distributed variable x that fall within a range of n standard deviations on both sides of the mean is the same for *any* normal distribution, as shown in Figure 9.3.

The properties of the distribution are such that:

■ 68.26 per cent of all x values fall within the range of *one* standard deviation on both sides of the mean (i.e. $\mu - 1\sigma$ to $\mu + 1\sigma$)

■ 95.44 per cent of all x values fall within the range of *two* standard deviations on both sides of the mean (i.e. $\mu - 2\sigma$ to $\mu + 2\sigma$)

■ 99.73 per cent of all x values fall within the range of *three* standard deviations on both sides of the mean (i.e. $\mu - 3\sigma$ to $\mu + 3\sigma$)

Thus, for a variable x which is normally distributed with a mean μ of, say, 15 and a standard deviation σ of, say, 4, one standard deviation on either side of the mean gives the range 15 ± 4 i.e. from 11 to 19 and we can say, therefore, that the probability of any individual value of x, selected at random, falling within the range from 11 to 19 is roughly 68 per cent. This can be expressed simply in notation form as $P[(15 - 4) \leq x \leq (15 + 4)] = 0.6826$. This is illustrated in Figure 9.4. Note that the required probability is shown as the shaded area under the normal distribution curve. Similarly, for ± 2 standard deviations and ± 3 standard deviations around the mean we have:

$$P[(15 - 8) \leq x \leq (15 + 8)] = 0.9544$$
$$P[(15 - 12) \leq x \leq (15 + 12)] = 0.9973$$

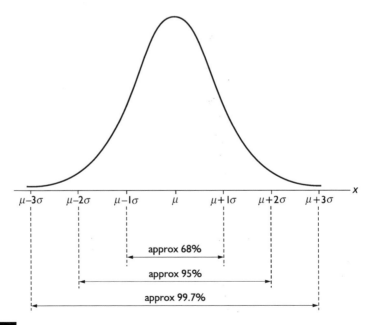

Figure 9.3 Properties of the normal distribution

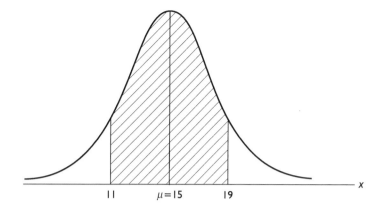

Figure 9.4 Area under the normal distribution curve: shaded area represents $P[(15 - 4) \leq x \leq (15 + 4)] = 0.6826$ (i.e. 68.26 per cent)

Thus, in general notation form,

$$P[(\mu - 1\sigma) \leq x \leq (\mu + 1\sigma)] = 0.6826$$
$$P[(\mu - 2\sigma) \leq x \leq (\mu + 2\sigma)] = 0.9544$$
$$P[(\mu - 3\sigma) \leq x \leq (\mu + 3\sigma)] = 0.9973$$

A key point to note is that the total area under any normal curve denotes the total probability and thus has the value of 1 (or 100 per cent). Areas under a normal curve represent the *cumulative density function* (cdf) and give the cumulative probability associated with a pdf – that is, a cdf gives the area under the pdf up to a specified value. It will be noted that the range of $\mu \pm 3\sigma$ encompasses virtually all possible values of x, i.e. 99.73 per cent of all observations. We have referred here, by way of example, to only three situations where x falls within ranges defined by one, two or three standard deviations precisely around the mean. The associated probability values are available in a standard table which also gives probabilities for all intermediate ranges. It will be appreciated that there can be an infinite number of normal distributions with different means and standard deviations, but fortunately it is possible to accommodate all of these in a single table known as the *standard normal distribution table*. The derivation of this, and the use of the table, are explained below.

The standard normal distribution table

Suppose a variable x is normally distributed with a given value of mean μ and standard deviation σ and we wish to calculate the probability of a particular value of the variable x falling within the range x_1 to x_2. This is equivalent to finding the area under the normal curve for the variable x between the specific values x_1 and x_2. We can read this area off directly from the *standard normal table*. Before showing how to use this table we must first discuss its derivation.

The standard normal table is based on the fact that any normal distribution can be transformed into one *standard* form. To do this, each value of the variable x is expressed as a deviation from its mean μ and then divided by its standard deviation σ, i.e.

$$\frac{x - \mu}{\sigma}$$

This transforms the variable x into what is referred to as a *standard normal variable* denoted by the letter Z, i.e.:

Standard normal variable (Z)

$$Z = \frac{x - \mu}{\sigma}$$

where x is any observation of the variable x, μ is the population mean of the variable x and σ is the population standard deviation of the variable x.

It further follows that Z must have a normal distribution with a mean of 0 ($\mu = 0$) and standard deviation of 1 ($\sigma = 1$), i.e. *a standard normal distribution*. These properties are illustrated in Figure 9.5 in which particular x values are transformed into Z (standard normal) values. As the distribution of Z is unique (in having a precise mean and standard deviation of 0 and 1 respectively), and virtually the entire distribution is encompassed

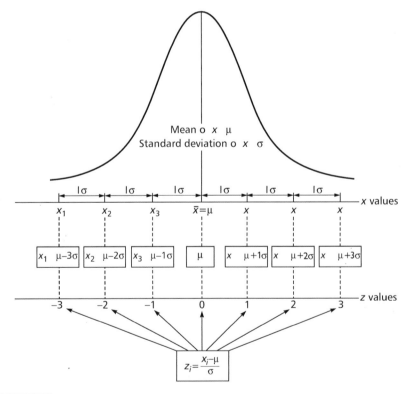

Figure 9.5 Standard normal distribution

within the range $\mu \pm 3\sigma$, a table corresponding to areas under the standard normal curve is very short. This table is given in Appendix A Table A3. Entries in the table are the areas which correspond to the probability of Z lying between 0 (the mean of Z) and one particular value, Z_1; i.e. $P(0 \leq Z \leq Z_1)$, as shown in Figure 9.6. The total area under the standard normal curve is equal to 1.0 (as noted earlier). So, for any particular value of $x = x_1$, we know that $P(0 \leq Z \leq Z_1)$ is equivalent to

$$P\left(0 \leq \frac{x - \mu}{\sigma} \leq \frac{x_1 - \mu}{\sigma}\right)$$

The shaded area in Figure 9.6 represents the probability that x will be within Z_1 standard deviations of its mean, or, given its standard deviation σ, that x will be x_1 and its mean μ. If, for example, Z is 1.96, the appropriate area found from the Appendix A Table A3 is 0.475, i.e. 47.5 per cent of the total area of 1.0. This means that the probability of a randomly selected x value falling within the range from μ to $\mu + 1.96\sigma$ is 0.475 (or 47.5 per cent). Note that the standard normal table in Appendix A Table A3 only gives values for half of the distribution (since it is perfectly symmetrical around the mean). Thus, it is also the case that the probability of x falling between μ and $\mu - 1.96\sigma$ is also 0.475. Hence, we can state that:

$$P[(\mu - 1.96\sigma) \leq x \leq (\mu + 1.96\sigma)] = 0.475 + 0.475 = 0.95$$

or, equivalently,

$$P[-1.96 \leq Z \leq +1.96] = 0.475 + 0.475 = 0.95$$

This is shown diagrammatically in Figure 9.7 as the sum of the two areas A and B. Naturally the standard normal table in Appendix A Table A3 gives precise probabilities and hence we now see that a probability of 95 per cent corresponds only approximately to $\mu \pm 2\sigma$ (the exact probability corresponding to $\mu \pm 2\sigma$ is 95.44 per cent as stated earlier).

Alternatively, we may wish to focus on the tail areas of the distribution, rather than areas in the body of the table, as above. It follows, therefore, that:

$$P(x > \mu + 1.96\sigma) = 0.025 = P(x < \mu - 1.96\sigma)$$

This is equivalent to

$$P(Z > +1.96) = 0.025 = P(Z < -1.96)$$

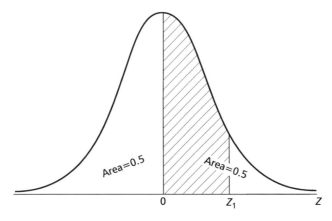

Figure 9.6 Area under the standard normal curve: shaded area $= P(0 \leq Z \leq Z_1)$

These probabilities and associated areas (denoted C) are also shown in Figure 9.7. Some textbooks include standard normal probability tables which show only the tail area rather than the area in the main body of the curve (as in Appendix A Table A3).

Use of the standard normal table

The table given in Appendix A Table A3 gives values of Z to two decimal places. The left-hand column of the table gives the first decimal place of Z while the top row gives the second decimal place. The entries in the body of the table give the probabilities that the standardized variable falls within the range between 0 and a particular value of Z (such as Z_1). Thus, for example, for $Z = 1.96$ we find, moving down the left-hand column to 1.9 and across to the column headed 0.06, the value of 0.4750. This gives the area under the curve for Z between 0 and 1.96. The table only gives values for half of the curve, the other half being obtained by symmetry. Thus the area of 0.4750 applies to $Z = +1.96$ and also to $Z = -1.96$.

Normal approximation to the binomial distribution

We noted earlier that under certain conditions the normal distribution provides a good approximation to the binomial distribution despite the fact that the latter is a discrete distribution while the former is a continuous one. There is a general rule of thumb to express these conditions as follows:

Rule of thumb for normal approximation to the binomial distribution

$$np \geq 5 \qquad \text{and} \qquad nq \geq 5$$

where n is the sample size (number of trials), p is the probability of 'success' and q is the probability of 'failure' $(= 1 - p)$.

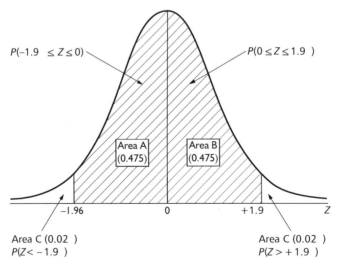

Figure 9.7 Area under the standard normal curve for $Z = \pm 1.96$

Consider a binomial case for the variable x and where $n = 20$ and $p = 0.4$. Then:

mean of $x = np = 20(0.4) = 8$

standard deviation of $x = \sqrt{(npq)} = \sqrt{[20(0.4)(0.6)]} = 2.19$

Using the binomial formula, we can calculate $P(x = x_i)$ for all values of x_i from 1 to 20. The distribution of these probabilities can be graphed as a histogram, as shown in Figure 9.8. Note, however, that in this case the normal approximation to the binomial can be applied because it satisfies the conditions set out above, namely $np = 20(0.4) = 8 > 5$ and $nq = 20(0.6) = 12 > 5$. A normal distribution with a mean of μ of 8 and a standard deviation σ of 2.19 is superimposed on the histogram in Figure 9.8. It will be seen that the fit between the normal curve and the binomial histogram is a good one.

Continuity correction factor

In applying the normal approximation it should be noted that a continuity correction factor needs to be applied. This arises because binomial variables are discrete while normally distributed variables are continuous. For example, the value of 12 in the case of a binomial variable is equivalent to an interval of 11.5–12.5 under the normal curve. Hence, if x is a discrete variable, we need to set up some correspondence between discrete values of x and the associated intervals along a continuous scale. Thus, when estimating $P(x \geq 12)$ using the normal approximation, this corresponds to the area covered by the intervals of 11.5–12.5, 12.5–13.5, 13.5–14.5, etc. This is equivalent to finding $P(x \geq 11.5)$ under the normal curve (see Worked example 9.3). The half distance added to or subtracted from the discrete value is called the *continuity correction factor*.

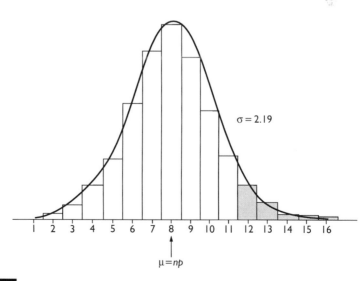

Figure 9.8 Normal approximation to the binomial

Applications and Worked Examples | **Worked example 9.1 Use of the standard normal (Z) table**

General descriptions of the MINITAB and Excel computer programs and their application are given in Appendices B and C respectively.

A random variable x, which is normally distributed, has a mean of 20 and a standard deviation of 5. Find the probability that a randomly selected observation of x will have a value:

(i) between 20 and 26

(ii) between 22 and 24

(iii) between 10 and 18

(iv) equal to 15 or more

(v) equal to 25 or less

(vi) more than 25

Worked solutions

The problems are expressed below in terms of x and Z along with the required probabilities shown as dark-shaded and hatched areas under the normal curve. These probabilities are then found with reference to the standard normal distribution table in Appendix A Table A3.

(i) $P(20 < x < 26)$

$$= P\left(\frac{20 - 20}{5} < Z < \frac{26 - 20}{5}\right)$$

$$= P(0 < Z < 1.2)$$

$$= 0.3849 \text{ (i.e. 38.49 per cent).}$$

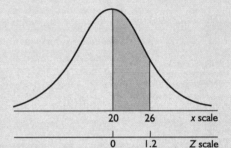

(ii) $P(22 < x < 24)$

$$= P\left(\frac{22 - 20}{5} < Z < \frac{24 - 20}{5}\right)$$

$$= P(0.4 < Z < 0.8).$$

The relevant area is the dark-shaded area which is obtained by subtracting $P(0.4 < Z)$ (the hatched area) from $P(0.8 < Z)$ (the total hatched and dark-shaded area).

$P(0.4 < Z < 0.8)$

$$= 0.2881 - 0.1554$$

$$= 0.1327 \text{ (i.e. 13.27 per cent).}$$

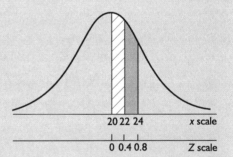

(iii) $P(10 < x < 18)$

$$= P\left(\frac{10 - 20}{5} < Z < \frac{18 - 20}{5}\right)$$

$$= P(-2.0 < Z < -0.4)$$

$$= 0.4772 - 0.1554$$

$$= 0.3218 \text{ (i.e. 32.18 per cent).}$$

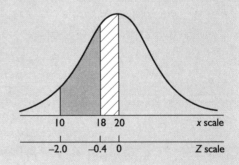

(iv) $P(x \geq 15)$

$$= P\left(Z \geq \frac{15 - 20}{5}\right)$$
$$= P(Z \geq -1)$$
$$= 0.3413 + 0.5$$
$$= 0.8413 \text{ (i.e. 84.13 per cent)}.$$

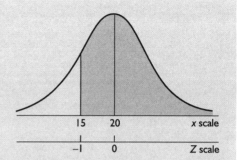

(v) $P(x \leq 25)$

$$= P\left(Z \leq \frac{25 - 20}{5}\right)$$
$$= P(Z \leq 1)$$
$$= 0.5 + 0.3413$$
$$= 0.8413 \text{ (i.e. 84.13 per cent)}.$$

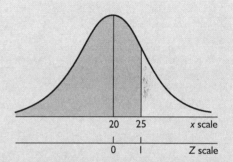

(vi) $P(x > 25)$

$$= 1 - P(x \leq 25)$$
$$= 1 - 0.8413$$
$$= 0.1587 \text{ (i.e. 15.87 per cent)}.$$

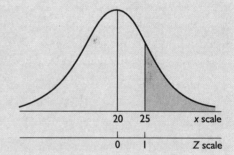

Computer solution using MINITAB

Ⓜ

Choose **Calc → Probability Distributions → Normal**
Check the **Cumulative probability** box
Enter the value of the mean ($\mu = 20$)

Enter the value of the standard deviation ($\sigma = 5$)
Check the **Input column** option and enter **C1**
Click **OK**

Worked example 9.1 Use of the standard normal (Z) table *(continued)*

The output is as follows:

Worksheet Output

x	Cumulative Distribution Function	
20		
26	Normal with mean = 20.0000 and	
22	standard deviation = 5.00000	
24	x	P(X <= x)
18	20.0000	0.5000
10	26.0000	0.8849
15	22.0000	0.6554
25	24.0000	0.7881
	18.0000	0.3446
	10.0000	0.0228
	15.0000	0.1587
	25.0000	0.8413

Authors' comment: The results in column 2 give total areas under the normal curve to the left of the x values specified, as shown in the diagram below, which permit the calculation of the required answers, as in the manual example above.

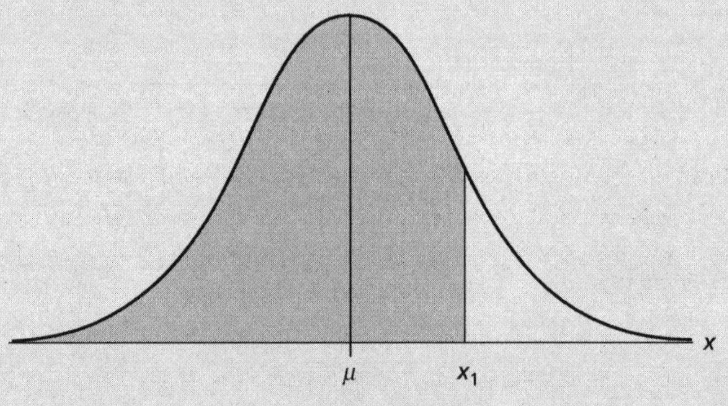

Computer solution using Excel

E

Use the **NORMDIST** function

Click = in the formula bar, select the **NORMDIST** function and enter:

In **X** the value for which the distribution is required (in this case, with the sequence of values: 20, 26, 22, 24, 18, 10, 15, 25, entered in column **A** of the worksheet, we enter **A1**)

In **Mean** the arithmetic mean of the distribution (20 in this case)

In **Standard dev** the standard deviation of the distribution (5 in this case)

In **Cumulative TRUE** for the cumulative distribution function or **FALSE** for the probability mass function (probability density function) **TRUE** is entered in this case

This returns, initially, the answer for $x = 20$ (the value in cell A1) – i.e. 0.5. Dragging the **fill handle**, on the cell in which the first result (0.5) is entered, down the column automatically inputs the probabilities for the remaining values of x, as shown below.

This permits the calculation of the required answers to the questions.

x	$P(X \leq x)$
20	0.5
26	0.88493
22	0.655422
24	0.788145
18	0.344578
10	0.02275
15	0.158655
25	0.841345

The required solutions to the problems can be readily obtained from either of these computer printouts as follows:

(i) $P(20 < x < 26) = 0.8849 - 0.5 = 0.3849$ (i.e. 38.49 per cent).

(ii) $P(22 < x < 24) = 0.7881 - 0.6554 = 0.1327$ (i.e. 13.27 per cent).

(iii) $P(10 < x < 18) = 0.3446 - 0.0228 = 0.3218$ (i.e. 32.18 per cent).

(iv) $P(x \geq 15) = 1.0 - 0.1587 = 0.8413$ (i.e. 84.13 per cent).

(v) $P(x \leq 25) = 0.8413$ (i.e. 84.13 per cent).

(vi) $P(x > 25) = 1.0 - 0.8413 = 0.1587$ (i.e. 15.87 per cent).

Worked example 9.2   The normal distribution

The times taken by doctors to examine patients during consultations at a hospital have been found to be normally distributed with a mean μ of ten minutes and a standard deviation σ of five minutes.

(i) What is the probability of a consultation requiring 20 minutes or more?

(ii) If 80 patients are examined in a day, how many will take, on average, between five and ten minutes each?

Worked solutions

(i) $P(x \geq 20)$ $= P\left(Z \geq \dfrac{20 - 10}{5}\right)$

$= P(Z \geq 2.0)$

$= 0.5 - 0.4772$ (from tables)

$= 0.0228$ (i.e. 2.28 per cent).

(ii) $P(5 \leq x \leq 10)$ $= P\left(\dfrac{5 - 10}{5} \leq Z \leq \dfrac{10 - 10}{5}\right)$

$= P(-1.0 \leq Z \leq 0.0)$

$= 0.3413$ (from tables).

Worked example 9.2 The normal distribution *(continued)*

Therefore, out of 80 patients, the number of consultations taking, on average, between five and ten minutes each will be 0.3413×80, i.e. probability $\times$ total number of patients $= 27.304$ (say 28 patients).

Computer solution using MINITAB

Choose **Calc → Probability Distributions → Normal**

Check the **Cumulative probability** box

Enter the value of the mean ($\mu = 10$)

Enter the value of the standard deviation ($\sigma = 5$)

Check the **Input column** option and enter **C1**

Click **OK**

The output is as follows:

Worksheet

20

5

10

Output

Cumulative Distribution Function

Normal with mean = 10.0000
and standard deviation = 5.00000

x	P(X <= x)
20.0000	0.9772
5.0000	0.1587
10.0000	0.5000

Authors' comment: As in Worked example 9.1, the results in column 2 give the required probabilities which allow the answers to be calculated as before.

Computer solution using Excel

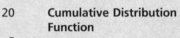

Use the **NORMDIST** function

Click = in the formula bar, select the **NORMDIST** function and enter:

In **X** – the value for which the distribution is required (in this case, with the sequence of values: 20, 5, 10 entered in column A of the worksheet, we enter **A1**)

In **Mean** – the arithmetic mean of the distribution (10 in this case)

In **Standard dev** – the standard deviation of the distribution (5 in this case)

In **Cumulative** – **TRUE** for the cumulative distribution function or **FALSE** for the probability mass function (probability density function) – **TRUE** is entered in this case

This returns, initially, the answer for $x = 20$ (the value in cell A1) – i.e. 0.97725. Dragging the **fill handle**, on the cell in which the first result (0. 97725) is entered, down the column automatically inputs the probabilities for the remaining values of x, as shown below, which permit the calculation of the required answers to the questions.

x	P(X <= x)
20	0.97725
5	0.158655
10	0.5

Required solutions:

(i) $P(x \geq 20) = 1 - P(x < 20) = 1.0 - 0.9772 = 0.0228$ (i.e. 2.28 per cent).

(ii) $P(5 \leq x \leq 10) = 0.5 - 0.1587 = 0.3413$.

Multiplying this probability by the total number of patients (80) gives the required answer of 28 patients (rounded up) as shown above.

Worked example 9.3 Normal approximation to the binomial

It is known that 40 per cent of candidates for a professional examination normally fail. Out of any group of 20 candidates selected at random, what is the probability that 12 or more will fail?

Worked solution

This question could be solved using the binomial formula considered in Chapter 7 but the calculation would be tedious. It is more convenient to use the normal approximation. The conditions necessary for the use of the normal approximation (i.e. that $np \geq 5$ and $nq \geq 5$) apply here as $np = 20(0.4) = 8$ and $nq = 20(0.6) = 12$. Thus we now know that $\mu = np = 20(0.4) = 8$ and $\sigma = \sqrt{(npq)} = \sqrt{[20(0.4)(0.6)]} = 2.1909$. Given that we are approximating a *discrete* variable using the *continuous* normal distribution, it is necessary to apply a continuity correction factor as explained earlier. Thus:

$$P(x \geq 12) = P(x \geq 11.5)$$
$$= P\left(Z \geq \frac{11.5 - 8}{2.1909}\right)$$
$$= P(Z \geq 1.5975)$$
$$= 0.5 - 0.4449$$
$$= 0.0551$$

(from the standard normal table in Appendix A Table A3, interpolating between the values for $Z = 1.59$ and $Z = 1.6$).

The exact binomial probability obtained by cumulation from the table of individual binomial probabilities (Appendix A Table A1) is 0.0566 (as shown below, using MINITAB and Excel the answer is 0.0565). Note, therefore, that the normal approximation provides an answer which is very close to the exact answer.

Computer solution using MINITAB

Applying the normal distribution	Choose **Calc** → **Probability Distributions** → **Normal**
$P(x \geq 12)$ is equivalent to $P(x \geq 11.5)$, thus: Enter 11.5 in column C1 of the MINITAB worksheet	Check the **Cumulative probability** box Enter the value of the mean ($\mu = 20$) Enter Standard deviation ($\sigma = 2.1909$)

Worked example 9.3   Normal approximation to the binomial *(continued)*

Check **Input column** and enter **C1**
Click **OK**

The output is as follows:

Cumulative distribution function

Normal with mean = 8.00000 and standard deviation = 2.19090

x	P(X <= x)
11.5000	0.9449

Authors' comment:
$P(x \geq 11.5) = 1.0 - 0.9449 = 0.0551$

Applying the binomial distribution

Enter 11.0 in column C1 of the MINITAB worksheet

Choose **Calc → Probability Distributions → Binomial**
Check the **Cumulative probability** box
Enter $n(= 20)$ in the **Number of trials** box
Enter $p(= 0.4)$ in the **Probability of success** box
Check **Input column** and enter **C1**
Click **OK**

The output is as follows:

Cumulative distribution function

Binomial with $n = 20$ and $p = 0.400000$

x	P(X <= x)
11.00	0.9435

Authors' comment: Therefore
$P(x \geq 12) = 1.0 - 0.9435 = 0.0565$

Computer solution using Excel

Using the normal approximation to the binomial

Use the **NORMDIST** function
Click = in the formula bar, select the **NORMDIST** function and enter:
In **X** – the value for which the distribution is required (11.5 in this case)
In **Mean** – the arithmetic mean of the distribution (8 in this case)
In **Standard dev** – the standard deviation of the distribution (2.19 in this case)
In **Cumulative** – **TRUE** for the cumulative distribution function or **FALSE** for the probability mass function (probability density function) – **TRUE** is entered in this case
This returns the probability: 0.944998

The required answer: $P(x \geq 11.5)$ is therefore: $1.0 - 0.944998 = 0.055002$

Using the binomial distribution

Use the **BINOMDIST** function
Click = in the formula bar, select the **BINOMDIST** function and enter:
In **Number_s**, the number of successes in trials (11 in this case)
In **Trials**, the number of independent trials (20 in this case)
In **Probability_s**, the probability of success on each trial (0.4 in this case)
In **Cumulative**, enter **TRUE** for the cumulative distribution function
This returns the answer: 0.943474

The required answer: $P(x \geq 12)$ is therefore: $1.0 - 0.943474 = 0.056526$

Key terms and concepts

Cumulative distribution function (cdf) This measures areas under the normal curve and gives the cumulative probability associated with a pdf up to a specified value.

Continuity correction factor A correction factor applied to the values of a discrete variable when using the normal distribution as an approximation to the binomial distribution (equal to half the distance between the discrete values which is added to, or subtracted from, each of them).

Normal distribution formula The probability density function, denoted $f(x)$, of a normally distributed random variable x expressed as:

$$f(x) = \frac{1}{\sqrt{(2\pi\sigma^2)}} e^{-\frac{1}{2}[(x-\mu)/\sigma]^2}$$

Normal probability distribution A continuous probability distribution whose form is a symmetrical, bell-shaped curve and is determined by the mean μ and standard deviation σ.

Properties of the normal distribution These are such that the proportion of all the observations of a normally distributed variable x that fall within a range of n standard deviations on both sides of the mean is the same for *any* normal distribution.

Probability density function (pdf) This determines the shape of a normally distributed variable x.

Standard normal distribution Distribution of the standard normal variable Z which follows a normal distribution with a mean of 0 and a standard deviation of 1.

Standard normal variable The transformation of the normally distributed variable x, expressed as Z, where:

$$Z = \frac{x - \mu}{\sigma}$$

This is also referred to as the *Z statistic*.

Z distribution Distribution of the standard normal variable Z which follows a normal distribution with a mean of 0 and a standard deviation of 1 (also referred to as the *standard normal distribution*).

Z statistic The transformation of the normally distributed variable x, expressed as Z (also referred to as the *standard normal variable*).

Self-study exercises

Normal distribution

9.1. What are the characteristics of a normal distribution?

9.2. What is meant by a 'standard normal distribution'? Why does a standard normal distribution have a mean of zero and a standard deviation equal to 1.0?

9.3. Find the area under the normal curve for each of the following:

    (a) $P(Z > 1.0)$

    (b) $P(0 < Z < 1.0)$

    (c) $P(Z < -1.0)$

    (d) $P(-1.0 < Z < +1.0)$

    (e) $P(Z > 1.96)$

(f) $P(Z < -1.5)$

(g) $P(-1 < Z < 3)$

9.4. A manufacturer of microchips knows that his products have a mean life of 3000 hours with a standard deviation of 850 hours. What percentage of his output will have lifetime performance of:

(a) less than 2000 hours?

(b) more than 3400 hours?

(c) 2500–3500 hours?

9.5. What are the upper and lower quartiles of the distribution given in question 9.4 above?

9.6. The treasurer of a local charity fund-raising committee collects data from a random sample of 100 jumble sales in the local area which show that the mean amount raised per sale was £100 with a standard deviation of £20. Assuming that the distribution of amounts is normal, what is the probability that the next jumble sale will raise more than £110?

9.7. The distribution of weeks of sickness leave in the construction industry is known to be normal, and a survey indicates that 33 per cent of workers had no more than $1\frac{1}{2}$ weeks of leave, while 16.6 per cent had at least $2\frac{1}{2}$ weeks of leave. Calculate the mean and standard deviation of leave in the industry.

9.8. The meat content of packed meat pies is normally distributed. Approximately one-third of the pies have a meat content of less than 61 per cent and a quarter above 62 per cent.

(a) What is the mean and standard deviation?

(b) What proportion of pies will have a meat content of more than 63 per cent?

(c) What can you say about the meat content of the best 10 per cent of the pies?

9.9. A firm makes a component with a nominal length of 14 cm. On inspection it is found that the length of the component is normally distributed with a standard deviation of 0.25 cm. For a component to be of use it must be between 13.6 cm and 14.5 cm.

(a) Estimate the proportions of components which are initially: (i) the correct size, (ii) oversized and (iii) undersized respectively.

(b) The cost of producing the component is 18p. Components of the correct size can be sold for 25p each. Those which are undersized are sold for scrap for 2p while those which are oversized can be made the correct size at an additional cost of 6p. Calculate the contribution to profit (or loss) of each 1,000 components.

9.10. A machine packs 'half-kilo' packets of butter. In fact the weights of the packets are normally distributed with mean 0.51 kg and standard deviation 0.008 kg.

(a) What percentage of packets is underweight?

(b) What percentage of packets weigh more than 0.505 kg?

(c) New trade regulations permit at most 5 per cent of packets to be underweight. If the machine is to comply with regulations, find the lowest value to which the firm could set the mean weight, assuming variability unchanged.

(d) To meet the above trade regulations, the company decides to adjust the machine so as to alter the variability of weights, while leaving the mean weight at 0.51 kg. What would be the maximum standard deviation of weight following this adjustment?

Normal approximation to the binomial distribution

9.11. Under what conditions is it appropriate to use the normal approximation to the binomial distribution?

9.12. Analysis of a bank's records shows that 8 per cent of its customers keep a minimum balance of £500 in their cheque book accounts. What is the probability that in a random sample of 100 customers a minimum balance of £500 is maintained by:

(a) exactly 6 customers?

(b) exactly 11 customers?

(c) 6 or fewer customers?

(d) 5 or more customers?

(e) 10 or fewer customers?

(f) 11 or fewer customers?

9.13. Suppose a garage examining cars for defective tyres finds a defective tyre in one out of every five cars examined. In a normal working week 80 cars are examined on average. What is the probability that 12 or more will be found to have a defective tyre?

9.14. A manufacturer guarantees to replace a new brand of long-life fluorescent tubes if they burn out within six months of the sale. Twenty per cent of such tubes burn out before the expiration of the guarantee. What is the probability that a dealer who has sold 50 tubes will have to replace:

(a) at least 15?

(b) exactly 15?

(c) at least 5 but not more than 10?

9.15. An examination consists of 100 multiple-choice questions. For each question there are four answers, of which only one is correct. A candidate enters the examination and chooses answers randomly. The pass mark is 35 per cent. What is the probability that the candidate will:

(a) obtain exactly 35 per cent?

(b) pass the examination?

(c) If all the candidates choose randomly, at what level would the pass mark have to be set to ensure that 10 per cent of candidates pass the examination?

Part III | Statistical inference

Ignorance gives one a large range of probabilities

GEORGE ELIOT

The following diagram provides an overview of the whole of this part of the book.

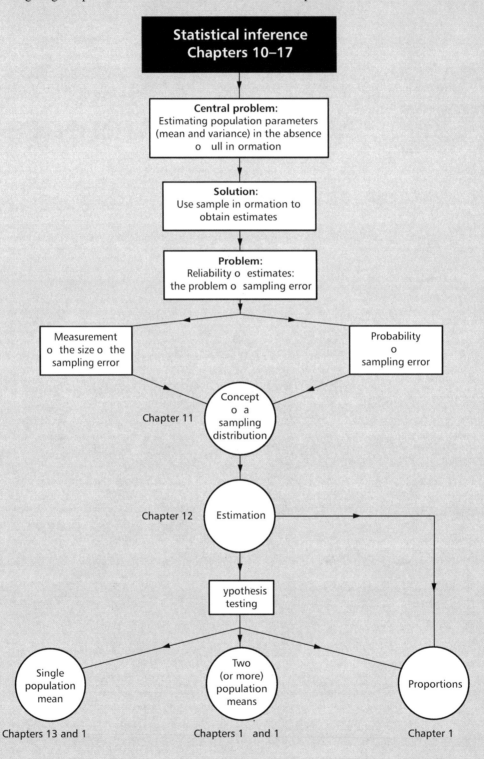

Chapter 10 | Statistical inference: an overview

Overview This part of the book deals with statistical inference; i.e. drawing inferences about a population parameter (such as the population mean) on the basis of information about samples drawn from the population (i.e. *sample statistics*). Sample information provides the basis of most decisions in practice for the simple reason that it is often too costly, or impracticable, to collect information for the whole population, i.e. to conduct a census.

In practice, the study of statistical inference falls into two main areas:

1 **estimation**   *the estimation of values of (unknown) population parameters.*

2 **hypothesis testing** *the testing of hypotheses about the values of the population parameters.*

As an example of the kind of problem addressed by the analysis of statistical inference, consider the following situation. A tyre manufacturer wishes to estimate the life expectancy of his tyres. Clearly all tyres cannot be tested to destruction before they are sold! It is necessary, therefore, to collect information for a sample of tyres which have been used in a variety of conditions on different vehicles. The results from this sample may then be used as the basis for estimating the average life of all tyres of the same type. Naturally, however, such an estimate will be subject to error as a result of *sampling error*. The question then arises of estimating the size of this error and the degree of confidence (i.e. probability of error) that can be attached to the estimate.

The basic idea of hypothesis testing can be conveyed by an extension of the example given above. Consider the situation whereby the manufacturer of tyres wishes to test whether or not a new production process produces tyres with a longer life expectancy than those produced hitherto. In particular, say that the normal life expectancy of a tyre using the old process is claimed to be 60 000 km. A sample of tyres produced using the new process gives an average life of 65 000 km. Is this sample evidence consistent with a hypothesis that the true (population) mean is still 60 000 km?

The statistical principles underlying the solution of these types of problems and the way they are applied are covered step by step in the next seven chapters which make up Part III of the book. A schematic overview of Part III is presented in the diagram at the beginning of this part of the book. It will be seen that this diagram summarizes the central problem and the organization of topics in the rest of this part of the book. In brief, the seven chapters are as follows:

Chapter 11 – Sampling Distributions This chapter deals with the concept of a sampling distribution and its properties. It provides the foundation for the rest of this part of the book.

Chapter 12 – Estimation This is concerned with the estimation of population parameters using sample information, the probability of error and the confidence levels that can be attached to these estimates.

Chapter 13 – Estimation and the _t_ Distribution This chapter extends the analysis of estimation principles set out in Chapter 12 and introduces a probability distribution known as the _t_ distribution.

Chapter 14 – Hypothesis Testing: Single Population Mean This chapter focuses on the testing of hypotheses about a population mean, i.e. tests involving a single population.

Chapter 15 – Hypothesis Testing: Two Population Means In this chapter the analysis of hypothesis testing in the context of a single population mean considered in Chapter 14 is extended to testing the significance of the difference between two means, i.e. tests involving two populations.

Chapter 16 – Estimation and Hypothesis Tests for Proportions The analysis of the preceding chapters concerning estimation and hypothesis tests for a single mean and the difference between two means is extended in this chapter to situations involving proportions.

Chapter 17 – Analysis of Variance (ANOVA) This extends the analysis of Chapter 15 to situations involving comparisons of the means of several populations simultaneously.

Chapter 11 | Sampling distributions

Aims   In this chapter we consider the concept of sampling distributions and their properties. Such distributions are fundamental to the whole theory of statistical inference. They provide the key to the measurement of sampling error and thus to the estimation of population parameters, the probability of error and the testing of hypotheses.

The topics dealt with in the next section are organized as follows:

- the concept of sampling distributions
- the relationship between sample statistics and population parameters
- the size and measurement of the standard error of the mean
- probability of sampling error – the *central limit theorem*

■ grasp the distinction between sample statistics and population parameters

■ understand the concept of a sampling distribution

■ appreciate the difference between sampling with and without replacement

■ show the derivation of a sampling distribution of the sample mean

■ understand what is meant by the standard error of the sample mean

■ recognize the relationship between sample statistics and population parameters

■ know what is meant by the finite population correction factor and when to apply it

■ know what is meant by the central limit population and its significance.

Principles

The concept of sampling distributions

Given a population of size N, samples of size n may be drawn from it (where n is smaller than N). There would generally be a large number of such samples and the value of a statistic, such as the arithmetic mean, computed for *each* sample will vary from one sample to another. The frequency distribution of *all* such sample statistics is an example of a *sampling distribution*.

The concept of a sampling distribution may be best illustrated by an example. Consider a population in which the variable can take a value of 0, 1, 2, 3 or 4 only. Such a situation could refer, for example, to one where a product could possess four possible faults and the values refer, therefore, to the number of faults any particular product is found to possess. Let us assume a population consists of five of the products, each of which possess 0, 1, 2, 3 and 4 faults respectively. Next, consider all possible samples of a given size n, say $n = 2$, that could be drawn from this population. These will depend on whether any one element of the population, once selected, can or cannot be selected again in the same sample, i.e. whether sampling is *with replacement* or *without replacement*. Sampling with replacement means that a value, having been selected, is 'replaced' and can then be selected again, whereas the opposite is true in the case of sampling without replacement. There are 25 possible samples in the case of sampling with replacement but only ten in the case of sampling without replacement in this example. These are shown in Table 11.1 along with all corresponding sample means $\bar{x}$. Also shown are the frequency distributions of these sample means, i.e. the *sampling distributions of means*.

Table 11.1 also shows the two summary measures – mean and variance – for each sampling distribution of means. Note that it is appropriate to use the symbols for population parameters (μ and σ^2) in referring to the mean and variance in this context because the sampling distribution defines the full set of possible observations (in this case the full set of possible sample means). Thus the mean of all the sample means can be denoted as $\mu_{\bar{x}}$ and the variance of all the sample means as $\sigma_{\bar{x}}^2$. These measures are required in subsequent exposition, but note at this stage that the means of the two distributions are the same ($\mu_{\bar{x}} = 2.0$) but the variances are different ($\sigma_{\bar{x}}^2 = 1.0$ and 0.75 respectively).

The next step, given that the objective is to use sample statistics to estimate population parameters, is to examine the relationship between the sample statistics we have just calculated and the corresponding population parameters.

Table 11.1 Sampling distributions with and without replacement (sample size $n = 2$; population $= 0, 1, 2, 3, 4$)

Sampling with replacement (25 samples)					Sampling without replacement (10 samples)			
(a) All possible samples								
0,0	1,0	2,0	3,0	4,0				
0,1	1,1	2,1	3,1	4,1	0,1	1,2	2,3	3,4
0,2	1,2	2,2	3,2	4,2	0,2	1,3	2,4	
0,3	1,3	2,3	3,3	4,3	0,3	1,4		
0,4	1,4	2,4	3,4	4,4	0,4			
(b) All possible sample means ($\bar{x}$)								
0.0	0.5	1.0	1.5	2.0				
0.5	1.0	1.5	2.0	2.5	0.5	1.5	2.5	3.5
1.0	1.5	2.0	2.5	3.0	1.0	2.0	3.0	
1.5	2.0	2.5	3.0	3.5	1.5	2.5		
2.0	2.5	3.0	3.5	4.0	2.0			

(c) Summary: the sampling distribution of means

Sample mean $\bar{x}$	Number of samples f	Probability of occurrence $f/\Sigma f$	Sample mean $\bar{x}$	Number of samples f	Probability of occurrence $f/\Sigma f$
0.0	1	1/25	0.5	1	1/10
0.5	2	2/25	1.0	1	1/10
1.0	3	3/25	1.5	2	2/10
1.5	4	4/25	2.0	2	2/10
2.0	5	5/25	2.5	2	2/10
2.5	4	4/25	3.0	1	1/10
3.0	3	3/25	3.5	1	1/10
3.5	2	2/25			
4.0	1	1/25			
Total	25	1.0	Total	10	1.0

(d) Mean and variance of sample means ($\mu_{\bar{x}}, \sigma_{\bar{x}}^2$)

$$\mu_{\bar{x}} = 2.0 \qquad \sigma_{\bar{x}}^2 = 1.0 \qquad\qquad\qquad \mu_{\bar{x}} = 2.0 \qquad \sigma_{\bar{x}}^2 = 0.75$$

Relationship between sample statistics and population parameters

To show the relationship between sample statistics and population parameters, we concentrate on the mean, rather than any other statistic, because estimation of the population mean is the most frequent objective of sampling. The population mean μ for the above example is:

$$\mu = (0 + 1 + 2 + 3 + 4) \div 5 = 2.0$$

Note that this is equal to the mean of the sampling distribution of means ($\mu_{\bar{x}}$) in each case, i.e.:

$$\mu_{\bar{x}} = \mu$$

It will be seen from Table 11.2 that this is also true when we take samples of size $n = 3$ and $n = 4$ from the same population. This is in fact always true for *all* sampling distributions and it is, therefore, an extremely important property of the sampling distribution of the mean.

It will also be noticed from Tables 11.1 and 11.2 that the sample means $\bar{x}$ are distributed symmetrically around the true population mean μ and that as the sample size n is increased, they become more and more closely clustered around μ. Clearly, therefore, the

Table 11.2 Sampling distributions of means – sampling with replacement (sample size $n = 3$ and $n = 4$)

Sample size $n = 3$			Sample size $n = 4$		
Sample mean $\bar{x}$	Number of samples f	Probability of occurrence $P(\bar{x})^a$	Sample mean $\bar{x}$	Number of samples f	Probability of occurrence $P(\bar{x})^a$
0.00	1	0.008	0.00	1	0.0016
0.33	3	0.024	0.25	4	0.0064
0.67	6	0.048	0.50	10	0.0160
1.00	10	0.080	0.75	20	0.0320
1.33	15	0.120	1.00	35	0.0560
1.67	18	0.144	1.25	52	0.0832
2.00	19	0.152	1.50	68	0.1088
2.33	18	0.144	1.75	80	0.1280
2.67	15	0.120	2.00	85	0.1360
3.00	10	0.080	2.25	80	0.1280
3.33	6	0.048	2.50	68	0.1088
3.67	3	0.024	2.75	52	0.0832
4.00	1	0.008	3.00	35	0.0560
			3.25	20	0.0320
Total	125	1.000	3.50	10	0.0160
			3.75	4	0.0064
			4.00	1	0.0016
			Total	625	1.0000
$\mu_{\bar{x}} = 2.0$ $\sigma_{\bar{x}}^2 = 0.66$			$\mu_{\bar{x}} = 2.0$ $\sigma_{\bar{x}}^2 = 0.5$		

^{a} Derived from relative frequencies ($f/\Sigma f$)

other summary statistics, the variance and standard deviation, for the population and sampling distribution are not the same. It is obvious that as sample size is *increased*, the variance and standard deviation of the sampling distribution *decrease*. Because the variation of sample means around the population mean varies as a result of sampling, a *sampling error* arises when any individual sample mean $\bar{x}_i$ is used as an estimate of the population mean μ. For this reason the term *standard error* is used rather than standard deviation to refer to the measurement of variation in a sampling distribution of means. We turn next to the measurement of the standard error of the sample mean and its relationship to the population variance and standard deviation.

The standard error of the sample mean

We have shown in Table 11.2 that as we increase sample size the variance of the sampling distribution of means declines. There is a close relationship between $\sigma_{\bar{x}}^2$ and σ^2 which depends on sample size and also on whether or not sampling is with or without replacement. The values for $\sigma_{\bar{x}}^2$ for $n = 2$, 3 or 4 for sampling with replacement are given in Table 11.1 and 11.2. The variance of the population values 0, 1, 2, 3 and 4 is $\sigma^2 = \Sigma(x_i - \bar{x})^2/N = 2.0$. Sampling with replacement, the relationship between $\sigma_{\bar{x}}^2$, σ^2 and n is:

$$\sigma_{\bar{x}}^2 = \frac{\sigma^2}{n}$$

This is illustrated for the three sample sizes, $n = 2$, 3 and 4, in the final column of Table 11.3.

Sampling *with* replacement, the population is in effect 'infinite' because the same member of the population can be selected again and again. Sampling *without* replacement, on the other hand, means that the population from which the samples are drawn becomes finite, and as a result the variance of the distribution of sample means ($\sigma_{\bar{x}}^2$) is reduced. The reason for this may be stated intuitively: sampling without replacement from a finite population means that ultimately the population is exhausted. This is intuitively obvious because the process of sampling *without* replacement from a finite population means that once an item is selected from the population and its value recorded, it cannot be selected again, whereas, in contrast, it may be selected again and again when sampling *with* replacement. Sampling without replacement is, therefore, a more efficient information-gathering process and the size of the variation is reduced. The magnitude of the reduction depends on the size of the sample (n) relative to the size of the population (N). This gives rise to a so-called *finite population correction* (fpc) factor given by:

Table 11.3 Variance of sampling distributions – sampling with replacement

Sample size (n)	Variance ($\sigma_{\bar{x}}^2$)	σ^2/n
2	1.0 (from Table 11.1)	= 2.0/2
3	0.66 (from Table 11.2)	= 2.0/3
4	0.5 (from Table 11.2)	= 2.0/4

Finite population correction factor (fpc)

$$\frac{N - n}{N - 1}$$

In these cases the relationship between $\sigma_{\bar{x}}^2$ and σ^2 is modified to:

$$\sigma_{\bar{x}}^2 = \frac{\sigma^2}{n}\left(\frac{N - n}{N - 1}\right)$$

Thus in the case of the example of sampling without replacement given in Table 11.1, we show that $\sigma_{\bar{x}}^2 = 0.75$. Note now that it is equal to:

$$\frac{\sigma^2}{n}\left(\frac{N - n}{N - 1}\right) = \frac{2.0}{2}\left(\frac{5 - 2}{5 - 1}\right) = \frac{3}{4} = 0.75$$

To compute the standard error requires, of course, that we take the square root of the expressions given above. To summarize, we define and measure the standard error of the mean in situations of sampling with and without replacement as follows.

Standard error of the mean ($\sigma_{\bar{x}}$)

    Sampling with replacement   Sampling without replacement

$$\sigma_{\bar{x}} = \frac{\sigma}{\sqrt{n}} \qquad\qquad \sigma_{\bar{x}} = \frac{\sigma}{\sqrt{n}}\sqrt{\left(\frac{N - n}{N - 1}\right)}$$

where σ is the population standard deviation, N is the population size and n is the sample size.

Finally, with regard to the application of the finite population correction factor, note that when sampling is carried out without replacement it is often from *large* populations of which the sample size n is a small fraction. In these situations the fpc factor may become very small and can thus be ignored. A rule of thumb regarding the use of the fpc factor is that if the sample size n is less than 5 per cent of the population size N (i.e. $n < 0.05N$) then it can be ignored.

Review

So far in this chapter we have defined the concept of a sampling distribution and established the relationship between the sampling distribution and the population from which it is derived in terms of the mean and standard deviation (standard error) of the distributions as follows:

$$\mu_{\bar{x}} = \mu$$
$$\sigma_{\bar{x}} = \frac{\sigma}{\sqrt{n}}$$

The final stage in laying the foundations of the principles of statistical inference is to consider how the *probability* of the errors that arise as a result of sampling can be measured. This is the subject of the next section. Having done so we shall then be in a position to consider the subject of estimation in Chapter 12.

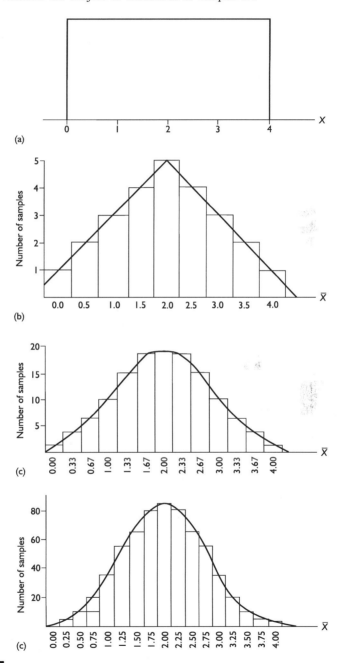

Figure 11.1 Sampling distributions of the mean from a population of values: 0, 1, 2, 3, 4; (a) population, a rectangular distribution; (b) sampling distribution of $\bar{x}$, $n = 2$; (c) sampling distribution of $\bar{x}$, $n = 3$; (d) sampling distribution of $\bar{x}$, $n = 4$

The probability of error – the central limit theorem

The key to establishing the probability of error is to examine the relationship between sampling distributions and their parent populations, not in terms of the summary descriptive statistics (considered above) but in terms of probability distributions. In the examples we used above we started with a uniform distribution of population values consisting of one value each of 0, 1, 2, 3, 4. A graph of the distribution is shown at the top of Figure 11.1 above along with graphs beneath it of the sampling distributions drawn from this population with sample sizes of $n = 2$, 3 and 4. It will be seen that the sampling distributions are symmetrical and tend to assume a shape which is similar to that of a normal distribution (considered in Chapter 9) as the sample size increases. This tendency is true for *all* sampling distributions regardless of the shape of the population distribution (i.e. whether or not it is symmetrical) as the sample size n is increased. Several examples of this are shown in Figure 11.2 below.

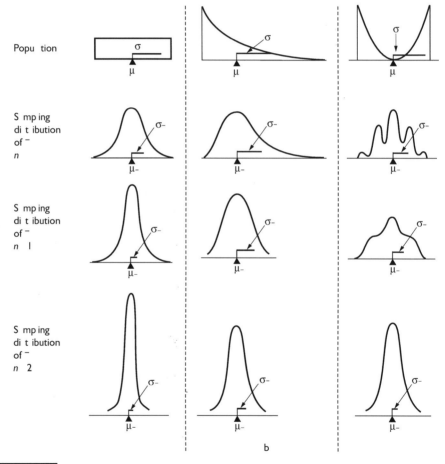

Figure 11.2 Illustration of the central limit theorem for three populations: (a) uniform population; (b) non-uniform population; (c) non-uniform population. *Source:* Example drawn from Tashman, L. J. and Lamborn, K. R. (1979) *The Ways and Means of Statistics*, New York: Harcourt Brace Jovanovich.

It will be seen from Figure 11.2 that the sampling distributions rapidly assume a symmetrical and increasingly bell-shaped form as the sample size *n* increases. In addition, the figure shows the size of the standard error $\sigma_{\bar{x}}$ declining as the sample size increases.

With regard to measuring the probability of an error of a given size occurring, the important point of the analysis here is that it has been found that as long as relatively large samples are taken, the sampling distribution closely approximates the *normal distribution* even if the distribution of the parent population itself is not normal. This property of sampling distributions, based on large samples from non-normal distributions, is referred to as the *central limit theorem*. Consequently, knowledge of the properties of the normal probability distribution, considered in Chapter 9, may be applied to measuring the probability of error that occurs due to sampling. For this reason, the normal distribution is the most important of all probability distributions. Strictly speaking, the central limit theorem is only true as *n* approaches infinity but, in practice, the distribution of $\bar{x}$ approaches a normal distribution very rapidly and becomes very close indeed for samples as small as 30. Thus the term 'large sample' is generally taken as being a sample size of 30 or more. In cases where the parent population is itself normal then the sampling distribution is normal for samples of *any* size.

These results of sampling from normal and non-normal populations are of the utmost importance. They provide theorems that constitute cornerstones of statistical inference. We therefore restate them formally here.

Sampling from a non-normal population – the central limit theorem

Where the parent population is not normally distributed, the distribution of the sample mean approaches a normal distribution as the sample size *n* becomes large ($n \geq 30$).

Sampling from a normal population

If the population distribution is normal then the distribution of the sample mean is also normal.

In both cases:

Mean $= \mu$

$$\text{Variance} = \begin{cases} \dfrac{\sigma^2}{n} & \text{for sampling } \textit{with replacement} \\[2ex] \dfrac{\sigma^2}{n}\left[\dfrac{N-n}{N-1}\right] & \text{for sampling } \textit{without replacement} \end{cases}$$

The application of the principles set out above is shown in the next section. Note, however, that these principles have been stated on the basis of a known value for σ, the standard deviation of a population. Normally, however, this will *not* be known and, consequently, it has to be estimated. The procedure in this case is considered in Chapter 12.

Applications and Worked Examples   **Worked example 11.1   Calculation of the standard error of the mean**

General descriptions of the MINITAB and Excel computer programs and their application are given in Appendices B and C respectively.

Calculate the standard error of the mean ($\sigma_{\bar{x}}$) in each of the following cases, where N is population size and n is sample size:

(i) $N = 100$, $\sigma = 8$, $n = 16$ and sampling is with replacement.

(ii) $N = 100$, $\sigma = 8$, $n = 16$ and sampling is without replacement.

(iii) N is large, $\sigma = 8$, $n = 64$.

Solution

(i) $\sigma_{\bar{x}} = \dfrac{\sigma}{\sqrt{n}} = \dfrac{8}{4} = 2.0$

(ii) $\sigma_{\bar{x}} = \dfrac{\sigma}{\sqrt{n}} \sqrt{\left(\dfrac{N-n}{N-1}\right)} = \dfrac{8}{\sqrt{16}} \sqrt{\left[\dfrac{(100-16)}{(100-1)}\right]} = 1.84$

   Note the finite population correction factor $(N-n)/(N-1)$ applies here because sampling is without replacement and because $n/N = 16/100 = 0.16 > 0.05$

(iii) $\sigma_{\bar{x}} = \dfrac{\sigma}{\sqrt{n}} = \dfrac{8}{\sqrt{64}} = 1.0$

Worked example 11.2   Probability of sampling error

The marks of 36 students in a statistics examination are normally distributed with a mean of 58 and a standard deviation of 18. What are the chances that:

(i) a random sample of nine students will have an average mark greater than 70?

(ii) an individual student, chosen at random, will have a mark greater than 70?

Solution

The solution here involves the use of the normal distribution, first in the context of sampling (as discussed in this chapter) and second in a non-sampling context in which the procedures set out in Chapter 9 are appropriate.

(i) It is first necessary to compute Z, the

standard normal variable. In this case, since we are dealing with the sampling distribution of means, $\bar{x}$, this is defined as:

$$Z = \frac{\bar{x} - \mu}{\sigma_{\bar{x}}}$$

Note the contrast with solution (ii) below where, in a non-sampling context,

$$Z = \frac{x - \mu}{\sigma}$$

Thus:

$$Z = \frac{\bar{x} - \mu}{\sigma_{\bar{x}}} = \frac{\bar{x} - \mu}{\sigma/\sqrt{n}} = \frac{70 - 58}{18/\sqrt{9}} = 2$$

Hence, from the standard normal distribution table (Appendix A Table A3),

$$P(0 \leq Z \leq 2) = 0.4772$$

Therefore, $P(\bar{x} > 70) = P(Z > 2) = 0.5 - 0.4772 = 0.0228$. That is, just over a 2 per cent chance. This is illustrated in the figure below along with the solution to (ii).

(ii) $P(x > 70) = P\left(Z > \dfrac{x - \mu}{\sigma}\right)$

$$= P\left(Z > \dfrac{70 - 58}{18}\right)$$

$$= P(Z > 0.67)$$

$$= 0.5 - 0.2486 \qquad \text{(from Appendix A Table A3)}$$

$$= 0.2514$$

That is, just over 25 per cent chance. Note the much greater chance of an individual scoring over 70 as opposed to a sample of nine students scoring 70 on average. This is due to the clustering of sample means around the population mean that occurs as a result of sampling.

The results for both (i) and (ii) are illustrated below.

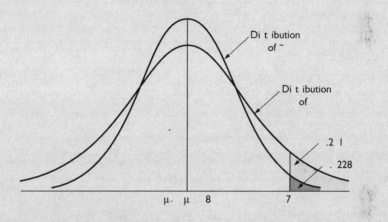

Worked example 11.3   Central limit theorem

The average life of TV tubes is known to be 20 000 hours with a standard deviation of 8000 hours. What is the probability that a sample of 33 tubes will have a mean life of more than 23 000 hours?

Solution

In this question we have no information about the nature of the population distribution but the central limit theorem can be applied because the sample is 'large' ($n \geq 30$).

$$Z = \dfrac{\bar{x} - \mu}{\sigma\sqrt{n}} = \dfrac{23\,000 - 20\,000}{8000/\sqrt{33}} = 2.15$$

Therefore

$P(\bar{x} > 23\,000)$

$= P(Z > 2.15)$

$= 0.5 - 0.4842$

(from Appendix A Table A3)

$= 0.0158$

Thus there is a less than 2 per cent chance that the sample will have a mean life of more than 23 000 hours.

Key terms and concepts

Central limit theorem A theorem that allows us to use the normal distribution to approximate the sampling distribution whenever the sample is large, even if the distribution of the parent population is not normal.

Finite population correction factor (fpc) Used to modify the expressions for $\sigma_{\bar{x}}$ when sampling without replacement from a finite population:

$$fpc = \sqrt{\frac{N-n}{N-1}} \left(\text{thus } \sigma_{\bar{x}} = \frac{\sigma}{\sqrt{n}} \sqrt{\frac{N-n}{N-1}} \right)$$

Population parameters Numerical summary measures used to describe a statistical population (such as the population average).

Sample statistics Numerical summary measures used to describe a sample such as the sample average (whose values vary according to the random sample collected).

Sampling distribution A probability distribution of all possible values of a sample statistic.

Sampling error The error which arises when any individual sample mean is used as estimate of the population mean.

Sampling with replacement A sampling procedure in which a value, having been selected, *is* replaced and therefore can then be selected again.

Sampling without replacement A sampling procedure in which a value, having been selected, *is not* replaced and therefore cannot then be selected again.

Standard error of the sample mean The standard deviation of the distribution of the sample mean:

Sampling With Replacement   *Sampling Without Replacement*

$$\sigma_{\bar{x}} = \frac{\sigma}{\sqrt{n}}$$ $$\sigma_{\bar{x}} = \frac{\sigma}{\sqrt{n}} \sqrt{\frac{N-n}{N-1}}$$

where: σ = population standard deviation

N = population size

n = sample size

Self-study exercises

11.1. Calculate the standard error of the mean ($\sigma_{\bar{x}}$) in each of the following cases, where N is population size and n is sample size.

(a) $N = 81$, $\sigma = 18$, $n = 9$, sampling with replacement.

(b) $N = 81$, $\sigma = 18$, $n = 9$, sampling without replacement.

(c) N is large, $\sigma = 16$, $n = 64$.

11.2. Given that a variable x is distributed with mean 25 and standard deviation 16:

(a) what is $\mu_{\bar{x}}$?

(b) what is $\sigma_{\bar{x}}$?

(c) what is the probability that the sample mean $\bar{x}$ based on 30 random observations will exceed 35?

(d) what is the probability that the sample mean $\bar{x}$ based on 30 random observations will be less than 18?

(e) what is the probability that the sample mean $\bar{x}$ based on 64 observations will be less than 18?

(f) what is the probability that the sample mean $\bar{x}$ based on 688 observations will be less than 24?

11.3. A machine packs 'half-kilo' packets of butter. In fact, the weights of the packets are normally distributed with mean 0.51 kg and standard deviation 0.008 kg. If a person buys four packets what is the chance that he or she obtains more than 2 kg?

11.4. A large normally distributed population has a mean of 10 and a variance of 16. What is the probability that a random sample would have a mean of less than 4, if the sample size is:

(a) 1?

(b) 4?

11.5. The lives of a certain type of battery are normally distributed with a mean of 100 hours and a standard deviation of 24 hours. What are the chances that:

(a) a random sample of 36 batteries will have an average life greater than 110 hours?

(b) a single battery, chosen at random, will last for more than 110 hours?

11.6. (Note: this question strictly requires the use of a continuity correction factor – see Chapter 9.)
A machine fills boxes with matches. On average, each box should contain 45. One day's production of 10 000 boxes includes 668 boxes containing less than 43 matches and 62 boxes containing more than 51 matches.

(a) On suitable assumptions, *which should be stated*, calculate the mean and standard deviation of the number of matches per box.

(b) What percentage of boxes are underfilled?

(c) What percentage of boxes contain exactly 45 matches?

(d) Do you think the manufacturer is meeting his specification?

(e) The boxes are wrapped in packs of 10 boxes for distribution. What is the probability that the mean content of the boxes in a pack does not fall below the nominal mean contents of 45?

(f) New trade regulations permit at most 5 per cent of boxes to be underfilled. If the machine is to comply with regulations, find the lowest value to which the firm could reduce the mean contents per box, assuming variability is unchanged.

11.7. (a) A manufacturer of car exhaust pipes wishes to check the achievement of a standard in which their average length of life is longer than one year (otherwise they have to be replaced under guarantee) but no longer than 18 months under normal use. From a pilot sample he computes the standard deviation as 30 months. What minimum size of sample will he require to check the achievement of the desired standard with 95 per cent confidence?

(b) If he is able to afford a sample only half the size of that indicated, what degree of confidence would be attached to the results?

11.8. The distribution of hourly earnings in a large firm was as follows:

Earnings (£)	Percentage of workers
Under 2.00	35.2
2.00–2.20	33.3
2.20 and over	31.5

The distribution is exactly normal. Find **(a)** the standard deviation, **(b)** the mean, **(c)** the percentage of workers earning between £2.20 and £2.30 per hour.

11.9. A factory is illuminated by 2000 fluorescent tubes. The lives of these tubes are normally distributed with a mean of 550 hours and a standard deviation of 50 hours. It is decided to replace all the tubes at such intervals of time that only about 20 tubes are likely to fail during each interval. How frequently should the tubes be changed?

Chapter 12 | Estimation

Aims

In this chapter we come to the first of the two main areas of statistical inference noted in Chapter 10, namely the estimation of population parameters from sample information and the determination of the degree of confidence that can be attached to the estimates. The two main topics to be considered are

■ point estimation of the population mean
■ interval estimation of the population mean

These topics are considered first in situations when σ, the population standard deviation (which permits the measurement of the standard error of the mean), is known. They are then considered in Chapter 13 in situations where σ is not known. Consideration of the latter involves the introduction of another probability distribution (closely allied to the normal distribution) known as the t distribution but involves no new principles.

After working through this chapter you will be able to:

- appreciate what is meant by estimation in the context of statistical inference
- grasp what is meant by the terms: unbiasedness, efficiency and consistency with respect to a sample statistic
- understand the meanings of point and interval estimation of a population mean
- make point estimates of a population mean
- construct confidence intervals for a population mean
- comprehend the distinction between standard error and sampling error
- understand the meaning of levels of confidence and confidence limits.

Principles

Point estimation of the population mean

In a point estimation we use the data from a sample to compute the value of a sample *statistic* (in this section we deal only with the mean). Thus we would refer to the numerical value of $\bar{x}$, the sample mean, as the point estimator of the population parameter – the population mean μ.

In order for a sample statistic to be a good estimator of a population parameter it should possess the following three key characteristics.

Unbiasedness   *An estimator is said to be unbiased if the mean of the sample statistic (in this case $\mu_{\bar{x}}$) is equal to the population parameter being estimated (μ in this case).*

Efficiency   *Efficiency requires that, for a given sample size, the standard error of the sample statistic is as small as possible.*

Consistency   *A sample estimator is said to be consistent if its value approaches that of the population parameter being estimated as the sample size increases.*

We showed earlier, in Chapter 11, that the sample mean satisfies the criteria of unbiasedness and consistency. It can also be shown that it is also efficient. Thus the sample mean $\bar{x}$ may be used as a *point estimator* of the population mean μ (often denoted $\hat{\mu}$ – pronounced 'mu-hat'). Since the value of the *estimator*, $\bar{x}$, computed from a single sample is a single value, we refer to it as a *point* estimate of the unknown population mean because it represents a single point on the scale of all possible values.

It should be borne in mind, however, that point estimates are nearly always wrong in the sense that they will frequently differ from the true population parameter. With point estimation the sample mean is taken as an estimate of the unknown population mean. Provided that the sample is obtained at random, we know that the sample mean provides a good estimator but it is, of course, unlikely that it will be exactly equal to the unknown population parameter. We know that it will be subject to sampling error and will thus give an estimate which may be too high or too low. For this reason *interval estimates*, which specify a range of values, are often used instead of point estimates.

Interval estimation of the population mean – σ known

Interval estimation uses the principles we have established in the last chapter to define an interval around our estimate, $\bar{x}$, within which we can state, with a given level of probability, say 95 per cent, that the true mean μ will lie. The probability that a specified interval contains the unknown population parameter is called *confidence*. The interval itself is called a *confidence interval* or simply an *interval estimate* and takes the following form:

Confidence interval for the population mean, μ

$\bar{\mu} = \bar{x} \pm (\text{a sampling error})$

---

where the horizontal bars above and below μ are notation used to denote a confidence interval for μ. In the exposition, measurement of sampling error assumes a known value of σ. When σ is unknown, the sample standard deviation s is used as an estimator, but if samples are small a different procedure is required (see Chapter 14).

The actual allowance for sampling error is computed by taking an appropriate number of standard errors ($\sigma_{\bar{x}}$) according to how confident we wish to be that our interval estimate will in fact include the true population value μ. It is common to choose a 95 per cent confidence level, i.e. an interval such that there is a 95 per cent chance that it will contain the population parameter. On the basis that the sampling distribution of $\bar{x}$ is normally distributed, we may obtain the confidence level of 95 per cent by taking the smallest range of $\bar{x}$ under the normal distribution that just encloses 95 per cent of the area. This is the *middle* section of the sampling distribution which leaves exactly 2.5 per cent of the area excluded in *each* tail – see Figure 12.1. From the table of areas under the normal curve (Appendix A Table A3) we know that we need to measure 1.96 standard deviations (in this case, standard errors) of $\bar{x}$ on either side of the mean.

Hence a confidence interval constructed as $\bar{\mu} \pm 1.96\sigma_{\bar{x}}$ will encompass 95 per cent of all the possible means in the sampling distribution. This is illustrated at the top of Figure 12.1. It follows that a similar interval (i.e. $\pm 1.96\sigma_{\bar{x}}$) constructed around *any single sample mean* $\bar{x}$ that falls *within* the range defined by the two vertical broken lines A and B in Figure 12.1 must also enclose the unknown population mean μ. There is only a 5 per cent chance that a sample mean would lie outside this range. Thus a 95 per cent confidence interval for the population mean is defined as

---

95 per cent confidence interval

$\bar{\mu} = \bar{x} \pm 1.96\sigma_{\bar{x}}$

where $\sigma_{\bar{x}} = \sigma/\sqrt{n}$

This is illustrated in the bottom half of Figure 12.1. Notice that confidence intervals for the first four samples illustrated do encompass μ but sample number 5 does not.

Normally, of course, we would only have one sample and thus only one confidence interval. We repeat, for emphasis, that given large samples so that the central limit theorem applies or, alternatively, sampling from a population which is itself known to be normal, the probability that a confidence interval constructed as above would *fail* to include the true population mean is only 5 per cent.

Levels of confidence and confidence limits

Levels of confidence other than 95 per cent may be obtained simply by substituting appropriate values of the standard normal variable Z. More generally, the desired level of confidence may be denoted as $100(1 - \alpha)$ per cent, where α is the sum of the two areas in the tails of the normal distribution. Thus, for 95 per cent confidence, $\alpha = 0.05$ and the area in each tail is given by $\alpha/2$ (i.e. 0.025). The general expression for a confidence interval ($\bar{\mu}$) using Z values is:

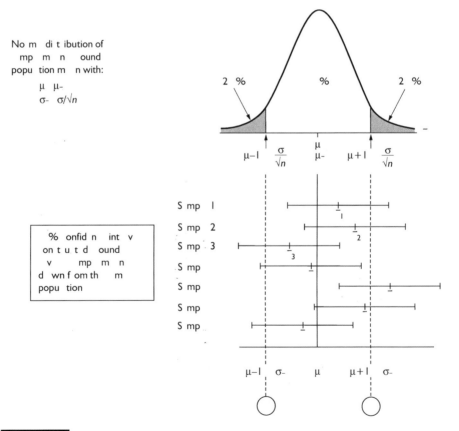

Figure 12.1   Interval estimation of the population mean

Confidence interval for population mean using Z values

$$\bar{\mu} = \bar{x} \pm Z_{\alpha/2}(\sigma_{\bar{x}})$$

where $100(1 - \alpha)$ is the desired level of confidence and $Z_{\alpha/2}$ denotes the value of Z which cuts off an area of $\alpha/2$ in each of the tails of the distribution.

The two values that together define a confidence interval are known as *confidence limits – a lower confidence limit* and an *upper confidence limit*. The most common confidence levels after 95 per cent are 90 per cent and 99 per cent. Obviously, the higher the degree of confidence required, the wider the confidence interval. The values of Z that correspond to these levels of confidence are shown in Table 12.1.

Table 12.1 Levels of confidence

Required level of confidence	Corresponding value of α	Value of $Z_{\alpha/2}$
90 per cent	0.10	1.65
95 per cent	0.05	1.96
99 per cent	0.01	2.58

Thus, the corresponding confidence intervals for the population mean μ are defined as follows:

90 per cent confidence interval: $\bar{\mu} = \bar{x} \pm 1.65\sigma_{\bar{x}}$

95 per cent confidence interval: $\bar{\mu} = \bar{x} \pm 1.96\sigma_{\bar{x}}$

99 per cent confidence interval: $\bar{\mu} = \bar{x} \pm 2.58\sigma_{\bar{x}}$

Applications and Worked Examples — Worked example 12.1   Point estimation of μ

General descriptions of the MINITAB and Excel computer programs and their application are given in Appendices B and C respectively.

The contents (in millilitres) of bottles of shampoo are known to be normally distributed with a standard deviation of 30 ml. A random sample of 15 bottles is found to have the following contents:

189, 204, 205, 234, 200, 198, 215, 178,

210, 212, 232, 210, 188, 201, 199.

(i) Make a point estimate of the mean content of the bottles.

(ii) Calculate the standard error of the mean.

Solution

(i) Let the point estimate of the mean be denoted by $\hat{\mu}$ (pronounced 'mu-hat'). The sample mean is found to be $\bar{x} = 205$ ml. Thus, a point estimate of the mean is:

$$\hat{\mu} = \bar{x} = 205 \text{ ml}$$

(ii) The standard error of the mean is given by:

$$\sigma_{\bar{x}} = \sigma/\sqrt{n} = 30/\sqrt{(15)} = 7.75 \text{ ml}$$

Worked example 12.2   Interval estimation of μ

Using the data in Worked example 12.1:

(i) calculate 95 per cent and 99 per cent confidence intervals for $\hat{\mu}$;

(ii) state the confidence limits.

Worked solution

(i) 95 per cent confidence interval: $\overline{\mu} = \bar{x} \pm Z_{0.05/2}(\sigma_{\bar{x}})$

$$= 205 \pm 1.96\,(7.75)$$
$$= 205 \pm 15.19$$
$$\text{i.e. from 189.8 to 220.2}$$

99 per cent confidence interval: $\overline{\mu} = \bar{x} \pm Z_{0.01/2}(\sigma_{\bar{x}})$

$$= 205 \pm 2.58\,(7.75)$$
$$= 205 \pm 20.0$$
$$\text{i.e. from 185.0 to 225.0}$$

(ii) 95 per cent confidence limits: 189.8 (lower) and 220.2 (upper)

99 per cent confidence limits: 185.0 (lower) and 225.0 (upper)

Computer solution using MINITAB

With the data entered to **Column C1** of the MINITAB worksheet, the procedure is:

Choose **Stat → Basic Statistics → 1-Sample Z. . .**
Select variable **(C1)**
In the **Confidence interval Level** box enter **95** or **99** etc. as appropriate
Enter the value of **Sigma**, σ (30 in this case)
Click **OK**

The output is as follows:

Z Confidence Intervals

The assumed sigma = 30.0

Variable	N	Mean	StDev	SE Mean	95.0 % CI
C1	15	205.00	15.05	7.75	(189.82, 220.18)

Standard error of the mean $(30/\sqrt{15})$

Lower and upper limits of 95% confidence interval respectively

Variable	N	Mean	St Dev	SE Mean	99.0 % CI
C1	15	205.00	15.05	7.75	(185.04, 224.96)

Lower and upper limits of 99% confidence interval respectively

Computer solution using Excel

Use function **CONFIDENCE**
Click = on the formula bar and select
    CONFIDENCE
In **Alpha**, enter significance level, thus for
    95% confidence enter 0.05; for 99% confi-
    dence enter 0.01 etc.
In **Standard_dev**, enter population standard

deviation (assumed to be known) – in this
    case enter 30
In **Mean**, enter sample size (in this case 15)

This returns: 15.18179 for 95% confidence
and 19.95233 for 99% confidence.

Authors' comment:

Thus 95% confidence interval = $\bar{x}$ (= 205) ± 15.18179
$\qquad\qquad\qquad\qquad\qquad$ = 189.8 to 220.2 to 1 decimal place – lower and upper limits

Thus 99% confidence interval = $\bar{x}$ (= 205) ± 19.95233
$\qquad\qquad\qquad\qquad\qquad$ = 185.0 to 225.0 to 1 decimal place – lower and upper limits

Interpretation

It will be seen that the 99 per cent confidence interval (from 185 to 225 roughly) is wider than the 95 per cent confidence interval (from 189.8 to 220.2) for the estimated value of μ (i.e. $\hat{\mu}$).

Worked example 12.3   Estimating confidence

A manufacturer of pads of writing paper knows that the number of sheets in each pad varies from batch to batch but that the standard deviation remains constant at a value of 20 sheets. He wishes to test the mean contents of the latest batch and, taking a random sample of 25 pads, finds the mean number of sheets per pad to be 103.

How accurate is the sample mean as a point estimate for the mean of the batch under test?

Solution

Allowing for the continuity correction, 103 is equivalent to 102.5 – 103.49. Thus, we need to find the probability of $102.5 \leq \bar{x} \leq 103.49$.

$$P(\bar{x} \leq 102.5) = P\left(Z \leq \frac{102.5 - 103}{20/\sqrt{(25)}}\right)$$
$$= P(Z \leq -0.5/4)$$
$$= P(Z \leq -0.125) = 0.45 \text{ (approximately)}$$
$$P(\bar{x} \geq 103.49) = P\left(Z \geq \frac{103.49 - 103}{20/\sqrt{(25)}}\right)$$
$$= P(Z \geq 0.49/4)$$
$$= P(Z \geq 0.1225) = 0.45 \text{ (approximately)}$$

Worked example 12.3 Estimating confidence *(continued)*

Thus,

$$P(102.5 \leq \bar{x} \leq 103.49) = 1 - 0.45 - 0.45$$
$$= 1 - 0.9 = 0.1$$

i.e. 10 per cent (approximately)

Authors' comment: Thus the probability of selecting a random sample with a mean of 103 is only 10 per cent. Thus the degree of accuracy (confidence) is low – there is a 90 per cent probability that the mean of the batch is greater or less than 103 sheets.

Worked example 12.4 Determining sample size

Suppose that the manufacturer in Worked example 12.3 is dissatisfied with the accuracy of his estimate based on a sample of 25. How large would this sample need to be so that he can be sure with a probability of 0.95 that his estimate will not be in error by more than two sheets either way?

Solution

Allowing for the continuity correction, an error exceeding two sheets is equivalent to an error of more than 2.49. Thus, for 95 per cent probability,

$$2.49 = 1.96 \, (20/\sqrt{n})$$

Hence $n = [1.96(20)/2.49]^2 \approx 248$. Note that, without the continuity correction, the solution becomes $2 = 1.96 \, (20\sqrt{n})$; therefore:

$$n = [1.96(20)/2]^2 = 384$$

Key terms and concepts

Confidence interval An interval estimate of the population parameter defined according to a specified level of confidence (probability).

Confidence interval of the population mean using Z values

$$\underline{\mu} = \bar{x} \pm Z_{\alpha/2}(\sigma_{\bar{x}})$$

Confidence interval for the population mean

$$\underline{\mu} = \bar{x} \pm \text{ (a sampling error)}$$

Confidence levels Degree of confidence associated with an interval estimate of the population parameter.

Confidence limits The upper and lower limits that together define an interval estimate.

Consistency A sample estimator is said to be consistent if its value approaches that of the population parameter being estimated as the sample size increases.

Efficiency For a given sample size, efficiency requires that the standard error of the sample statistic is as small as possible.

Interval estimate An estimate of a population parameter that provides an interval of values believed to contain the value of the parameter being estimated. The interval estimate is also referred to as a confidence interval.

Point estimate A single numerical value used as an estimate of a population parameter.

Unbiasedness An estimator is said to be unbiased if the mean of the sample statistic is equal to the population parameter being estimated.

Self-study exercises

12.1. Measurement of the IQ of 202 people selected at random produced the following results:

mean IQ = 105   standard deviation = 10

   (a) Calculate the standard error of the mean.
   (b) Estimate the confidence limits of the mean IQ:
      (i)   with 95 per cent confidence
      (ii) with 99 per cent confidence.
   (c) Without making further calculations, state how in general the confidence interval would be affected:
      (i)   if the sample were smaller?
      (ii)   if the sample were larger?
      (iii) if the sample size were unchanged but the standard deviation increased?

12.2. Tests carried out by a Trading Standards authority on the meat content of premium sausages produced by a local firm show that a random sample of 36 sausages had a mean meat content of 62.3 per cent with a standard deviation of 1.2 per cent.

   (a) Give the point estimate for the mean meat content of the firm's premium sausages.
   (b) Compute 95 per cent confidence limits for the meat content of the firm's premium sausages.

12.3. An automatic machine, used to fill cartons with 1000 ml (i.e. 1 litre) of fruit juice, does so with a standard deviation of 50 ml. A random sample of 100 cartons taken by the Trading Standards Inspectorate reveals a mean contents of 1005 ml.

   (a) What is the probability that the population mean is less than 1 litre?

(b) How much juice per carton should the machine be set to dispense on average in order to ensure with 99% confidence that the mean contents do not fall below 1 litre?

12.4. An investigation is to be mounted into the mean duration of unemployment. There are 1000 people on the local register of unemployed. A random sample of 100 is taken *without replacement* giving a mean and variance of 87 and 256 days respectively.

(a) State an interval estimate of μ that allows a margin of error of five days.

(b) How confident can you be about this estimate?

(c) How large a sample would be necessary to give 99 per cent confidence?

12.5. A sample survey is planned to determine the average amount of overtime pay received by nurses working in hospitals. A pilot investigation of nurses reveals a standard deviation of £500. How large a sample is needed in order to have:

(a) an estimate of the population mean within $\pm£100$ of the sample mean with 95 per cent confidence?

(b) an estimate within the same limits with 99.7 per cent confidence?

12.6. (a) Explain the importance of the distinction between sampling *with* or *without* replacement and the use of the finite population correction factor.

(b) As a result of a survey of parental contribution to student maintenance covering a random sample of 400 students (obtained without replacement) out of a total population of 10 000 the following results are obtained:

$$\Sigma x_i = £120,000 \quad \Sigma(x_i - \bar{x})^2 = £4,398,975.$$

(i) Estimate the mean and variance of the population.

(ii) Compute 90 per cent and 95 per cent interval estimates of the population mean.

(iii) What level of confidence can be attached to the interval estimate $\bar{x} \pm 15$?

(c) How would the limits be affected if:

(i) the sample had been four times larger?

(ii) the population had been four times larger?

12.7. An annual random sample of 36 manufacturing establishments is taken from the hosiery industry in order to obtain information about planned investment. The 95 per cent confidence interval of mean planned investment is estimated from the sample to be £10 million to £11 million.

(a) What is the value of the sample standard deviation?

(b) If it is known that the number of establishments in the industry is 100, find a more accurate 95 per cent confidence interval.

| Chapter 13 | # Estimation and the *t* distribution |

Aims

In the previous chapter we showed how to derive a confidence interval for the population mean μ, when the population standard deviation σ is known. In this chapter we now set out the equivalent procedures for calculating a confidence interval for the population mean when σ is *unknown* and introduce the distribution known as the *t distribution*. This distribution is used not only in this context but also in a wide range of other applications which are discussed in subsequent chapters.

The chapter covers the following topics:

- the standard error of the mean when σ is unknown
- the *t* statistic and the *t* distribution
- interval estimation using the *t* distribution

Principles

The standard error of the mean when σ is unknown

The principles of estimation set out in Chapter 12 were stated on the basis of a known value for σ – the standard deviation of the population. However, this will not usually be known and, consequently, it has to be estimated. The only information normally available comes from the sample. Just as we showed (in Chapter 11) that a sample mean $\bar{x}$ is an unbiased estimator of the population mean μ, so too it can be shown that the sample variance s^2 is an unbiased estimator of the population variance σ^2, provided that s^2 is computed using a denominator of $n - 1$. It was for this reason that the *sample* variance was defined in this way in Chapter 4. So we have:

Unbiased estimator of σ^2

$$s^2 = \frac{\Sigma(x_i - \bar{x})^2}{n - 1}$$

The standard error of the mean is then estimated as $s/\sqrt{n}$ and is now denoted $s_{\bar{x}}$. It should be noted that, as long as the sample size n is large, s is a very good estimator of σ. However, if n is small then s may badly underestimate σ, so the estimated standard error of the mean, $s/\sqrt{n}$, may be greatly understated also. It follows that in constructing a confidence interval using data from small samples, the use of $s/\sqrt{n}$ will greatly narrow the confidence interval and hence deprive us of the desired level of confidence. To maintain the desired confidence level requires compensation to expand the width of the interval. This is achieved by switching from the use of Z values from the standard normal distribution, as in Chapter 12, to the use of the *t statistic* from the *t distribution* (often referred to as *Student's t* distribution). The *t* distribution is wider and flatter than the normal distribution.

The t statistic and *t* distribution

It will be recalled that the standard normal variable Z is given by

$$Z = \frac{\bar{x} - \mu}{\sigma/\sqrt{n}}$$

and that the distribution of Z is symmetrical with a mean of 0 and a variance of 1. The t statistic is defined by a similar expression, i.e.

---

t statistic

$$t = \frac{\bar{x} - \mu}{s/\sqrt{n}}$$

The t distribution is also symmetrical with a mean of 0 but has a variance greater than 1. In fact, unlike the Z distribution, the t distribution is actually a whole family of distributions whose variances gets larger – the tails of the distribution get larger and more spread out – as n gets smaller. An illustration is given in Figure 13.1. The actual shape of a t distribution depends on sample size (as noted above) or, more strictly, *degrees of freedom* (df).

Degrees of freedom

The number of degrees of freedom is given by $n - 1$ where n is the sample size. This concept is used in many areas of statistical analysis. Essentially, the number of degrees of freedom indicates the number of values that are free to vary in a random sample of a given

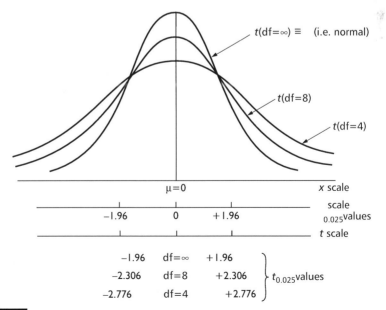

Figure 13.1 *Z* and *t* distributions

size. In general, the number of degrees of freedom lost is equal to the number of population parameters estimated as the basis for a statistical inference procedure. For example, if s (the sample standard deviation) is used to estimate σ (the population standard deviation), because σ is unknown, one degree of freedom is lost. This is because, in calculating s one population parameter (μ) is estimated using $\bar{x}$ (the sample mean). Thus, the number of degrees of freedom, df, is equal to $n - 1$.

Tables of t: format and explanation

A table of areas under the curve for the t distribution which allows us to make probability statements is available, just as in the case of the standard normal curve – see Appendix A Table A4. Note, however, that as the t distribution is not a single distribution but a whole family of distributions, the table is laid out differently from the table for the standard normal curve (shown in Appendix A Table A3). Whereas the latter gives areas under the curve (i.e. probabilities) for specified values of Z, the t table gives t values denoted t_α corrresponding to specified probabilities (α): each row of the table indicating a separate t distribution according to the degrees of freedom. Degrees of freedom are shown in the first column and probabilities are shown for *one tail* of the distribution along the top row ranging from 0.100 down to 0.005. Thus, for ten degrees of freedom, the values of t which leaves an area of 0.025 in the tail of the distribution (i.e. $t_{0.025}$) is 2.228 from the t table. Note that, unlike the Z distribution, the probabilities now relate to only one tail of the distribution rather than the main body of the distribution – compare Appendix A Table A3 with Appendix A Table A4.

It will be seen from the table of t that in the final row (where degrees of freedom are very large and strictly equal to infinity) the t statistics are identical to the Z statistics for corresponding levels of probability. In fact the t statistic is very close to the Z statistic when $n \geq 30$. Consequently, the Z statistic is often used in place of t, even in situations where t is appropriate (i.e. when $n \geq 30$), and, in practice, the t statistic therefore tends to be used only when $n < 30$. It is for this reason that t is sometimes referred to as the *small-sample statistic*.

Finally, it should be noted that the use of the t distribution requires that the population from which the samples are drawn does not depart markedly from normality, especially if the sample size is particularly small (less than 15). If this is not the case, the use of t as a measure of probability becomes unreliable. So, in the case of small samples drawn from a markedly skewed (non-normal) population, neither the Z nor t statistics are appropriate descriptive statistics. Some more advanced texts recommend a procedure based on 'Chebyshev's inequality' in these cases but this is beyond the scope of a basic statistics text and hence it is not dealt with here. An alternative procedure, of course, is to try to increase sample size.

To summarize, the t distribution is appropriate when all of the following conditions exist:

Conditions for the use of the t distribution

1. the sample standard deviation, s, is used to estimate the unknown population standard deviation, σ;

2. the sample size is small ($n < 30$);

3. the population is approximately normally distributed.

Interval estimation using the t distribution

Interval estimation using the t distribution follows the same principles as those involved in using the normal distribution using Z set out in Chapter 12. It will be recalled that by using Z, a confidence interval for the population mean μ may be defined as

$$\underline{\mu} = \bar{x} \pm Z_{\alpha/2}(\sigma_{\bar{x}})$$

where the level of confidence is $100(1 - \alpha)$ per cent as defined earlier. Correspondingly, where the t distribution applies, the formula for calculating a $100(1 - \alpha)$ per cent confidence interval for the population mean may be expressed as follows:

Formula for confidence interval for population mean using the t statistic

$$\underline{\mu} = \bar{x} \pm t_{(\alpha/2, n-1df)}(s_{\bar{x}}).$$

An illustration of the use of this formula is given in the applications section below.

| **Applications and Worked Examples** | **Worked example 13.1   Use of the t table – determining t values** |

General descriptions of the MINITAB and Excel computer programs and their application are given in Appendices B and C respectively.

What is the value of t, with ten degrees of freedom, that leaves:

(i) 10 per cent of the distribution in the right-hand tail of the t distribution?

(ii) 2.5 per cent of the distribution in the right-hand tail of the t distribution?

(iii) 2.5 per cent of the distribution in the left-hand tail of the distribution?

Solution

Using the t table in Appendix A Table A4 we can read off the corresponding t values directly:

(i) $t_{0.10,10} = 1.372$

(ii) $t_{0.025,10} = 2.228$

(iii) $t_{0.025,10} = -2.228$

Worked example 13.2   Use of the t table – determining probabilities

With 18 degrees of freedom, use the t table to find the probability that:

(i) t exceeds 1.734

(ii) t is less than -1.734

(iii) $-1.330 < t < 2.552$

(iv) $-1.330 > t$ or $t > 2.552$

Solution

(i) $P(t > 1.734) = 0.05 = \alpha$, i.e. 5 per cent

(ii) $P(t < -1.734) = 0.05 = \alpha$, i.e. 5 per cent

(iii) $P(-1.330 < t < 2.552)$
$= 1 - P(-1.330 > t > 2.552)$
$= 1 - (0.10 + 0.01)$
$= 0.89$, i.e. 89 per cent

(iv) $P(-1.330 > t$ or $t > 2.552) = 0.10 + 0.01$
$= 0.11$, i.e. 11 per cent

Worked Example 13.3   Confidence interval estimation of μ using t

The weights of persons using a lift are normally distributed. A random sample of eight people gives the following weights (in kg):

71, 85, 68, 72, 58, 76, 74, 80.

Estimate a 95 per cent confidence interval for the population mean.

Worked solution

Confidence interval: $\overline{\mu} = \bar{x} \pm t_{(\alpha/2, n-1df)}(s_{\bar{x}})$

The value of t, for 95 per cent confidence, with $8 - 1 = 7$ degrees of freedom, is:

$t_{(\alpha/2, n-1)} = t_{(0.05/2, 8-1df)} = 2.365$

The sample mean $\bar{x} = 73\,\text{kg}$ and the sample standard deviation $s = 8.09\,\text{kg}$. Thus:

$$s_{\bar{x}} = \frac{s}{\sqrt{n}} = \frac{8.09}{\sqrt{8}} = 2.86$$

Therefore, a 95 per cent confidence interval for the population is given by

$$\overline{\mu} = 73 \pm (2.365 \times 2.86)$$
$$= 73 \pm 6.76$$
$$= 66.24 \text{ to } 79.76\,\text{kg}$$

Computer solution using MINITAB

With the data entered to Column **C1** of the MINITAB worksheet, the procedure is:

Choose **Stat → Basic Statistics → 1-Sample t**...

Select variable **(C1)**

In the **Confidence interval Level** box enter 95

Click **OK**

The output is as follows:

T Confidence Intervals

Variable	N	Mean	StDev	SE Mean	95.0 % CI
CI	8	73.00	8.09	2.86	(66.23, 79.77)

Standard error of the mean $(8.09/\sqrt{8})$

Lower and upper limit respectively of 95% confidence interval

---

Computer solution using Excel

There is no pre-defined function in Excel for computing a confidence interval using the t distribution.

> **Worked example 13.4 Confidence interval estimation of μ with a finite population correction factor**

Forty children attend a nursery school. A random sample of eight children is selected and their heights measured (in cm) with the following results:

110, 112, 85, 117, 100, 98, 104, 90.

Construct a 95 per cent confidence interval for the population mean height.

Solution

Note that the sample size n is greater than 5 per cent of the population N, i.e. $n/N = 8/40 = 0.20 > 0.05$. In this situation, a finite population correction factor has to be applied in calculating the standard error (see Chapter 11). The confidence interval is now given by:

$$\underline{\mu} = \bar{x} \pm t_{(\alpha/2,n-1df)}(s_{\bar{x}})\sqrt{\left(\frac{N-n}{N-1}\right)}$$

For 95 per cent confidence,

$$t_{(\alpha/2,n-1df)} = t_{(0.05/2,8-1df)} = 2.365$$

from the t table. The sample mean $\bar{x} = 102$ cm and the sample standard deviation $s = 10.994$ cm. Thus,

$$s_{\bar{x}} = \frac{s}{\sqrt{n}} = \frac{10.994}{\sqrt{8}} = 3.89$$

Therefore, a 95 per cent confidence interval for the population mean is given by:

$$\underline{\mu} = 102 \pm (2.365 \times 3.89)\sqrt{\left(\frac{40-8}{40-1}\right)}$$
$$= 102 \pm 8.33$$
$$= 93.67 - 110.33 \text{ cm}$$

Note: there are no pre-defined tools or functions in MINITAB or Excel for solving this problem.

Key terms and concepts

Interval estimation of the mean using the t distribution Construction of an interval around a sample mean to give an estimate of the lower and upper limits within which a population mean lies with a degree of confidence using the t distribution, denoted:

$$\underline{\mu} = \bar{x} \pm t_{(\alpha/2)}(s_{\bar{x}})$$

where t has $n-1$ degrees of freedom.

Standard error of the mean when σ is unknown Standard deviation of the sampling distribution of means when the population standard deviation (σ) is not known, thus:

$$s_{\bar{x}} = \frac{s}{\sqrt{n}}$$

t **distribution** A family of probability distributions which can be used to develop interval estimates of a population mean and test statistical hypotheses whenever the population standard deviation is unknown and the population has a normal or near-normal probability distribution.

t **statistic** The difference between a sample mean and a population mean divided by the standard error computed from a small sample, thus:

$$t = \frac{\bar{x} - \mu}{s/\sqrt{n}}$$

Unbiased estimator of σ^2 An estimate of the population variance using sample information, denoted s^2:

$$s^2 = \frac{\Sigma(x_i - \bar{x})^2}{n - 1}$$

Self-study exercises

13.1. What is the value of *t*, with 15 degrees of freedom, that leaves:

    (a) 5 per cent of the distribution in the right-hand tail?

    (b) 1 per cent of the distribution in the left-hand tail?

    (c) 10 per cent of the distribution in the left-hand tail?

13.2. With 21 degrees of freedom, use the *t* table to find the probability that:

    (a) *t* exceeds 2.08

    (b) *t* is less than −2.518

    (c) −1.721 < *t* < 2.518

13.3. A consumer association, conducting a survey of torch batteries, tests random samples of each brand to determine how long they last. The results for one brand (in hours) were:

    9.0   9.2   9.5   10.1   9.8   9.3   9.7   9.6   10.0

Calculate a 95 per cent confidence interval for the population mean life.

13.4. In an industry consisting of 40 firms, a survey of 10 firms showed that the mean number of injuries per thousand man hours was 2.35 with a standard deviation of 0.5. Calculate a 95 per cent confidence interval for the mean for all firms.

| Chapter 14 | # Hypothesis testing: single population mean |

The testing of hypotheses represents the second main application of the principles of statistical inference (estimation being the first) noted in Chapter 10. In Chapters 12 and 13 we were concerned with the estimation of population parameters in terms of point estimates and confidence intervals. In this chapter we use sample statistics to test hypothesized values of the population parameters. In particular, we now consider the application of the principles of statistical inference in the context of hypotheses about population means. The topic is of central importance in statistics because hypothesis tests are very widely used as the basis for making decisions in industry and government and in research more generally. Some examples of the kinds of decisions to which the principles of hypothesis testing may be applied are as follows.

- Does a survey of television viewing hours show that there has been a significant increase in the number of hours watched by children under the age of 16 each week since the previous survey when the average was shown to be 25 hours per week?

- Is a manufacturing process running 'under control' in the sense that it performs its operation within pre-specified limits (e.g. machining a component to given size limits or filling containers with predetermined amounts of material) or has the machine drifted out of control and thus requires adjustment or repair?

- Has a new method for producing television tubes significantly increased their life expectancy over the previous method which produced tubes with an average life of 15 000 hours of operational time?

In each of these examples we start with some predetermined standard or knowledge of the previous value of a population parameter (e.g. the population mean) and wish to decide whether the new information, drawn from a sample, is or is not consistent with having been drawn from a population with, in this case, the same population mean.

The principles and procedures involved in hypothesis tests of these kinds are set out in this chapter as follows:

■ the basic concepts of hypothesis testing

■ conducting hypothesis tests

■ errors in hypothesis testing

■ use of hypothesis tests in quality control situations

The aim here is to convey the essence of the principles in such a way that their application across a wider range of problems can be appreciated. Consequently, we focus attention mainly on the application of the principles to the most important area – namely hypothesis testing of means. This chapter deals with hypothesis testing in the case of the mean of a *single* population while the next chapter is concerned with hypothesis testing involving *two*-population means, using tests based on Z or t as appropriate.

Learning Outcomes   After working through this chapter you will be able to:

■ understand the principles of statistical inference in the context of hypotheses about a single population mean

■ grasp the basic concepts of statistical significance and the terminology used in conducting tests of statistical hypotheses

■ understand what is meant by hypothesis testing as well as null and alternative hypotheses

■ comprehend the meaning of two-tailed and one-tailed hypothesis tests

■ understand what is meant by levels of significance and critical values

■ make comparisons between hypothesis testing and confidence interval estimation

■ conduct hypothesis tests

■ appreciate the meaning of Type I and Type II errors in hypothesis testing and their practical importance in decision making

■ understand the use of hypothesis tests in quality control procedures.

Principles

The basic concepts of hypothesis testing

Null and alternative hypotheses

The basic idea behind hypothesis testing is that we begin by making a tentative assumption about the value of a population parameter, such as the population mean μ, and then use sample information to 'test' the hypothesis. This tentative assumption is called the *null hypothesis* and is denoted by H_0. For example, the null hypothesis may be of the form that μ is equal to a hypothesized value μ_0, i.e. $H_0: \mu = \mu_0$. We then define another hypothesis, called the *alternative hypothesis*, which is denoted by H_1. The alternative hypothesis may take one of three possible forms which are in contradiction to the null hypothesis; i.e. $H_1: \mu \neq \mu_0$, $H_1: \mu > \mu_0$ or $H_1: \mu < \mu_0$. The hypothesis testing procedure consists of comparing the appropriate sample statistic (i.e. a sample mean denoted $\bar{x}$) with the corresponding parameter value hypothesized under H_0 (i.e. μ_0). If there is no difference between the two values, the null hypothesis may be accepted. However, it is more likely, of course, to find some difference either as a result of chance (i.e. sampling error – see Chapter 11), the hypothesis itself being true, or because it is in fact untrue. Thus the procedure involved in hypothesis testing is to reject the null hypothesis and accept the alternative only when the difference is so large that the probability of it occurring by chance is very small.

The rationale for the procedure is best understood by comparing it with that for making interval estimates of the population mean μ (as described in Chapters 12 and 13) because there is a direct parallel between the two. In making interval estimates of population means, upper and lower confidence limits are determined with reference to a desired level of confidence or probability. We can have no certainty that the true population mean does not fall outside these confidence limits but, as we saw, the probability of it doing so can be made very small by choosing large values of Z or t (as appropriate) in calculating these limits and/or increasing the sample size. Likewise in hypothesis testing, a hypothesis may be accepted or rejected not with certainty but with confidence that the likelihood of error in making the decision is small.

In hypothesis testing if the hypothesized mean, denoted by μ_0, is regarded as the true mean lying at the centre of a sampling distribution of $\bar{x}$ – shown in Figure 14.1(a) on the next page – then there is a 5 per cent probability of any single sample mean falling outside the limits defined by:

$$\mu_0 \pm 1.96 \text{ (standard error)}$$

assuming that the sampling distributed may be regarded as normal. As in Chapters 12 and 13, the level of probability (significance), in this case 5 per cent, is denoted by the symbol α. If the 5 per cent probability of error is acceptable, then we regard only sample means coming within this range as acceptable evidence that the sample has been drawn from a population with a mean that is *no different* from the hypothesized value μ_0. Thus the region within the limits becomes an *acceptance region* for the hypothesis and the regions outside the limits are then *rejection regions* for the hypothesis – see Figure 14.1. The probability level itself is known in this context as the *significance level* of the test. Note, therefore, that a 95 per cent confidence interval corresponds to a 5 per cent level of significance. Similarly, a 99 per cent confidence interval corresponds to a 1 per cent significance level. The level of significance represents the probability of making a mistake by rejecting a null hypothesis because of sampling error when it is in fact true. It is set independently by the person conducting the test.

Two-tailed and one-tailed tests

It will be seen in Figure 14.1(a) that there are two rejection regions, one in each tail. The test for 'no difference' is therefore known as a *two-tailed test*. Consequently, the 5 per cent rejection region is split into two equal halves with $2\frac{1}{2}$ per cent of the total area under the curve in each tail, i.e. $\alpha/2$. In contrast, so-called *one-tailed tests* deal with the situation

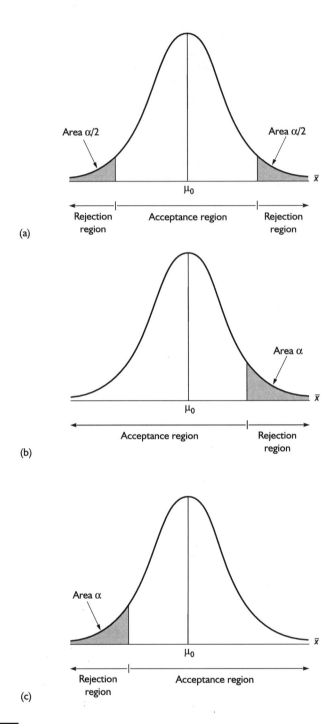

Figure 14.1 Acceptance and rejection regions in hypothesis testing for two-tailed and one-tailed tests: (a) two-tailed test; (b) one-tailed test; (c) one-tailed test

where we are interested in testing hypotheses that the population mean is specifically *greater than* or *less than* some hypothesized value – see Figures 14.1(b) and 14.1(c). In such cases there is only *one* rejection region – in only one tail or the other – the whole area of which equals α (and not $\alpha/2$ as above).

Comparison between hypothesis testing and confidence interval estimation

The similarly between the setting of confidence intervals and the testing of hypotheses is clear. The essential difference between the two lies in the objectives. With interval estimation we start with no assumption (hypothesis) about the value of the population parameter before sample data are collected, since the parameter value is unknown and the statistical objective is to estimate it with a given level of confidence. In hypothesis testing, on the other hand, we *do* start with an assumption about the value of the population parameter and then collect sample data to test the hypothesis. In an interval estimate, therefore, the sample mean $\bar{x}$ is placed at the centre of the interval (see Figure 12.1 in Chapter 12) whereas in the directly comparable hypothesis test it is the hypothesized mean μ_0 that is placed at the centre (see Figure 14.1).

Conducting hypothesis tests

As in the calculation of confidence intervals, a distinction has to be made between situations where the Z statistic may be used to determine the required level of significance (α) for the test from situations where the t statistic has to be used. We deal with each in turn in the context of two-tailed and one-tailed tests.

In the conduct of any hypothesis test, there are always five steps to be followed. These are explained in turn below.

Step 1: formulation of null (H_0) and alternative (H_1) hypotheses

The null hypothesis specifies the value of the parameter to be tested, e.g. that the population mean μ (unknown) is equal to some hypothesized value μ_0. The form of the alternative hypothesis, however, differs for two-tailed and one-tailed tests. In the case of two-tailed tests we are concerned with testing whether or not the true population parameter μ simply *differs* from the hypothesized value μ_0, whereas in the case of one-tailed tests the *direction of difference* is of concern and thus whether the true population parameter μ is *bigger* or *smaller* than the hypothesized value μ_0.

Thus the hypotheses are formulated as follows:

Two-tailed tests: $H_0: \mu = \mu_0$
                       $H_1: \mu \neq \mu_0$
One-tailed tests: *either* (a) $H_0: \mu = \mu_0$; $H_1: \mu > \mu_0$
               or   (b) $H_0: \mu = \mu_0$; $H_1: \mu < \mu_0$

Step 2: specification of the level of significance (α)

Having formulated the hypotheses to be tested and thus consequently determined whether a two-tailed or a one-tailed test is required, the next step is to fix the dividing line between the regions of acceptance and rejection. This simply depends upon the level of significance (α) at which it is decided to conduct the test and whether the sampling distribution of the statistic $\bar{x}$ can be regarded as being normal or whether the t distribution needs to be employed.

It is most common to work with 5 per cent or 1 per cent significance levels but any level may be chosen. Having chosen a significance level, the point of division between the acceptance and rejection regions marks the critical value of the test statistic. It is a critical value in the sense that it is the value against which the actual sample value is compared. For a two-tailed test there are, of course, two critical values (one positive and one negative) and for a one-tailed test, one critical value. These are shown in Figure 14.1 above. Following this terminology, rejection regions are themselves sometimes called 'critical regions'.

Step 3: selection of the test statistic and its critical value

There are two ways of stating the test statistic and thus of carrying out the test. One way is to calculate a critical value of the statistic itself ($\bar{x}$). The other is to work in terms of the standardized normal variable Z (or, where appropriate, the t value). Both ways, of course, are equivalent to each other. In practice, statisticians generally prefer the use of Z or t values (rather than $\bar{x}$ values) because they allow the level of significance of any test situation to be readily identified. Thus, the critical values in each case for testing a hypothesized mean μ_0 are as shown in Table 14.1.

Step 4: determination of actual value of the test statistic

If the test is being conducted using $\bar{x}$ then, of course, the actual value (as opposed to the critical value) of the test statistic is simply the actual value of the particular sample mean, $\bar{x}$. On the other hand, if the test is conducted using Z or t, then the actual value of the test statistic is obtained by transforming the value of the sample mean, $\bar{x}$, into its equivalent standardized form (Z or t). Thus,

$$Z = \frac{\bar{x} - \mu_0}{\sigma_{\bar{x}}} \quad \text{or} \quad t = \frac{\bar{x} - \mu_0}{s_{\bar{x}}}$$

Step 5: decision rule

Once the critical values have been determined then we are in a position to decide whether to accept or reject the null hypothesis in the two-tailed and one-tailed situations. It is essential to note that we focus upon acceptance or rejection of the *null* hypothesis – rejection of H_0 implies acceptance of H_1. The decision rule in each case is as follows.

Table 14.1 Critical values for testing a hypothesized mean μ_0

| Type of test | Hypotheses | | Critical values of test statistic using | | |
	H_0	H_1	$\bar{x}$	Z	t
Two-tailed	$\mu = \mu_0$	$\mu \neq \mu_0$	$\mu_0 \pm Z_{\alpha/2}(\sigma_{\bar{x}})$	$\pm Z_{\alpha/2}$	$\pm t_{(\alpha/2, df)}$
One-tailed	$\mu = \mu_0$	$\mu > \mu_0$	$\mu_0 + Z_\alpha \sigma_{\bar{x}}$	$+Z_\alpha$	$+t_{(\alpha, df)}$
One-tailed	$\mu = \mu_0$	$\mu < \mu_0$	$\mu_0 - Z_\alpha \sigma_{\bar{x}}$	$-Z_\alpha$	$-t_{(\alpha, df)}$

Note: df denotes degrees of freedom.

Decision rule: two-tailed test

Using $\bar{x}$ as the test statistic:

Accept H_0 *if the actual value of the sample mean $\bar{x}$ falls between the two critical values of $\bar{x}$*

Reject H_0 *otherwise*

Alternatively, using Z or t as the test statistic:

Accept H_0 *if the actual value of Z or t corresponding to $\bar{x}$ falls between the two critical values of Z or t*

Reject H_0 *otherwise*

Decision rule: one-tailed test

Using $\bar{x}$ as the test statistic:

Accept H_0 *if the actual value of $\bar{x}$ falls above or below the critical value of $\bar{x}$ depending on whether the lower or upper tail is the appropriate rejection region*

Reject H_0 *otherwise*

Alternatively, using Z or t as the test statistic:

Accept H_0 *if the actual value of Z or t corresponding to $\bar{x}$ falls above or below the critical value of Z or t depending on whether the lower or upper tail is appropriate*

Reject H_0 *otherwise*

All of these cases are shown in Figure 14.1. Examples of the application of both two-tailed and one-tailed tests, using $\bar{x}$, Z and t as the test statistics, are given in the Applications and Worked Examples section below.

Summary of procedural steps in hypothesis testing

We give below a summary of the five procedural steps in hypothesis testing which have been set out above.

Step 1   *Formulate the null (H_0) and alternative (H_1) hypotheses (thus determining whether a two-tailed or a one-tailed test is required).*

Step 2   *Specify the level of significance (α) to be used.*

Step 3   *Select the test statistic (either $\bar{x}$ or the corresponding Z or t value) and determine its critical value.*

Step 4   *Determine the actual value of the test statistic.*

Step 5   *Compare the actual value of the test statistic against its critical value and state the conclusion, i.e. accept or reject H_0 and state the decision clearly.*

Statistical significance and the *p* value approach to hypothesis testing

An alternative approach to using tests based on critical values of $\bar{x}$ or the standardized variables, Z or t, is to use the *probability value* or *p value*. This value is simply the probability of observing a sample outcome even more extreme than the observed value when the null hypothesis is true. So, for example, in a case where $Z = 2.1$, the p value would be (using Appendix A Table A3) $0.5 - 0.4821 = 0.0179$. This is the tail area of $Z > 2.1$. For any specified value of α exceeding this p value, we reject the null hypothesis. Figure 14.2 illustrates the relationship between the values of p and the critical values of Z for a one-tailed test situation.

The p value approach has the advantage of allowing us to specify the level of risk of rejecting a null hypothesis (which is in fact true) *after* the sample results are obtained. The approach is becoming very popular because many statistical software packages (such as MINITAB and Excel) automatically print out the p value of the test statistic and the user has no need to determine a critical value of the test statistic from the corresponding tables. Illustratioins of the use of p values are given in the Applications and Worked Examples section below.

Errors in hypothesis testing

It is important to understand that the conclusions of hypothesis tests do not provide *proof* of the hypotheses accepted or rejected. They are merely probability statements, not certainties. This follows from the fact that we are dealing with sampling data which are inevitably subject to sampling error. Thus, for example, acceptance of the null hypothesis (H_0) indicates *only* that the sample evidence is *insufficient to reject* at the chosen level of significance, *not* that it is necessarily true.

It will be appreciated that it is possible to reject a null hypothesis when it is in fact true and, conversely, to accept it when it is in fact false. Errors such as these are referred to as *type I* and *type II* errors respectively. The occurrence of these errors can be summarized as shown in Table 14.2. The importance of recognizing the possibility of errors in hypothesis testing is that it is possible to *control* them, to some extent, when setting up the sample scheme.

The probability of committing a type I error can be controlled by specifying α; i.e. the significance level chosen specifies the *risk* we are willing to take of rejecting a true null hypothesis. A type II error, however, cannot be controlled simultaneously, although it is possible to compute it. The two types of error are interdependent and it is more usual to control type I error. The only way to reduce both types of error simultaneously is to

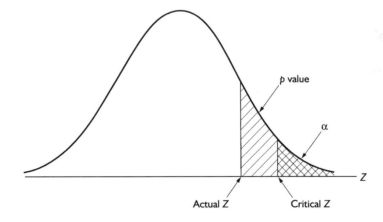

Figure 14.2 Comparison of p values and critical values of Z in a one-tailed test

Table 14.2 Errors in hypothesis testing

	Situation in the population	
Test decision	H_0 true	H_0 false
Accept H_0	Correct decision	Type II error
Reject H_0	Type I error	Correct decision

increase the sample size, though, of course, this may not be practicable. The techniques required to control type I and type II errors are beyond the scope of this book: they involve the development of *operating characteristic curves* or corresponding *power curves* which allow the judgement of how well a test performs in minimizing type II error in various possible test situations.

Operating characteristic curves and power curves are especially useful in industry in, for example, acceptance sampling. In this context the management will have in mind some values of the probability of making type I and type II errors, which it is willing to tolerate, and also some deviation from the hypothesized value of the parameter which it considers of practical importance and wishes to detect.

Practical importance of type I and type II errors

The balance between type I and type II errors can be important in many practical decision-making situations. A standard example may be taken from the pharmaceutical industry. Consider a company faced with making a decision concerning whether or not to introduce a newly-developed vaccine. There is a danger of the company concluding that the new vaccine is *not* better than another one *when in fact it is* (a type II error). This means that the vaccine will not be supplied and that people may suffer or die unnecessarily. Conversely, the conclusion that it *is* better *when in fact it is not* (a type I error) will mean that the vaccine is supplied even though it is, at best, no better than a cheaper alternative and, more seriously, it could even be a dangerous vaccine.

Use of hypothesis tests in quality control situations

Another common application of the principles of hypothesis testing involves quality control in production processes. For example, it is not unusual for variations to arise in the dimensions, tensile strength, finish, etc. of repetitive items of production despite the apparently identical conditions of production. Management face the problem of deciding whether any observed variations are due to chance fluctuations which reflect the inherent variability of the process or are caused by faults in the machine or materials or by operative error.

Given a required mean value of the item being produced (e.g. mean length, diameter, weight), regular samples can be taken and their mean values compared against a range of values indicating acceptable performance: sample means falling within the range indicate that the production process is 'under control', mean values falling outside the range indicate that it is out of control and that remedial action is required. These limits are set by measuring so many standard errors $\sigma_{\bar{x}}$ on either side of the required mean μ_0. It is common to *set warning limits* as $\mu_0 \pm 2\sigma_{\bar{x}}$ and *control limits* as $\mu_0 \pm 3\sigma_{\bar{x}}$. These are often drawn on a *control chart* on which the results of repeated sampling are recorded. A worked example in the next section makes the procedure clear.

Applications and Worked Examples   Worked example 14.1   Two-tailed Z test – σ known

General desciptions of the MINITAB and Excel computer programs and their applications are given in Appendices B and C respectively.

A census of households living in urban areas shows an average family size of 3.9 persons with a standard deviation of 0.6. A random sample of 100 families in rural areas showed an average family size of 4.2. Test the hypothesis that the family size in rural areas is no different from that found in urban areas. The null hypothesis is that the sample of 100 families comes from a population in which the mean is 3.9.

We set out the test formally below, following the steps set our earlier. For this purpose we shall conduct the test at the 5 per cent level of significance i.e. $\alpha = 0.05$.

Solution

Note that the *normal* distribution and thus a Z test may be used here because the value of σ is known.

Given: $\mu = 3.9; \sigma = 0.6; n = 100$

Hypotheses: $H_0: \mu_0 = 3.9$
                             $H_1: \mu_0 \neq 3.9$ (i.e. two-tailed test)

Specification of significance level : $\alpha = 0.05$, hence for a two-tailed test we specify

$$Z_{\alpha/2} = Z_{0.025} = \pm 1.96$$

Standard error of the mean: $\sigma_{\bar{x}} = \dfrac{\sigma}{\sqrt{n}} = \dfrac{0.6}{\sqrt{(100)}} = 0.06$

Critical values: Using $\bar{x}: \mu_0 \pm Z_{\alpha/2}\sigma_{\bar{x}} = 3.9 \pm 1.96\,(0.06)$
                                 $= 3.78$ and 4.02
              or $Z: \pm Z_{\alpha/2} = -1.96$ and $+1.96$

The diagram below illustrates the equivalence of the two ways of defining critical values.

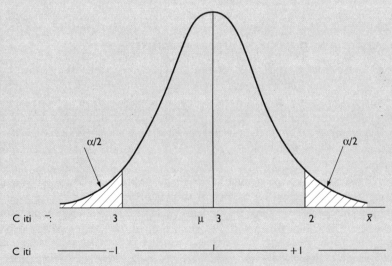

Test:

(a) Using critical $\bar{x}$ values:

actual $\bar{x}$ $(= 4.2)$ falls outside the critical value (i.e. $4.02 < 4.2$)

(b) Alternatively, using critical Z values:

$$\text{actual } Z = \frac{\bar{x} - \mu_0}{\sigma_{\bar{x}}} = \frac{4.2 - 3.9}{0.06} = 5$$

which falls outside the critical value (i.e. $1.96 < 5$)

Conclusion

Since the actual values of $\bar{x}$ or Z fall in the rejection region, we reject the null hypothesis (at the 5 per cent level of significance) that the family size in rural areas is no different from that found in urban areas. It can be concluded, therefore, that the average family size in rural areas is *different* from that in urban areas (note that we cannot conclude that it is greater because that would require a one-tailed test – see Worked example 14.2 below).

Computer solutions to Worked examples 14.1 and 14.2 (in which σ is known) are not given as the procedure is the same as in Worked examples 14.3 and 14.4 (in which σ is not known) and the latter represent the usual situation in practice.

Worked example 14.2   One-tailed Z test – σ known

Using the same data as in Worked example 14.1, we may wish to test whether or not average family size in rural areas is *greater than* that in urban areas, again at the 5 per cent level of significance.

Solution

Hypotheses: H_0: $\mu_0 = 3.9$
H_1: $\mu_0 > 3.9$ (i.e. a one-tailed test)

Specification of significance level: $\alpha = 0.05$, hence for a one-tailed test we specify

$$Z_\alpha = + Z_{0.05} = + 1.645$$

Standard error of the mean: $\sigma_{\bar{x}} = 0.06$ (from Worked example 14.1 above)

Critical values: Using $\bar{x}$: $\mu_0 + Z_\alpha(\sigma_{\bar{x}}) = 3.9 + 1.645\,(0.06)$
$= 4.0$
or Z: $+ Z_\alpha = +1.645$

Test:

(a) Using the critical $\bar{x}$ value:

actual $\bar{x}$ $(= 4.2)$ is greater than the critical value (i.e. $4.0 < 4.2$)

(b) Alternatively, using the critical Z value:

$$\text{actual } Z = \frac{\bar{x} - \mu_0}{\sigma_{\bar{x}}} = 5$$

(from Worked example 14.1) which is greater than the critical value (i.e. $1.645 < 5$)

Worked example 14.2   One-tailed Z – σ known *(continued)*

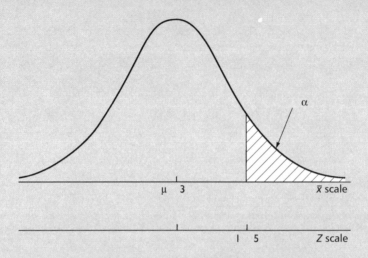

μ   3   x̄ scale

I   5   Z scale

Conclusion

As before, since the actual values of $\bar{x}$ and Z fall in the rejection region, we reject the null hypothesis that $\mu_0 = 3.9$ and conclude, at the 5 per cent level of significance, that family size in rural areas is greater than that for urban areas.

Worked example 14.3   Two-tailed Z test – σ unknown

A battery manufacturer claims that his batteries have an average life of 55 hours (i.e. $\mu_0 = 55$). A batch of 40 batteries is tested, giving a mean life $\bar{x}$ of 50 hours with a standard deviation s of 11.734 hours. Does this show that the manufacturer's claim is justified at the 1 per cent level of significance?

Worked solution

Given:	$\mu_0 = 55; \bar{x} = 50; s = 11.734; n = 40$
Hypotheses:	$H_0: \mu_0 = 55$
	$H_1: \mu_0 \neq 55$ (i.e. a two-tailed test)
Specification of significance level:	$\alpha = 0.01$
Standard error of the mean:	$s_{\bar{x}} = \dfrac{s}{\sqrt{n}} = \dfrac{11.734}{\sqrt{(40)}} = 1.86$
Critical values:	$\pm Z_{\alpha/2} = \pm 2.58$

In the interests of brevity we only use Z values. The corresponding critical values of $\bar{x}$ would be:

$$\bar{x} = \mu_0 \pm Z_{\alpha/2}(s_{\bar{x}}) = 55 \pm 2.58\,(1.86) = 50.2 \text{ and } 59.8$$

It is normal practice to focus on Z values rather than $\bar{x}$ values since the results are identical and, as explained above, there are advantages in conducting the test in this way.

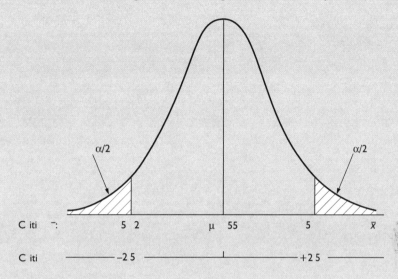

Test:

$$\text{actual}\quad Z = \frac{\bar{x} - \mu_0}{s_{\bar{x}}} = \frac{50 - 55}{1.86}$$
$$= -2.69$$

In *absolute* terms, this exceeds the critical Z value of -2.58 and hence falls in the rejection region.

Conclusion

Stating the conclusion formally:

$$|Z_{\text{actual}}| > |Z_{\text{critical}}|$$

(The vertical bars signify that absolute values (i.e. ignoring negative signs) are being compared.) Thus we must reject the null hypothesis and conclude that the manufacturer's claim of an average battery life of 55 hours is unjustified (at the 1 per cent level of significance).

Computer solution using MINITAB

With the data entered in column **C1** of the MINITAB worksheet, the procedure is as follows:

Choose **Stat** → **Basic Statistics** →
   1-Sample Z...
Enter **Variable** (C1)

Check **Test Mean** and enter value of the hypothesized mean (55.0)
Select **Alternative** (i.e. the alternative hypothesis): 'Not equal' in this case
Enter the value of **Sigma** (11.734)
Click **OK**

Worked example 14.3: Two-tailed *Z* test–σ unknown *(continued)*

The output is as follows:

Z-Test

Test of mu = 55.00 vs mu not = 55.00
The assumed sigma = 11.7

Variable	N	Mean	StDev	SE Mean	Z	P
C1	40	50.00	11.73	1.86	−2.69	0.0072

Authors' comment: In line with the manual worked solution, because the calculated value of *Z* (= 2.69 in absolute terms) is greater than the critical value of *Z* from tables (= 2.58 in absolute terms), we arrive at the same conclusion as above and reject the null hypothesis. Note, however, that the MINITAB results give a 'P' value alongside the calculated *Z* value. This gives the precise probability of the result. Thus rather than looking up tables of *Z*, we can compare this 'P' value with the required significance level (α = 0.01). As 0.0072 < 0.01, we reject the null hypothesis at the 1% level of significance, as before. This 'P' value can be used, of course, to test at any pre-specified α level – as long as the 'P' value is less than the α level, the null hypothesis can be rejected.

Should it be desired to obtain the critical value of *Z*, note that this may be derived directly using MINITAB, rather than looking up tables, by using the *inverse cumulative probability* – this is illustrated below.

*Using MINITAB to derive critical values of *Z**

MINITAB may be used to derive critical values for various probability distributions. We illustrate the procedure here for the Normal distribution.

Choose **Calc → Probability Distributions → Normal...**
Select **Inverse cumulative probability**
In **Mean** enter 0 and in **Standard deviation** enter 1 (in order to give values for the standard normal distribution)
In **Input constant** enter the significance level

(α) desired, i.e. tail area (here, as we are interested in a 1% level of significance in a two-tailed test we may enter 0.005, which will give the critical value of *Z* for the left-hand tail). Alternatively, entering the complement of 0.005 (0.995) gives the critical value of *Z* for the right-hand tail. Both of these are shown below
Click **OK**

The output is as follows:

Inverse Cumulative Distribution Function

Normal with mean = 0 and standard deviation = 1.00000

P(X <= x)	x
0.0050	−2.5758

 Critical value of Z (left-hand tail)

Inverse Cumulative Distribution Function

Normal with mean = 0 and standard deviation = 1.00000

P(X <= x)	x
0.9950	−2.5758 ← Critical value of Z (right-hand tail)

Computer solution using Excel

E

There is no pre-defined tool in Excel for solving this problem. The Excel function ZTEST returns the two-tailed P-value of a Z-test by generating a standard score for x with respect to a *data set (array)*, and returning the two-tailed probability for the normal distribution. In this example the raw data set is not given and thus ZTEST cannot be used.

Worked example 14.4   One-tailed Z test – σ unknown

Using the same data as in Worked example 14.3, we may wish to test specifically whether the average battery life is significantly *less* than the manufacturer's claim of 55 hours at the 1 per cent level of significance.

Worked solution

Given: $\mu_0 = 55; \bar{x} = 50; s = 11.734; n = 40$

Hypotheses: $H_0: \mu_0 = 55$
  $H_1: \mu_0 < 55$ (i.e. a one-tailed test)

Significance level: $\alpha = 0.01$

Standard error of the mean: $s_{\bar{x}} = \dfrac{s}{\sqrt{n}} = \dfrac{11.734}{\sqrt{(40)}} = 1.86$ (as before)

Critical value: $-Z_\alpha = -2.33$ from Appendix A Table A3

(Note that since we are testing whether the average battery life is *less* than the manufacturer's claim of 55, we assign a negative sign to the critical Z value.)

Test:

$$\text{actual } Z = \frac{\bar{x} - \mu_0}{s_{\bar{x}}} = \frac{50 - 55}{1.86}$$
$$= -2.69$$
$$-2.69 < -2.33$$

Conclusion: Since the actual Z value is less than the critical Z value, it falls in the rejection region. Therefore we must reject the null hypothesis and conclude that the average battery life is less than 55 hours (at the 1 per cent level of significance).

Worked example 14.4: One-tailed *Z* test – σ unknown *(continued)*

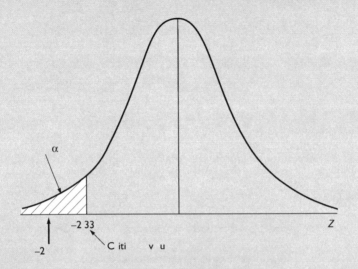

Computer solution using MINITAB

With the data entered to column **C1** of the MINITAB worksheet, the procedure is:

Choose **Stat → Basic Statistics → 1-Sample Z**
Enter **Variables** (C1)
Enter **Test Mean** (55)

Enter **Alternative**, i.e. Alternative hypothesis (less than)
Enter **Sigma** (11.734)
Click **OK**

The output is as follows:

Z-Test

Test of mu = 55.00 vs mu < 55.00
The assumed sigma = 11.7

Variable	N	Mean	StDev	SE Mean	Z	P
C1	40	50.00	11.73	1.86	−2.69	0.0036

Authors' comment: This gives a calculated *Z* value of −2.69, (as above) which can be compared with the critical value of −2.33 (from tables or using the MINITAB *inverse cumulative probability* – see MINITAB solution to Worked example 14.3) as before, and thus the null hypothesis is rejected and we conclude that the mean battery life is less than 55 hours at the 1% level of significance. Alternatively, the 'P' value may be compared with the desired significance level for the test: as 0.0036 < 0.01, the null hypothesis is rejected at the 1% level of significance.

Computer solution using Excel **E**

There is no pre-defined tool in Excel for solving this problem.

Worked example 14.5   Two-tailed *t* test – σ unknown

A survey of a country's workforce found that the number of days lost each year through sickness was 15 days per worker on average. Suppose that a researcher wants to test this survey result by monitoring the records of a sample of 25 workers. The following data show the number of days absent for each of the 25 workers during a particular year.

5	25	10	0	3	50	12	14	40
12	32	8	4	47	20	14	16	10
1	22	58	5	23	18	9		

Use these data to determine whether the claim of 15 days lost per worker each year should be rejected by the researcher's sample survey results (at the 5 per cent level of significance).

Worked solution

Given:

$\mu_0 = 15; n = 25$
By calculation: $\bar{x} = 18.32, s = 15.845$

Hypotheses:

$H_0: \mu_0 = 15$
$H_1: \mu_0 \neq 15$ (i.e. a two-tailed test)

Significance level:

$\alpha = 0.05$

Standard error of the mean: $s_{\bar{x}} = \dfrac{s}{\sqrt{n}} = \dfrac{15.845}{\sqrt{(25)}} = 3.169$

Critical values: $\pm t_{\alpha/2} = \pm 2.064$ from Appendix A Table A4 with $(25 - 1)$ df

Test: actual $t = \dfrac{\bar{x} - \mu_0}{s_{\bar{x}}} = \dfrac{18.32 - 15}{3.169} = 1.05$

Thus,

critical $t(= +2.064) >$ actual $t(= 1.05)$

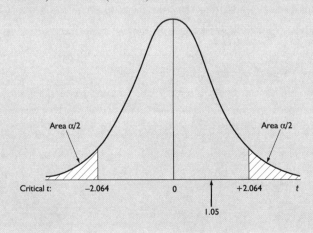

Worked example 14.5   Two-tailed *t* test – σ unknown *(continued)*

Conclusion: Since the actual *t* value is less than the critical +*t* value, we accept the null hypothesis at the 5 per cent level of significance and conclude that the number of days of sickness is not significantly different from 15.

Computer solution using MINITAB

With the data entered to column **C1** of the MINITAB worksheet, the procedure is as follows:

Choose **Stat → Basic Statistics →**
  1-Sample t...
Enter **Variable** (C1)

Check **Test Mean** and enter value of the hypothesized mean (15.0)
Select **Alternative**, i.e. the alternative hypothesis: ('Not equal' in this case)
Click **OK**

The output is as follows:

T-Test of the Mean

Test of mu = 15.00 vs mu not = 15.00

Variable	N	Mean	StDev	SE Mean	T	P
C1	25	18.32	15.84	3.17	1.05	0.31

Authors' comment: Note that, unlike the Z test above (Worked solution 14.3) there is no requirement to specify sigma (σ) in the commands; this is because sigma is unknown and the test uses the sample standard deviation *s* in calculating the standard error of the mean and this is computed by MINITAB directly. At the 5% level of significance the critical value of *t* with 24 degrees of freedom (i.e. $n - 1 = 25 - 1$) is ±2.064. Therefore, as the actual (absolute) value of *t* (= 1.05) is less than + 2.064, we accept the null hypothesis and conclude that the number of days of sickness is not significantly different from 15. If desired, the critical value of *t* may be derived using the MINITAB *inverse cumulative probability* function – see MINITAB solution to Worked example 14.3 which illustrates the procedure using *Z*.

Alternatively, using the 'P' value, as $0.31 > \alpha (= 0.05)$ we accept the null hypothesis, as before, and reach the same conclusion.

Computer solution using Excel

There is no pre-defined tool in Excel for solving this problem.

Worked example 14.6   One-tailed *t* test – σ unknown

Using the information given in Worked example 14.5, conduct one-tailed *t* tests:

(a) that the number of days lost, on average, *exceeds* the claim of 15 and
(b) that the number of days lost, on average, is *less than* the claim of 15 at the 5 per cent level of significance.

Worked solutions

(a) **Alternative hypothesis H$_1$: μ$_0$ > 15**
Actual *t* (from Worked example 14.5) = 1.05
Critical *t* with (25–1) df (from the tables) = 1.711
Thus, critical *t* (= 1.711) > actual *t* (= 1.05) and we conclude that the null hypothesis cannot be rejected, i.e. that the number of days lost *does not exceed* 15.

(b) **Alternative hypothesis H$_1$: μ$_0$ < 15**
Actual *t* (from Worked example 14.5) = 1.05
Critical *t* with (25–1) df (from the tables) = −1.711
Thus, critical *t* (= −1.711) < actual *t* (= +1.05) and we conclude that the null hypothesis cannot be rejected, i.e. that the number of days lost is *not less than* 15.

Computer solution using MINITAB

(a) **Alternative hypothesis H$_1$: μ$_0$ > 15**
With the data entered in column **C1** of the MINITAB worksheet, the procedure is as follows:

Choose **Stat → Basic Statistics →
    1-Sample t...**
Enter **Variable** (C1)

Check **Test Mean** and enter value of the hypothesized mean (15.0)
Select **Alternative** (i.e. the alternative hypothesis): 'greater than' in this case
Click **OK**

The output is as follows:

T-Test of the Mean

Test of mu = 15.00 vs mu > 15.00

Variable	N	Mean	StDev	SE Mean	T	P
C1	25	18.32	15.84	3.17	1.05	0.15

Authors' comment: Without determining the critical value of *t*, as $P = 0.15 > \alpha$ (= 0.05) we may conclude that the null hypothesis cannot be rejected – as above.

Worked example 14.6   One-tailed *t* test–σ unknown *(continued)*

Computer solution using MINITAB

(b)   Alternative hypothesis H₁: μ₀ < 15
With the data entered in column **C1** of the MINITAB worksheet, the procedure is as follows:

Choose **Stat → Basic Statistics →**
    1-Sample t...
Enter **Variable** (C1)

Check **Test Mean** and enter value of the hypothesized mean (15.0)
Select **Alternative** (i.e. the alternative hypothesis): 'less than' in this case
Click **OK**

The output is as follows:

T-Test of the Mean

Test of mu = 15.00 vs mu < 15.00

Variable	N	Mean	StDev	SE Mean	T	P
C1	25	18.32	15.84	3.17	1.05	0.85

Authors' comment: Again, without determining the critical value of *t*, as P = 0.85 > α (= 0.05) we may conclude that the null hypothesis cannot be rejected – as above.

In both cases ((a) and (b)), since the *p* value exceeds α(= 0.05) we cannot reject the null hypothesis that $\mu_0 = 15$. It may be useful to note that, as a general rule of thumb, for extremely small *p* values relative to the chosen value of α, we would tend to reject the null hypothesis while for relatively larger *p* values (relative to α) we would tend to accept H₀. The figures below summarize these findings.

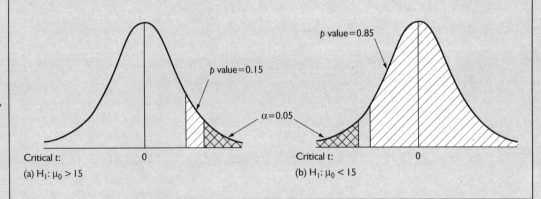

(a) H₁: μ₀ > 15   (b) H₁: μ₀ < 15

Computer solution using Excel

There is no pre-defined tool in Excel for solving this problem.

Worked example 14.7   Quality control charts

A manufacturing company produces metal rods for reinforcing concrete. The 'target' length specified for the rods is 3 metres. The usual variability of the process gives a standard deviation σ of 0.05 metres. From time to time samples of ten rods are examined and the mean length of each sample is calculated. These sample means are then compared with 'warning limits' of $\mu \pm 2\sigma_{\bar{x}}$ and 'control limits' of $\mu \pm 3\sigma_{\bar{x}}$, to assess whether the production process remains under control. If the following eight sample mean results are obtained, what actions should be taken?

Sample No.	1	2	3	4	5	6	7	8
Mean	2.98	3.01	3.06	3.02	2.96	2.99	3.00	3.07

Worked solution

Standard error of the mean $(\sigma_{\bar{x}}) = \dfrac{\sigma}{\sqrt{n}} = \dfrac{0.05}{\sqrt{(10)}} = 0.0158$

Hence

Warning limits: $(\mu \pm 2\sigma_{\bar{x}}) = 3 \pm (2 \times 0.0158) = 2.97$ to 3.03

Control limits: $(\mu \pm 3\sigma_{\bar{x}}) = 3 \pm (3 \times 0.0158) = 2.95$ to 3.05

The fifth sample mean (2.96 metres) falls outside the warning limits but is inside the control limits; this in itself is not serious and no action is required. But the means of the third and eighth samples (3.06 and 3.07 respectively) fall outside both the warning limits and the control limits, suggesting that remedial action is required for this process. The control chart for this production process is shown in the MINITAB printout below.

Worked example 14.7   Quality control charts *(continued)*

Computer solution using MINITAB

Choose **Stat → Control Charts → Xbar...**

Complete **Data are arranged as** boxes as follows:

Enter C1 in **Single column box** and '1' in **Subgroup size** box

Enter **Historical mean** (3)

Enter **Historical sigma** (.0158)

Click on **S Limits...**

Enter 2 3 in **Sigma limit positions** box [this is optional but produces upper and lower warning and control limit lines at 2 and 3 standard errors above and below the specified (historical) mean]

Select **Attributes of Sigma Limits** as desired (line type, colour and size)

Click **OK** twice

The output is as follows:

X-bar Chart for C1

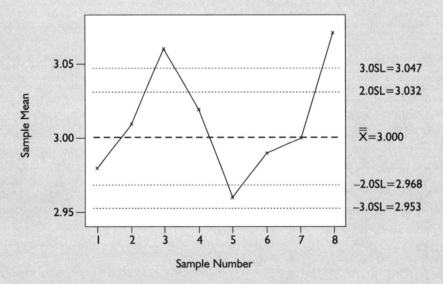

Authors' comment: The MINITAB program has been used here to produce a basic control chart. The program also contains a wide range of options for the production of different types of control charts and the conduct of tests designed to highlight automatically situations which merit further investigation. These are beyond the scope of this book but are fully documented in MINITAB itself as well as more specialized texts.

Computer solution using Excel

There is no tool in Excel for constructing control charts.

Key terms and concepts

Acceptance region Region in which the null hypothesis cannot be rejected.

Alternative hypothesis Specifies the hypothesis assumed true if the null hypothesis is rejected.

Control chart A diagram used for quality control purposes in which the results of repeated sampling are recorded.

Control limits Limits on a control chart defined as $\mu_0 \pm 3\sigma_{\bar{x}}$

Critical region Region in which the null hypothesis cannot be accepted (sometimes referred to as rejection region).

Critical value A value that is compared with the test statistic to determine whether H_0 is to be accepted or rejected.

Errors in hypothesis testing Errors relating to the possibility of rejecting a null hypothesis when it is in fact true and, conversely, to accept it when it is in fact false.

Hypothesis testing Procedures using sample statistics to test hypothesized values of population parameters.

Level of significance The maximum probability of a Type I error that the user will tolerate in the hypothesis-testing procedure.

Null hypothesis Specifies the hypothesized value of the parameter to be tested.

One-tailed test An hypothesis test in which rejection of the null hypothesis occurs for values of the test statistic in one tail of the sampling distribution.

p value The probability of observing a sample outcome even more extreme than the observed value when the null hypothesis is true.

Rejection region Region in which the null hypothesis cannot be accepted (sometimes referred to as critical region).

Statistical inference Drawing conclusions about statistical populations based on information obtained from sample data.

Test statistic Statistic, such as Z or t, which defines the acceptance and rejection regions according to defined levels of probability.

Two-tailed test An hypothesis test in which rejection of the null hypothesis occurs for values of the test statistic in either tail of the sampling distribution.

Type I error The error of rejecting H_0 when it is true.

Type II error The error of accepting H_0 when it is false.

Warning limits Limits on a control chart defined as $\mu_0 \pm 2\sigma_{\bar{x}}$

Self-study exercises

14.1. In what circumstances should t rather than Z be used in hypothesis testing?

14.2. For each of the following test situations, each involving a test of a hypothesized mean, state the critical value of $\bar{x}$, given $\mu_0 = 80$, where α denotes the level of significance and μ_0 denotes the hypothesized mean:

(a) Two-tailed test, $\alpha = 0.05$, $n = 35$, $\sigma = 4$

(b) One-tailed test, $\alpha = 0.05$, $n = 16$, $\sigma = 4$

(c) One-tailed test, $\alpha = 0.01$, $n = 20$, $s = 9$

(d) $H_0: \mu = \mu_0$; $H_1: \mu \neq \mu_0$; $\alpha = 0.01$; $n = 90$, $s = 19$

(e) $H_0: \mu = \mu_0$; $H_1: \mu < \mu_0$; $\alpha = 0.05$; $n = 18$, $\sigma = 9$

14.3. Stating any necessary assumptions, indicate whether the Z or the t distribution applies in hypothesis testing of the mean in the following situations:

(a) $n = 21$, $\sigma = 3$

(b) $n = 26$, $s = 15$

(c) $n = 38$, $s = 10$

(d) $n = 45$, $\sigma = 5$

14.4. The mean number of accidents experienced by a group of 200 coal miners over an observation period was 2.0 with a standard deviation of 1.487. Test the hypothesis that the mean number of accidents amongst all miners represented by this sample group is 1.75.

14.5. Electric motors for washing machines are known to have an expected life of 2000 hours of use. If a sample of 100 of a new model of motor has a mean life of 2500 hours and a standard deviaiton of 1600 hours, can one conclude with 95 per cent confidence that the new model is more reliable?

14.6. A random sample of 25 unemployed persons in a village showed that the mean duration of unemployment was 87 days with a variance of 256 days. A census conducted ten years previously when the register was much smaller showed that the mean duration then was 64 days. Can we safely conclude that the duration of unemployment has increased?

14.7. The lives of six candles are found to be 8.1, 8.7, 9.2, 7.8, 8.4 and 9.4 hours giving a mean of 8.6 and a variance of 0.388. The manufacturer claims that the average life is 9.5 hours. Making a suitable assumption concerning the nature of the distribution of the life of a candle, carry out a statistical test of the manufacturer's claim.

14.8. A company guarantees that its torch batteries will last an average of 9.75 hours. After complaints, the company takes a random sample of batteries from stock and tests their lifetime with the following results:

9.0   9.2   9.5   10.1   9.8   9.3   9.7   9.6   10.0

Stating any assumption made, test the guarantee at the 0.05 significance level.

Chapter 15 | Hypothesis testing: two population means

Aims   In the last chapter we were concerned with tests of hypotheses about a single population mean using information from a single sample. In practice, the situation often arises where information is available from two samples and we wish to test whether or not the difference between the two sample means is a statistically significant difference. For example, a supermarket chain that is considering opening a branch in one of two cities may want to know whether there is a significant difference between average household income in the two cities. To compare the two cities, information may be obtained by selecting random samples from each city. As with tests of a simple mean, the hypothesis to be tested is one concerning the means of the populations from which the samples have been drawn. In this case, the null hypothesis – the hypothesis of no difference – is that the mean of the parent population from which the first sample is drawn is equal to the mean of the parent population from which the second sample is drawn. Formally, therefore, the null hypothesis to be tested is:

$$H_0: \mu_1 = \mu_2 \quad \text{(or equivalently } \mu_1 - \mu_2 = 0\text{)}$$

where μ_1 and μ_2 denote the unknown means of the two parent populations respectively.
  The alternative hypothesis may take one of three forms, as before:

$$H_1: \mu_1 \neq \mu_2 \quad \text{(or equivalently } \mu_1 - \mu_2 \neq 0\text{)}$$
$$H_1: \mu_1 > \mu_2 \quad \text{(or equivalently } \mu_1 - \mu_2 > 0\text{)}$$
$$H_1: \mu_1 < \mu_2 \quad \text{(or equivalently } \mu_1 - \mu_2 < 0\text{)}$$

An important distinction to be made is between *matched* and *independent* samples. For example, one may wish to make a comparison of the effect of, say, a training programme in improving the speed with which operatives carry out a particular task. Measurements may be taken of the times taken by the *same* group of operatives before and after the training programme. This is a *matched sample*. Obviously it would be foolish to try to investigate the impact of the training programme by taking samples of different operatives. The latter is known as *independent sampling*. Matched sampling is more efficient than independent sampling because, by observing the *same* group before and after, an important source of variation is removed – i.e. the variation between different people which is not related to the training. Matched sampling is appropriate where repeated measurement is required and feasible. On the other hand, independent sampling is appropriate in situations involving one-off, or non-recurrent, studies.

The basic principles involved in hypothesis tests involving matched and independent samples remain the same but the procedure in the case of matched samples requires adaptation and is, in fact, simpler to apply. In both cases both large and small samples may be involved and need to be treated separately. As in the case of tests for a single mean (see Chapter 14), this involves the use of Z and t tests.

The principles and applications of hypothesis tests are organized below according to the nature of the sample, i.e. using:

■ *large* independent samples

■ *small* independent samples

■ *large* matched samples

■ *small* matched samples

Learning Outcomes   After working through this chapter you will be able to:

■ understand the principles of statistical inference in the context of hypotheses about two population means and the terminology used

■ comprehend the concept of the sampling distribution of the difference between two population means

■ recognize the distinction between matched and independent sampling and samples

■ conduct hypothesis tests of the difference between two population means using the standard normal and t distributions as appropriate

■ know when to use the standard normal and t distributions in conducting such tests

■ conduct tests of hypotheses concerning matched samples.

Principles

Hypothesis testing using large independent samples

As we are dealing with large samples here (i.e. samples n_1 and n_2 both greater than or equal to 30), the Z test is appropriate (the central limit theorem may be invoked if the population standard deviations σ_1 and σ_2 are not known – see Chapter 11). The test procedures involve the same steps as set out in Chapter 14 involving tests of a single mean. There is only one additional factor to consider. This involves the computation of the test statistic (actual Z). The problem we are faced with is one of deciding how large the difference between the two sample means, $\bar{x}_1$ and $\bar{x}_2$, needs to be to allow us to conclude that they are likely to have come from two different populations, and thus that the difference between the sample means is statistically significant at some defined level of probability. To answer this question demands information about the sampling distribution of the difference between *all possible pairs of sample means*, i.e. the distribution of $\bar{x}_1 - \bar{x}_2$ for all possible values of $\bar{x}_1$ and $\bar{x}_2$ given the populations and sample sizes.

Fortunately, as long as the sample sizes n_1 and n_2 are both large, the distribution of $\bar{x}_1 - \bar{x}_2$ may be approximated by the normal probability distribution. Furthermore, the mean of this distribution ($\mu_{\bar{x}_1 - \bar{x}_2}$) is *zero* when the samples are drawn under the null hypothesis assumption that $\mu_1 = \mu_2$.

In order to conduct significance tests on the hypothesis that $\mu_1 = \mu_2$, we now need to establish the standard error of this sampling distribution. It can be shown that this is equal to:

Standard error of the difference between two means

$$\sigma_{\bar{x}_1 - \bar{x}_2} = \sqrt{\frac{\sigma_1^2}{n_1} + \frac{\sigma_2^2}{n_2}}$$

where σ_1^2 is the variance of population 1, σ_2^2 is the variance of population 2, n_1 is the size of the sample from population 1 and n_2 is the size of the sample from population 2.

Note that it is the variances of each population (relative to sample sizes) and not standard deviations that are included in this expression.

When σ_1^2 and σ_2^2 are *unknown*, we must use the values of the *sample* variances, i.e. s_1^2 and s_2^2. In such situations, provided that the sample sizes are sufficiently large (30 or more), an estimate for $\sigma_{\bar{x}_1 - \bar{x}_2}$ is:

Estimate of $\sigma_{\bar{x}_1 - \bar{x}_2}$

$$s_{\bar{x}_1 - \bar{x}_2} = \sqrt{\frac{s_1^2}{n_1} + \frac{s_2^2}{n_2}}$$

The steps to be followed in conducting the test may now be stated formally.

Step 1   *State the null and alternative hypotheses:*

$H_0: \mu_1 - \mu_2 = 0$
$H_1: \mu_1 - \mu_2 \neq 0$ *(requiring a two-tailed test)*

or

$H_1: \mu_1 - \mu_2 > 0$ *(requiring a one-tailed test in the right-hand tail)*

or

$H_1: \mu_1 - \mu_2 < 0$ *(requiring a one-tailed test in the left-hand tail)*

Step 2   *State the level of significance (α) at which the test is to be conducted.*

Step 3   *Compute the test statistic Z corresponding to the difference $\bar{x}_1 - \bar{x}_2$ as follows (where σ_1 and σ_2 are unknown):*

$$Z = \frac{(\bar{x}_1 - \bar{x}_2) - (\mu_1 - \mu_2)}{s_{\bar{x}_1 - \bar{x}_2}}$$

Under the null hypothesis, $\mu_1 - \mu_2 = 0$, and therefore the above expression may be written as

$$Z = \frac{\bar{x}_1 - \bar{x}_2}{s_{\bar{x}_1 - \bar{x}_2}}$$

Step 4   *Compare the test statistic with its critical value (obtained from the standard normal table) at the chosen level of significance.*

Step 5   *Accept or reject the null hypothesis on the basis of the comparison in Step 4. Acceptance of the null hypothesis means that any difference between the two sample means must be regarded as arising by chance (sampling error) rather than providing evidence in favour of a true difference in the corresponding populations.*

An illustration of the test procedure is given in Worked example 15.1 below.

Hypothesis testing using small independent samples

When dealing with situations where σ_1 and σ_2 are not known and sample sizes are small ($n_1 < 30$ and $n_2 < 30$), the test procedure makes use of the t distribution. The procedural steps are identical to those specified above for large independent samples subject, however, to a special case noted below. Under the null hypothesis $H_0: \mu_1 - \mu_2 = 0$, the test statistic t is calculated as

$$t = \frac{\bar{x}_1 - \bar{x}_2}{s_{\bar{x}_1 - \bar{x}_2}}$$

where

$$s_{\bar{x}_1 - \bar{x}_2} = \sqrt{\frac{s_1^2}{n_1} + \frac{s_2^2}{n_2}}$$

It is important to note that the t statistic in this case only approximately follows the t distribution. The degrees of freedom (df) are given by the expression

$$df = \frac{(s_1^2/n_1 + s_2^2/n_2)^2}{[(s_1^2/n_1)^2/(n_1 - 1)] + [(s_2^2/n_2)^2/(n_2 - 1)]}$$

Admittedly, this expression looks cumbersome but a good calculator or computer package (such as MINITAB or Excel) makes the calculation relatively painless. To be on the conservative side, if df as calculated is not an integer $(1, 2, 3, \ldots)$ it should be rounded *down* to the nearest integer.

Special case of equal population variances

The general procedure outlined above may be modified (producing a more powerful test) for those situations in which it is reasonable to assume that the two population variances (σ_1^2 and σ_2^2) are equal. This situation is common, for example, in many long-running production processes for which, based on past experience, we are convinced that the variation within population 1 is the same as the variation within population 2. The test procedure here differs in that the two sample variances (s_1^2 and s_2^2) are now 'pooled' to provide the best estimate of the *common* population variance (if equal population variances cannot be assumed then this pooling procedure is not appropriate). For example, for two samples of sizes n_1 and n_2 and variances s_1^2 and s_2^2 respectively, the pooled variance of the samples (s_p^2) is computed as the weighted average of the separate variances (using degrees of freedom as the weights), i.e.

Pooled sample variance

$$s_p^2 = \frac{(n_1 - 1)s_1^2 + (n_2 - 1)s_2^2}{n_1 + n_2 - 2}$$

Hence, the standard error of the difference between two sample means is then estimated as

$$s_{\bar{x}_1 - \bar{x}_2} = \sqrt{\left(\frac{s_p^2}{n_1} + \frac{s_p^2}{n_2}\right)}$$

The t statistic under the null hypothesis $H_0: \mu_1 - \mu_2 = 0$ is calculated as before, i.e.

$$t = \frac{\bar{x}_1 - \bar{x}_2}{s_{\bar{x}_1 - \bar{x}_2}}$$

The degrees of freedom for the t statistic in this special case is much easier to derive and are simply given by $(n_1 - 1) + (n_2 - 1) = n_1 + n_2 - 2$.

Summary of hypothesis testing using independent samples

A summary of the test procedures for both large and small independent samples is presented in Table 15.1.

Table 15.1 Summary of hypothesis tests using independent samples

	Large samples	Small samples
(a) TWO-TAILED TEST		
Hypotheses	$H_0: \mu_1 - \mu_2 = 0$ $H_1: \mu_1 - \mu_2 \neq 0$	$H_0: \mu_1 - \mu_2 = 0$ $H_1: \mu_1 - \mu_2 \neq 0$
Significance level	α	α
Standard error	σ known σ unknown $\sigma_{\bar{x}_1 - \bar{x}_2} = \sqrt{\left(\dfrac{\sigma_1^2}{n_1} + \dfrac{\sigma_2^2}{n_2}\right)}$ $s_{\bar{x}_1 - \bar{x}_2} = \sqrt{\left(\dfrac{s_1^2}{n_1} + \dfrac{s_2^2}{n_2}\right)}$	*General case* $s_{\bar{x}_1 - \bar{x}_2} = \sqrt{\left(\dfrac{s_1^2}{n_1} + \dfrac{s_2^2}{n_2}\right)}$ *Special case (pooled variance)* $\sqrt{\left(\dfrac{s_p^2}{n_1} + \dfrac{s_p^2}{n_2}\right)}$ where $s_p^2 = \dfrac{(n_1 - 1)s_1^2 + (n_2 - 1)s_2^2}{n_1 + n_2 - 2}$
Test statistic Actual value	$Z = \dfrac{(\bar{x}_1 - \bar{x}_2) - (\mu_1 - \mu_2)}{\text{standard error}} = \dfrac{(\bar{x}_1 - \bar{x}_2)}{\text{standard error}}$	$t = \dfrac{(\bar{x}_1 - \bar{x}_2) - (\mu_1 - \mu_2)}{\text{standard error}} = \dfrac{(\bar{x}_1 - \bar{x}_2)}{\text{standard error}}$
Critical value	$\pm Z_{\alpha/2}$	$\pm t_{\alpha/2}$ with $df = \dfrac{(s_1^2/n_1 + s_2^2/n_2)^2}{[(s_1^2/n_1)^2/(n_1 - 1)] + [(s_2^2/n_2)^2/(n_2 - 1)]}$ $\pm t_{\alpha/2}$ with $df = (n_1 - 1) + (n_2 - 1)$ $= (n_1 + n_2 - 2)$
Decision	Reject H_0 if $|Z| > |Z_{\alpha/2}|$ Accept H_0 otherwise	Reject H_0 if $|t| > |t_{\alpha/2}|$ Accept H_0 otherwise
Conclusion	Rejection of H_0 indicates that the difference in the means of the two independent samples cannot be regarded as being equal to 0 at the α level of significance	

	Large samples (continued)	*Small samples (continued)*
(b) ONE-TAILED TEST		
Hypotheses	$H_0: \mu_1 - \mu_2 = 0$ $H_1: \mu_1 - \mu_2 > 0$ or $\mu_1 - \mu_2 < 0$	$H_0: \mu_1 - \mu_2 = 0$ $H_1: \mu_1 - \mu_2 > 0$ or $\mu_1 - \mu_2 < 0$
Significance level	α	α
Standard error	As above	As above
Test statistic Actual value	As above	As above
Critical value	$+Z_\alpha$ or $-Z_\alpha$	$+t_\alpha$ or $-t_\alpha$ General case: df as above for two-tailed test Special case: df as above for two-tailed test
Decision	Reject H_0 if $Z > +Z_\alpha$ or $Z < -Z_\alpha$ Accept H_0 otherwise	Reject H_0 if $t > +t_\alpha$ or $t < -t_\alpha$ Accept H_0 otherwise
Conclusion	Rejection of H_0 indicates that the difference between the means of the two independent samples is either greater than zero (right-hand tail test) or less than zero (left-hand tail test) at the α level of significance	

209

Examples of the application of the test procedures for both the general case and special cases involving small independent samples are given in the Applications and Worked Examples section.

Hypothesis testing using large matched samples

So far in this chapter we have assumed that the two samples are independent. We now examine the situation where the samples are *matched*, first for large samples and then for small samples.

As noted above, matched samples are obtained by taking samples of matched pairs so that each observation from the first population is matched with a twin observation from the second population. Matching could be achieved, for example, by comparing data relating to the performance of a set of workers at one point in time to the performance of the *same* set of workers at another point in time. In this way, observations are matched in the sense that the before and after measurements are taken on the same individuals. The importance of this is that such a sample design leads to smaller sample error than the use of samples which are independent of each other – i.e. two *separate* groups of workers altogether. This method thus minimizes the variability due to extraneous factors and increases sample efficiency.

The test of the difference between the two samples may now be conducted, not by focusing on the two sample means $\bar{x}_1$ and $\bar{x}_2$, but by analysis of the difference D between *each* matched pair of observations, i.e. $D = x_1 - x_2$. The test is then conducted as a test of a single mean $\overline{D}$ (i.e. the mean of the differences of the matched pairs). Thus:

Mean of matched paired differences

$$\overline{D} = \frac{\Sigma D}{n}$$

where $D = x_1 - x_2$ for *each* matched pair and n is the number of matched pairs in the sample.

The $\overline{D}$ statistic will have an approximately normal distribution (because of the central limit theorem) as long as the sample size n is sufficiently large. The mean of the sampling distribution of $\overline{D}$ (i.e. μ_D) will be equal to $\mu_1 - \mu_2$; i.e. it provides an unbiased estimate of the population mean difference. The test procedure may then be conducted in the same way as for large independent samples (as above) where Z may now be defined as

$$Z = \frac{\overline{D} - \mu_D}{s_{\overline{D}}}$$

where $\overline{D}$ is the sample mean difference, μ_D is the population mean difference (under the null hypothesis this equals 0) and $s_{\overline{D}}$ is the standard error of the mean difference defined as

$$s_{\overline{D}} = s_D/\sqrt{n}$$

where

$$s_D = \sqrt{\left[\frac{\Sigma(D - \overline{D})^2}{n - 1}\right]} = \sqrt{\left[\frac{\Sigma D^2 - (\Sigma D)^2/n}{n - 1}\right]}$$

Hypothesis testing using small matched samples

Where we are dealing with *small* matched samples ($n < 30$) the t distribution, rather than Z, is appropriate. Assuming that the two populations are approximately normal, the t statistic is then defined exactly as for Z above:

$$t = \frac{\overline{D} - \mu_D}{s_{\overline{D}}}$$

and the critical value of t from the t tables is obtained with $n - 1$ degrees of freedom.

Summary of hypothesis testing using matched samples

A summary of the test procedures for both large and small matched samples is presented in Table 15.2 overleaf.

Applications and Worked Examples	Worked example 15.1 Hypothesis testing using large independent samples

General descriptions of the MINITAB and Excel computer programs and their application are given in Appendices B and C respectively.

A manufacturer of golf balls introduces a new type of ball (brand 2) which he claims is superior to the previous brand (brand 1) in terms of the distance it will carry under identical playing conditions. The best test of this claim would be to use a matched sample of players using each type of ball in turn (as in Worked Example 15.4 below) but, for purposes of illustration, we use independent samples here. A golfer, selected at random, hits 35 shots using the old brand (brand 1) while another randomly selected golfer hits 40 shots with the new brand (brand 2). The distances in metres for each brand are recorded and summarized as follows:

Brand 1   Brand 2

$n_1 = 35$ $n_2 = 40$

$\bar{x}_1 = 202.9$ metres $\bar{x}_2 = 223.7$ metres

$s_1 = 29.9$ metres $s_2 = 35.0$ metres

Table 15.2 Summary of hypothesis tests using matched samples

	Large samples	Small samples
(a) TWO-TAILED TEST		
Hypotheses	$H_0: \mu_D = 0$ $H_1: \mu_D \neq 0$	
Significance level	α	
Standard error	$s_{\bar{D}} = s_D/\sqrt{n}$ where $s_D = \sqrt{\dfrac{\Sigma(D-\bar{D})^2}{n-1}} = \sqrt{\dfrac{\Sigma D^2 - (\Sigma D)^2/n}{n-1}}$	
Test statistic		
Actual value	$Z = \dfrac{\bar{D} - \mu_D}{s_{\bar{D}}} = \dfrac{\bar{D}}{s_{\bar{D}}}$	$t = \dfrac{\bar{D} - \mu_D}{s_{\bar{D}}} = \dfrac{\bar{D}}{s_{\bar{D}}}$
Critical value	$\pm Z_{\alpha/2}$	$\pm t_{\alpha/2}$ with $n-1$ df
Decision	Reject H_0 if $|Z| > |Z_{\alpha/2}|$ Accept H_0 otherwise	Reject H_0 if $|t| > |t_{\alpha/2}|$ Accept H_0 otherwise
Conclusion	Rejection of H_0 indicates that the mean difference between the matched values of the two samples cannot be regarded as being equal to 0 at the α level of significance	
(b) ONE-TAILED TEST		
Hypotheses	$H_0: \mu_D = 0$ $H_1: \mu_D > 0$ or $\mu_D < 0$	
Significance level	α	
Standard error	As above	As above
Test statistic		
Actual value	As above	As above
Critical value	$+Z_\alpha$ or $-Z_\alpha$	$+t_\alpha$ or $-t_\alpha$ with $n-1$ df
Decision	Reject H_0 if $Z > +Z_\alpha$ or $Z < -Z_\alpha$ Accept H_0 otherwise	Reject H_0 if $t > +t_\alpha$ or $t < -t_\alpha$ Accept H_0 otherwise
Conclusion	Rejection of H_0 indicates that the mean difference between the matched values of the two samples is either greater than zero (right-hand tail test) or less than zero (left-hand tail test) at the α level of significance	

Worked example 15.1   Hypothesis testing using large independent samples *(continued)*

Test the hypotheses:

(a) that there is no difference in the performances of the two types of ball;

(b) that brand 1 is inferior to brand 2.

Worked Solution

(a) The test requires a two-tailed Z test.

Hypotheses: $\quad\quad\quad\quad\quad\quad$ H_0: $\mu_1 - \mu_2 = 0$

$\quad\quad\quad\quad\quad\quad\quad\quad\quad\quad$ H_1: $\mu_1 - \mu_2 \neq 0$

Significance level: $\quad\quad\quad$ $\alpha = 0.05$

Standard error of the mean: $\quad s_{\bar{x}_1 - \bar{x}_2} = \sqrt{\left(\dfrac{s_1^2}{n_1} + \dfrac{s_2^2}{n_2} \right)}$

$$= \sqrt{\left[\frac{(29.9)^2}{35} + \frac{(35.0)^2}{40} \right]} = 7.49$$

Critical values: $\quad\quad\quad\quad$ $\pm Z_{\alpha/2} = \pm 1.96$

Test: $\quad\quad\quad\quad\quad\quad\quad\quad$ $Z = \dfrac{(\bar{x}_1 - \bar{x}_2) - (\mu_1 - \mu_2)}{s_{\bar{x}_1 - \bar{x}_2}}$

$$= \frac{(202.9 - 223.7) - 0}{7.49} = -2.77$$

Conclusion: $|Z| > |Z_{\alpha/2}|$, i.e. $2.77 > 1.96$; therefore H_0 is rejected. We conclude that there is a significant difference in the mean distances achieved by the two types of ball at the 5 per cent level of significance. Note that this test provides no conclusion about the superiority of one type of ball over another – this requires a one-tailed test, as in (b) below.

(b) We specifically test here the claim that, on average, brand 1 balls are inferior to brand 2 balls. This requires a one-tailed Z test.

Hypotheses: $\quad\quad\quad\quad\quad\quad$ H_0: $\mu_1 - \mu_2 = 0$

$\quad\quad\quad\quad\quad\quad\quad\quad\quad\quad$ H_1: $\mu_1 - \mu_2 < 0$

Significance level: $\quad\quad\quad$ $\alpha = 0.05$

Standard error of the mean: $\quad s_{\bar{x}_1 - \bar{x}_2} = 7.49$ (as above)

Critical values: $\quad\quad\quad\quad$ $-Z_\alpha = -1.645$

Test: $\quad\quad\quad\quad\quad\quad\quad\quad$ $Z = -2.77$ (as above).

Conclusion: $Z < -Z_\alpha$, i.e. $-2.77 < -1.645$. Therefore H_0 is rejected at the 5 per cent level of significance and we conclude that brand 1 is inferior to brand 2. Thus the manufacturer's claim is justified.

Worked example 15.1   Hypothesis testing using large independent samples *(continued)*

Computer solution using MINITAB

(a) Two-tailed Z test

With the data for Brand 1 and Brand 2 entered in C1 and C2 respectively of the MINITAB worksheet, the procedure is as follows:

Choose **Stat → Basic statistics → 2-sample t…**

Check **Samples in different columns** and enter C1 and C2 in the **First** and **Second** boxes respectively

Select **Alternative**, i.e. alternative hypothesis: 'not equal' in this case

Click **OK**

Authors' comment: This procedure conducts a *t* test but the computed *t* value is the same as *Z* and thus may be compared directly with the critical *Z* value for the test. Note that the procedure also generates a 95% confidence interval. Note that MINITAB shows '*t*' as '*T*'

The output is as follows:

Two Sample T-Test and Confidence Interval

Two sample T for Brand 1 vs Brand 2

	N	Mean	StDev	SE Mean
Brand 1	35	202.9	29.9	5.1
Brand 2	40	223.7	35.0	5.5

95% CI for mu Brand 1− mu Brand 2: (− 35.7, − 5.8)
T-Test mu Brand 1 = mu Brand 2 (vs not =): T = − 2.77 P = 0.0071 DF = 72

Authors' comment: Actual $Z(\equiv T) = -2.77$. The decision, as before, is therefore that one must reject H_0 as $|Z| > |Z_{\alpha/2}|$, i.e. 2.77 > 1.96. The same conclusion is reached using the P value as $P = (0.0071) < \alpha (= 0.05)$.

(b) One-tailed Z test

With the data for Brand 1 and Brand 2 entered in C1 and C2 respectively of the MINITAB worksheet, the procedure is as follows:

Choose **Stat → Basic statistics → 2-sample t…**

Check **Samples in different columns** and enter C1 and C2 in the **First** and **Second** boxes respectively

Select **Alternative** i.e. alternative hypothesis: 'less than' in this case

Click **OK**

Authors' comment: This procedure conducts a *t* test but the computed *t* value is the same as *Z* and thus may be compared directly with the critical *Z* value for the test. Note that the procedure also generates a 95% confidence interval.

The output is as follows:

Two Sample T-Test and Confidence Interval

Two sample T for Brand 1 vs Brand 2

	N	Mean	StDev	SE Mean
Brand 1	35	202.9	29.9	5.1
Brand 2	40	223.7	35.0	5.5

95% CI for mu Brand 1− mu Brand 2: (−35.7, −5.8)
T-Test mu Brand 1 = mu Brand 2 (vs <): T = −2.77 P = 0.0035 DF = 72

Authors' comment: actual $Z(\equiv T) = -2.77$. The decision, as above, is therefore, that one must reject H_0 as $Z < -Z_\alpha$, i.e. $-2.77 < -1.645$. The same conclusion is reached using the P value as P = $(0.0035) < \alpha(= 0.05)$.

Computer solution using Excel

There is no pre-defined tool in Excel for solving this problem. Excel does contain an analysis tool, called **Z-Test: Two Sample for Means**, which performs a two-sample Z-test for means (with known variances) but it requires the data arrays for each sample to be input not just the summary statistics available in this example.

Worked example 15.2   Hypothesis testing using small independent samples

The data below show the typing rates (words per minute) achieved by each of ten secretaries working in two branches of an insurance company which use two different types of word processing programs.

Branch 1:   64   55   58   55   52   68   72   60   72   49
Branch 2:   57   64   55   64   60   66   75   82   48   54

Summary statistics for the branches are as follows:

Branch 1	Branch 2
$n_1 = 10$	$n_2 = 10$
$\bar{x}_1 = 60.5$	$\bar{x}_2 = 62.5$
$s_1 = 8.2$	$s_2 = 10.2$

Conduct hypothesis tests to determine whether or not

(a) there is a significant difference in performance in the two branches;

(b) the use of one type of program is associated with *better* performance than the other.

Worked example 15.2 Hypothesis testing using small independent samples *(continued)*

Worked solution

First, it is necessary to decide whether or not it is reasonable to assume that the two population variances are equal. As we are dealing with the performance of two different groups of typists using different equipment, there is no reason to assume equality of population variances and we conduct the tests on this basis.

(a) A two-tailed test is required using t, since population variances are not known and the sample sizes are small ($n_1 < 30$ and $n_2 < 30$).

Hypotheses:	$H_0: \mu_1 - \mu_2 = 0$
	$H_1: \mu_1 - \mu_2 \neq 0$
Significance level:	$\alpha = 0.05$

Standard error of the mean:
$$s_{\bar{x}_1 - \bar{x}_2} = \sqrt{\left(\frac{s_1^2}{n_1} + \frac{s_2^2}{n_2}\right)}$$

$$= \sqrt{\left[\frac{(8.2)^2}{10} + \frac{(10.2)^2}{10}\right]} = 4.14$$

Critical values: $\pm t_{\alpha/2} = \pm t_{0.025} = \pm 2.11$ with 17 df

where df ($= 17$) is given by

$$\frac{(s_1^2/n_1 + s_2^2/n_2)^2}{[(s_1^2/n_1)^2/(n_1 - 1)] + [(s_2^2/n_2)^2/(n_2 - 1)]}$$

$$= \frac{[(8.2)^2/10 + (10.2)^2/10]^2}{[(8.2)^2/10]^2/(10 - 1) + [(10.2)^2/10]^2/(10 - 1)}$$

$$= 17.2 \text{ (i.e. 17 rounded to nearest } lowest \text{ integer)}$$

Test:
$$t = \frac{(\bar{x}_1 - \bar{x}_2) - (\mu_1 - \mu_2)}{s_{\bar{x}_1 - \bar{x}_2}}$$

$$= \frac{(60.5 - 62.5) - 0}{4.14} = -0.48$$

Conclusion: As $|t| < |t_{\alpha/2}|$, i.e. $0.48 < 2.11$, we cannot reject H_0. We conclude that there is not a significant difference in the performance of the two branches at the 5 per cent level of significance.

(b) This involves a one-tailed test as follows.

Hypotheses:	$H_0: \mu_1 - \mu_2 = 0$
	$H_1: \mu_1 - \mu_2 > 0$
Significance level:	$\alpha = 0.05$
Standard error of the mean:	$s_{\bar{x}_1 - \bar{x}_2} = 4.14$ (as above)
Critical values:	$t_\alpha = + t_{0.05} = +1.74$, with 17 df (as above)
Test:	$t = -0.48$ (as above).

Conclusion: As $t < t_\alpha$, i.e. $-0.48 < +1.74$, we cannot reject H_0 at the 5 per cent level of significance. We conclude that the performance of the typists using one type of word processing program is no better than their performance using the other type.

Computer solution using MINITAB

(a) Two-tailed *t* test

With the data for Branch 1 and Branch 2 entered in C1 and C2 respectively of the MINITAB worksheet, the procedure is as follows:

Authors' comment: This procedure conducts a *t* test at the 5% level of significance and also generates a 95% confidence interval for the difference between two means.

Choose **Stat → Basic statistics →
   2-sample t...**
Check **Samples in different columns** and enter C1 and C2 in the **First** and **Second** boxes respectively
Select **Alternative** i.e. alternative hypothesis ('not equal' in this case)
Click **OK**

The output is as follows:

Two Sample T-Test and Confidence Interval

Two sample T for Brand 1 vs Brand 2

	N	Mean	StDev	SE Mean
BRANCH 1	10	60.50	8.20	2.6
BRANCH 2	10	62.5	10.2	3.2

95% CI for mu BRANCH 1 − mu BRANCH 2: $(-10.7, 6.7)$
T-Test mu BRANCH 1 = mu BRANCH 2 (vs not =): T = − 0.48   P = 0.63   DF = 17

Authors' comment: The decision, as before, is that one cannot reject H_0 at the 5% level of significance, as $|t| < |t_{\alpha/2}|$, i.e. $0.48 < 2.11$. The same conclusion is reached using the P value as $P(= 0.63) > \alpha(= 0.05)$.

(b) One-tailed t test

With the data for Branch 1 and Branch 2 entered in C1 and C2 respectively of the MINITAB worksheet, the procedure is as follows:

Choose **Stat → Basic statistics →
   2-sample t...**

Worked example 15.2   Hypothesis testing using small independent samples *(continued)*

Check **Samples in different columns** and enter C1 and C2 in the **First** and **Second** boxes respectively

Select **Alternative** i.e. alternative hypothesis ('greater than' in this case)

Click **OK**

Authors' comment: This procedure conducts a one-tailed *t* test at the 5% level of significance and also generates a 95% confidence interval for the difference between two means.

The output is as follows:

Two Sample T-Test and Confidence Interval

Two sample T for BRANCH 1 vs BRANCH 2

	N	Mean	StDev	SE Mean
BRANCH 1	10	60.50	8.20	2.6
BRANCH 2	10	62.5	10.2	3.2

95% CI for mu BRANCH 1 − mu BRANCH 2: (−10.7,6.7)
T-Test mu BRANCH 1 = mu BRANCH 2 (vs >): T = − 0.48   P = 0.68   DF = 17

Authors' comment: As before, as P(= 0.68) > α(= 0.05), we conclude that the null hypothesis cannot be rejected.

Computer solution using Excel

With the two data series entered in columns A and B of the worksheet, as shown below:

Choose **Tools → Data Analysis... → *t*-Test: Two-Sample Assuming Unequal Variances**

In **Variable 1 Range** select cells A1 : A10 and in **Variable 2 Range** select cells B1 : B10
In **Hypothesized Mean Difference** enter 0
Click **OK**
The input and output tables are shown overleaf

Authors' comment: Note that the output gives the actual *t* statistic along with its probability for both two-tailed and one-tailed tests and the critical value of *t* at the 5 per cent level of significance in both cases.

Two-tailed test conclusion: H_0 cannot be rejected at the 5% level of significance, as $|t| < |t_{\alpha/2}|$, i.e. 0.48 < 2.11. The same conclusion is reached using the P value as P(= 0.63) > α(= 0.05).

One-tailed test conclusion: H_0 cannot be rejected at the 5% level of significance, as $t < +t_{\alpha}$, i.e. −0.48 < +1.74. The same conclusion is reached using the P value as P(= 0.32) > α(= 0.05).

Input table		Output table		
A	B			
64	57	t-Test: Two-Sample Assuming Unequal Variances		
55	64			
58	55		Variable 1	Variable 2
55	64	Mean	60.5	62.5
52	60	Variance	67.166667	103.16667
68	66	Observations	10	10
72	75	Hypothesized Mean Difference	0	
60	82	df	17	
72	48	t Stat	−0.4845964	
49	54	P(T <=t) one-tail	0.3170737	
		t Critical one-tail	1.7396064	
		P(T <=t) two-tail	0.6341474	
		t Critical two-tail	2.1098185	

Worked example 15.3 Hypothesis testing using small independent samples – special case of pooled variances

Identical equipment is used in two manufacturing plants to produce car components. The number of defects per day is recorded in each plant in a seven-day period as follows:

Plant 1:   12   6   3   8   12   7   9
Plant 2:   11   13   15   4   10   8   7

Summary statistics for the plants are as follows:

Plant 1	Plant 2
$n_1 = 7$	$n_2 = 7$
$\bar{x}_1 = 8.14$	$\bar{x}_2 = 9.71$
$s_1 = 3.24$	$s_2 = 3.73$

It is reasonable to assume in this case that, as identical production processes are being used, the (unknown) population variances are equal, i.e. $\sigma_1^2 = \sigma_2^2$. Test the hypotheses that:

(a) there is no significant difference between the two plants;

(b) plant 2 has more defects than plant 1.

Worked example 15.3 Hypothesis testing using small independent samples – special case of pooled variances (continued)

Worked solution

(a) This is a two-tailed test, conducted as follows.

Hypotheses: $H_0: \mu_1 - \mu_2 = 0$

$H_1: \mu_1 - \mu_2 \neq 0$

Significance level: $\alpha = 0.05$

Standard error of the mean: $s_{\bar{x}_1 - \bar{x}_2} = \sqrt{\left(\dfrac{s_p^2}{n_1} + \dfrac{s_p^2}{n_2}\right)}$

where

$$s_p^2 = \frac{(n_1 - 1)s_1^2 + (n_2 - 1)s_2^2}{n_1 + n_2 - 2}$$

$$= \frac{(7 - 1)(3.24)^2 + (7 - 1)(3.73)^2}{7 + 7 - 2}$$

$$= 12.205$$

Thus

$$s_{\bar{x}_1 - \bar{x}_2} = \sqrt{\left(\frac{12.205}{7} + \frac{12.205}{7}\right)} = 1.867$$

Critical values: $t_{\alpha/2} = \pm t_{0.025} = \pm 2.179$ with

$n_1 + n_2 - 2 = 12$ df

Test: $t = \dfrac{(\bar{x}_1 - \bar{x}_2) - (\mu_1 - \mu_2)}{s_{\bar{x}_1 - \bar{x}_2}}$

$$= \frac{(8.14 - 9.71) - 0}{1.867} = -0.84$$

Conclusion: As $|t| < |t_{\alpha/2}|$, i.e. $0.84 < 2.179$, we cannot reject H_0 at the 5 per cent level of significance. Thus we cannot conclude that there is a significant difference between the two plants.

(b) This is a one-tailed test, conducted as follows.

Hypotheses:	$H_0: \mu_1 - \mu_2 = 0$
	$H_1: \mu_1 - \mu_2 < 0$
Significance level:	$\alpha = 0.05$
Standard error of the mean:	$s_{\bar{x}_1 - \bar{x}_2} = 1.867$ (as above)
Critical values:	$t_\alpha = -t_{0.05} = -1.782$, with 12 df
Test:	$t = -0.84$ (as above).

Conclusion: As $t > -t_\alpha$, i.e. $-0.84 > -1.782$, we cannot reject H_0 at the 5 per cent level of significance. We cannot conclude that plant 2 produces significantly more defects than plant 1.

Computer solution using MINITAB

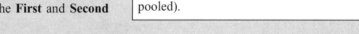

(a) Two-tailed *t* test

With the data for Plant 1 and Plant 2 entered to C1 and C2 respectively of the MINITAB worksheet, the procedure is as follows:

Choose **Stat → Basic statistics → 2-sample t...**

Check **Samples in different columns** and enter C1 and C2 in the **First** and **Second** boxes respectively.

Select **Alternative** i.e. alternative hypothesis ('not equal' in this case)

Check the **Assume equal variances** box

Click **OK**

Authors' comment: this procedure generates a 95% confidence interval for the difference between two means and also conducts a two-tailed *t*-test at the 5% level of significance on the basis that the population variances may be assumed to be equal (hence the sample variances are pooled).

The output is as follows:

Two Sample T-Test and Confidence Interval

Two sample T for Plant 1 vs Plant 2

	N	Mean	StDev	SE Mean
Plant 1	7	8.14	3.24	1.2
Plant 2	7	9.71	3.73	1.4

95% CI for mu Plant 1 − mu Plant 2: (−5.6, 2.5)
T-Test mu Plant 1 = mu Plant 2 (vs not =): T= −0.84 P = 0.42 DF = 12
Both use Pooled St Dev = 3.49

Authors' comment: Actual $t = -0.84$ (shown as T above). The decision, as before, is that one cannot reject H_0 at the 5% level of significance, as $|t| < |t_{\alpha/2}|$, i.e. $0.84 < 2.179$. The same conclusion is reached using the P value as P($= 0.42$) $> \alpha (= 0.05)$.

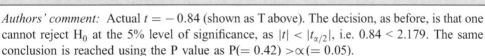

Worked example 15.3   Hypothesis testing using small independent samples – special case of pooled variances *(continued)*

(b) One-tailed *t* test

With the data for Plant 1 and Plant 2 entered in C1 and C2 respectively of the MINITAB worksheet, the procedure is as follows:

Choose **Stat → Basic statistics → 2-sample t...**

Check **Samples in different columns** and enter C1 and C2 in the **First** and **Second** boxes respectively

Select **Alternative** i.e. alternative hypothesis ('less than' in this case)

Check the **Assume equal variances** box
Click **OK**

Authors' comment: This procedure generates a 95% confidence interval for the difference between two means and also conducts a one-tailed *t*-test at the 5% level of significance on the basis that the population variances may be assumed to be equal (hence the sample variances are pooled).

The output is as follows:

Two Sample T-Test and Confidence Interval

Two sample T for Plant 1 vs Plant 2

	N	Mean	StDev	SE Mean
Plant 1	7	8.14	3.24	1.2
Plant 2	7	9.71	3.73	1.4

95% CI for mu Plant 1 − mu Plant 2: (−5.6, 2.5)
T-Test mu Brand 1 = mu Plant 2 (vs <): T= −0.84   P = 0.21   DF = 12
Both use Pooled StDev = 3.49

Authors' comment: Actual *t* (shown as T above) = −0.84. The decision, as before, is that one cannot reject H_0 at the 5% level of significance, as $t > -t_\alpha$; i.e. −0.84 > −1.782. The same conclusion is reached using the P value as P(= 0.21) > α(= 0.05).

Computer solution using Excel

With the two data series entered in columns A and B of the worksheet, as shown below:

Choose **Tools → Data Analysis... →**
   ***t*-Test: Two-Sample Assuming Equal**

Variances

In **Variable 1 Range** select cells A1 : A7 and in **Variable 2 Range** select cells B1 : B7

In **Hypothesized Mean Difference** enter 0
Click **OK**

The input and output tables are shown below:

Input table		Output table
A	B	t-Test: Two-Sample Assuming Equal Variances

			Variable 1	Variable 2
12	11			
6	13	Mean	8.1428571	9.714286
3	15	Variance	10.47619	13.90476
8	4	Observations	7	7
12	10	Pooled Variance	12.190476	
7	8	Hypothesized Mean Difference	0	
9	7	df	12	
		t Stat	−0.842012	
		P(T <= t) one-tail	0.2081235	
		t Critical one-tail	1.7822867	
		P(T <= t) two-tail	0.4162469	
		t Critical two-tail	2.1788128	

Authors' comment: Note that, as in Worked example 15.2 above, the output gives the actual t statistic along with its probability for both two-tailed and one-tailed tests and the critical value of t at the 5 per cent level of significance in both cases.

Two-tailed test conclusion: H_0 cannot be rejected at the 5% level of significance, as $|t| < |t_{\alpha/2}|$, i.e. $0.84 < 2.179$. The same conclusion is reached using the P value as $P(= 0.416) > \alpha(= 0.05)$.

One-tailed test conclusion: H_0 cannot be rejected at the 5% level of significance, as $t > -t_{\alpha}$, i.e. $-0.84 > -1.782$. The same conclusion is reached using the P value as $P(= 0.21) > \alpha(= 0.05)$.

Worked example 15.4   Hypothesis testing using large matched samples

A random sample of 35 golfers test two brands of golf ball (brands 1 and 2), each golfer driving each brand of ball once, and the distance is measured in metres. This provides a large matched sample of 35 paired observations. A summary of the results is shown below, where D is the difference between each pair of distances for each golfer, i.e. $x_2 - x_1$.

$$\overline{D} = 20.886 \qquad s_D = 42.866 \qquad n = 35$$

Test the following hypotheses:

(a) that there is no difference between the two brands of golf ball;

(b) that brand 2 is superior to brand 1.

Worked example 15.4   Hypothesis testing using large matched samples *(continued)*

Worked solution

(a) This is a two-tailed Z test conducted as follows.

Hypotheses:	$H_0: \mu_D = 0$
	$H_1: \mu_D \neq 0$
Significance level:	$\alpha = 0.05$
Standard error of the mean:	$s_{\overline{D}} = \dfrac{s_D}{\sqrt{n}} = \dfrac{42.866}{\sqrt{35}} = 7.246$
Critical values:	$\pm Z_{\alpha/2} = \pm Z_{0.025} = \pm 1.96$
Test:	$Z = \dfrac{\overline{D} - \mu_D}{s_{\overline{D}}} = \dfrac{20.866 - 0}{7.246} = 2.88.$

Conclusion: $|Z| > |Z_{\alpha/2}|$, i.e. $2.88 > 1.96$, and therefore H_0 is rejected at the 5 per cent level of significance. We conclude that the two brands of golf ball are different in performance.

(b) This is a one-tailed Z test conducted as follows.

Hypotheses:	$H_0: \mu_D = 0$
	$H_1: \mu_D > 0$
Significance level:	$\alpha = 0.05$
Standard error of the mean:	$s_{\overline{D}} = 7.246$ (as above)
Critical values:	$Z_\alpha = +Z_{0.05} = +1.64$
Test:	$Z = 2.88$ (as above).

Conclusion: $Z > +Z_\alpha$, i.e. $2.88 > +1.64$, and therefore H_0 is rejected at the 5 per cent level of significance. We conclude that brand 2 ball is superior to brand 1.

Computer solution using MINITAB

(a) Two-tailed Z test

Enter the data for Brand 1 and Brand 2 in columns C1 and C2 respectively of the MINITAB worksheet.

Calculate the difference ('D') between C2 and C1 (C2 – C1) and enter the result in C3 (This may be done as follows:

Choose **Calc → Calculator**; Enter C3 in **Store result in variable** box; Enter C2 – C1 in **Expression** box; Click **OK**, name C3 'D')

The procedure is then as follows:

Choose **Stat → Basic Statistics → 1-Sample Z**...

Enter **Variable** (C3)

Check **Test Mean** and enter value of the hypothesized mean (0.0)

Select **Alternative** (i.e. the alternative hypothesis): 'Not equal' in this case

Enter the value of **Sigma** (42.866)

Click **OK**

The output is as follows:

Z-Test

Test of mu = 0.00 vs mu not = 0.00
The assumed sigma = 42.9

Variable	N	Mean	StDev	SE Mean	Z	P
D	35	20.89	42.87	7.25	2.88	0.0040

Authors' comment: Actual $Z = 2.88$. The decision, as before, is therefore that one must reject H_0 at the 5% level of significance as $|Z| > |Z_{\alpha/2}|$, i.e. $2.88 > 1.96$. The same conclusion is reached using the P value as $P(= 0.0040) < \alpha(= 0.05)$.

(b) One-tailed *Z* test

With data for the difference ('D') between Brand 1 and Brand 2 (C2 – C1) entered in column C3 of the MINITAB worksheet (see two-tailed Z test example in (a) above regarding calculation in MINITAB), the procedure is as follows:

Choose **Calc → Calculator**;
Choose **Stat → Basic Statistics → 1-Sample Z**...

Enter **Variable** (C3)
Check **Test Mean** and enter value of the hypothesized mean (0.0)
Select **Alternative** i.e. the alternative hypothesis ('greater than' in this case)
Enter the value of **Sigma** (42.866)
Click **OK**

The output is as follows:

Z-Test

Test of mu = 0.00 vs mu > 0.00
The assumed sigma = 42.9

Variable	N	Mean	StDev	SE Mean	Z	P
D	35	20.89	42.87	7.25	2.88	0.0020

Authors' comment: Actual $Z = 2.88$. The decision, as before, is therefore that one must reject H_0 as $Z > Z_{\alpha}$, i.e. $2.88 > 1.64$. The same conclusion is reached using the P value as $P(= 0.0020) < \alpha(= 0.05)$.

Worked example 15.4   Hypothesis testing using large matched samples *(continued)*

> ### Computer solution using Excel
> **E**
>
> There is no pre-defined tool in Excel for solving this problem. Excel does contain an analysis tool, called **t-Test: Paired Two Sample for Means**, which performs a paired two-sample *t*-test but it requires the *data arrays* for each sample to be input, not just the summary statistics available in this example. The procedure for using this tool is the same as that set out in the Excel solutions for Worked examples 15.2 and 15.3.

Worked example 15.5   Hypothesis testing using small matched samples

The times taken by seven workers to perform a task are measured in minutes, first before training and then after training. Test the hypotheses that:

(a) the training programme has made no difference to performance;

(b) the training programme has in fact improved performance.

This is a small matched sample analysis of seven workers and thus the test can be conducted as a test of the single mean $\overline{D}$ using the *t* test. The raw data and computations are as follows:

	Time (minutes)		Worksheet		
Worker	Before training (x_1)	After training (x_2)	D $(x_2 - x_1)$	$D - \overline{D}$	$(D - \overline{D})^2$
1	7	8	1	−2.71	7.34
2	6	6	0	−3.71	13.76
3	2	10	8	4.29	18.40
4	5	12	7	3.29	10.82
5	4	9	5	1.29	1.66
6	6	10	4	0.29	0.08
7	8	9	1	−2.71	7.34
			$\Sigma D = 26$		$\Sigma(D - \overline{D})^2 = 59.4$

From this table, we can compute

$$\overline{D} = \Sigma D / 7$$
$$= 3.71$$

$$s_D = \sqrt{\left[\frac{\Sigma(D - \overline{D})^2}{n - 1}\right]} = \sqrt{\left(\frac{59.4}{7 - 1}\right)} = 3.15$$

Worked solution

(a) This is a two-tailed t test conducted as follows.

Hypotheses:	$H_0: \mu_D = 0$
	$H_1: \mu_D \neq 0$
Significance level:	$\alpha = 0.05$
Standard error of the mean:	$s_{\overline{D}} = \dfrac{s_D}{\sqrt{n}} = \dfrac{3.15}{\sqrt{7}} = 1.19$
Critical values:	$t = \pm t_{\alpha/2} = \pm 2.447$, with $n - 1 = 6$ df
Test:	$t = \dfrac{\overline{D} - \mu_D}{s_{\overline{D}}} = \dfrac{\overline{D} - 0}{s_{\overline{D}}} = \dfrac{3.71}{1.19} = 3.12$

Conclusion: $|t| > |t_{\alpha/2}|$, i.e. $3.12 > 2.447$, and therefore we must reject H_0 at the 5 per cent level of significance. We conclude that the training programme makes a difference to performance (note, however, that to conclude that performance has *improved* requires a one-tailed test – see (b) below).

(b) This is a one-tailed t test conducted as follows.

Hypotheses:	$H_0: \mu_D = 0$
	$H_1: \mu_D > 0$
Significance level:	$\alpha = 0.05$
Standard error of the mean:	$s_{\overline{D}} = 1.19$ (as before)
Critical values:	$t_{\alpha} = +t_{0.05} = +1.943$, with 6 df
Test:	$t = 3.12$ (as above)

Conclusion: As $t > +t_{\alpha}$, i.e. $3.12 > +1.943$, we must reject H_0 at the 5 per cent level of significance. We conclude that the training programme in fact improves performance.

Computer solution using MINITAB

(a) Two-tailed t test

Enter the data for 'Before training' and 'After training' in columns C1 and C2 respectively of the MINITAB worksheet. Calculate the difference ('D') between C2 and C1 (C2 – C1) and enter the result in C3 (see the two-tailed Z test computer solution in Worked example 15.4 above regarding the calculation of this within MINITAB). The procedure is then as follows:

Choose **Stat** → **Basic Statistics** →
    1-Sample t...
Enter **Variable** (C3)
Check **Test Mean** and enter value of the hypothesized mean (0.0)
Select **Alternative** (i.e. the alternative hypothesis): 'Not equal' in this case
Click **OK**

Worked example 15.5   Hypothesis testing using small matched samples *(continued)*

The output is as follows:

T-Test of the Mean

Test of mu $= 0.00$ vs mu not $= 0.00$

Variable	N	Mean	StDev	SE Mean	T	P
D	7	3.71	3.15	1.19	3.12	0.021

Authors' comment: Actual $t = 3.12$ (shown as T by MINITAB). The decision, as before, is that one must reject H_0 at the 5% level of significance as $|t| > |t_{\alpha/2}|$, i.e. $3.12 > 2.447$ with df $= n - 1 = 6$. The same conclusion is reached using the P value as $P(= 0.021) < \alpha(= 0.05)$.

(b) One-tailed *t* test

With data for the difference ('D') between 'Before training' and 'After training' (C2 − C1) entered in column C3 of the MINITAB worksheet (see the two-tailed Z test computer solution in Worked example 15.4 above regarding the calculation of this within MINITAB). The procedure is then as follows:

Choose **Stat → Basic Statistics → 1-Sample t**...
Enter **Variable** (C3)
Check **Test Mean** and enter value of the hypothesized mean (0.0)
Select **Alternative** i.e. the alternative hypothesis ('greater than' in this case)
Click **OK**

The output is as follows:

T-Test of the Mean

Test of mu $= 0.00$ vs mu > 0.00

Variable	N	Mean	StDev	SE Mean	T	P
D	7	3.71	3.15	1.19	3.12	0.010

Authors' comment: Actual $t = 3.12$ (shown as T by MINITAB). The decision, as before, is that one must reject H_0 at the 5% level of significance as $t > +t_{\alpha}$, i.e. $3.12 > +1.943$ with df $= n - 1 = 6$. The same conclusion is reached using the P value as $P(= 0.010) < \alpha(= 0.05)$.

E

Computer solution using Excel

With the two data series entered in columns A and B of the worksheet, as shown below (note that X_2 and X_1 are put in columns A and B respectively because Excel calculates on the basis of column A – column B):

Choose **Tools** → **Data Analysis**... →
    t-**Test: Paired Two-Sample for Means**

In **Variable 1 Range** select cells A1 : A7 and in **Variable 2 Range** select cells B1 – B7

In **Hypothesized Mean Difference** enter 0

Click **OK**

The output table is shown below:

Input table		Output table		
A	B			
X_2	X_1	t-Test: Paired Two Sample for Means		
8	7			
6	6		Variable 1	Variable 2
10	2	Mean	9.142857	5.428571
12	5	Variance	3.47619	3.952381
9	4	Observations	7	7
10	6	Pearson Correlation	−0.33402	
9	8	Hypothesized Mean Difference	0	
		df	6	
		t Stat	3.122499	
		P(T <= t) one-tail	0.01026	
		t Critical one-tail	1.943181	
		P(T <= t) two-tail	0.02052	
		t Critical two-tail	2.446914	

Authors' comment: Note that, as in Worked examples 15.2 and 15.3 above, the output gives the actual *t* statistic along with its probability for both two-tailed and one-tailed tests and the critical value of *t* at the 5 per cent level of significance in both cases.

Two-tailed test conclusion: as before, H_0 is rejected at the 5% level of significance, as $|t| > |t_{\alpha/2}|$, i.e. $3.12 > 2.447$. The same conclusion is reached using the P value as $P(= 0.021) < \alpha(= 0.05)$.

One-tailed test conclusion: as before, H_0 is rejected at the 5% level of significance, as $t > +t_\alpha$, i.e. $3.12 > +1.943$. The same conclusion is reached using the P value as $P(= 0.010) < \alpha(= 0.05)$.

Key terms and concepts

Hypothesis testing of the difference between two population means Procedures using sample statistics to test hypotheses about the difference between two population means.

Independent samples Samples in which the observations are taken from different objects.

Matched samples Samples in which the observations are taken from the same objects.

Pooled sample variance s_p^2 The weighted average of the separate variances of two samples where the weights used are the respective degrees of freedom:

$$s_p^2 = \frac{(n_1 - 1)s_1^2 + (n_2 - 1)s_2^2}{n_1 + n_2 - 2}$$

Sampling distribution of the difference between two population means The distribution representing the differences between all possible sample means which could be obtained by sampling from two populations.

Standard error of the difference between two means

$$\sigma_{\bar{x}_1 - \bar{x}_2} = \sqrt{\frac{\sigma_1^2}{n_1} + \frac{\sigma_2^2}{n_2}} \quad \text{when } \sigma_1 \text{ and } \sigma_2 \text{ are known}$$

$$s_{\bar{x}_1 - \bar{x}_2} = \sqrt{\frac{s_1^2}{n_1} + \frac{s_2^2}{n_2}} \quad \text{when } \sigma_1 \text{ and } \sigma_2 \text{ are unknown}$$

Self-study exercises

Independent samples

15.1. A random survey of 300 houses in region A showed that the average number of rooms per dwelling was 7 with a standard deviation of 3 and a comparable survey of 400 houses in region B showed that the average number of rooms per dwelling was 5 with a standard deviation of 2. Test the hypotheses at the 5 per cent level of significance that

(a) the number of rooms per dwelling in A is greater than in B;

(b) the number of rooms per dwelling in A differs from that in B.

15.2. A survey of weekly industrial earnings in one region of the country covering 400 manual workers showed that the mean earnings were £30 with a standard deviation of £6, while in another region a comparable survey covering 100 workers showed that mean earnings were £26 with a standard deviation of £4. Did this indicate that earnings in the two regions were significantly different?

15.3. A manufacturer of electric light bulbs produces two types: a 'short life' (SL) bulb, which is claimed to have a mean life of 1000 hours, and a 'long life' (LL) bulb which is claimed to have a mean life of 1500 hours. Samples of 100 of each type produce the results given in the table below.

(a) Do these results confirm that the bulbs are of a different type?

(b) Are the manufacturer's claims justified?

(c) How large a sample would the manufacturer need to take to be sure with 99 per cent confidence that the mean life of SL bulbs was not less than 1000 hours?

	SL Type	LL Type
Mean life (hours)	1100	1400
Standard deviation (hours)	500	520

15.4. Men's heights are normally distributed with mean 1.73 metres and standard deviation 0.076 metres and women's heights are normally distributed with mean 1.65 metres and standard deviation 0.064 metres. If, in a random sample of 100 married couples, 0.05 metres is the average value of the difference between the husband's height and the wife's height, is the choice of partner in marriage influenced by consideration of height?

15.5. In a random sample survey of family expenditure it was found that the amount spent on meat by one income group consisting of 20 families was £10 per week with a standard deviation of 90p. For another income group consisting of 30 families the amount spent was £9 per week with a standard deviation of 75p. Discuss the proposition that there is no difference in the expenditure on meat of these two groups of families. (*Note*: assume equal population variances.)

15.6. Two processes on an assembly line are performed manually: operation A and operation B. A random sample of 15 different assemblies using operation A shows that the mean time per assembly is 8.68 minutes with a standard deviation of 1.84 minutes. A random sample of 24 different assemblies using operation B shows that the sample average time per assembly is 9.46 minutes with a standard deviation of 1.56 minutes. Does the evidence support the view at the 10 per cent level of significance that operation B takes significantly longer to perform than operation A? (*Note:* assume unequal population variances.)

Matched samples

15.7. A motor cycle manufacturer is interested in testing the fuel economy of its motor cycles using leaded and unleaded petrol and collects the information shown below on fuel efficiency measured in kilometres per litre (km/l) achieved by ten identical motor cycles using leaded and unleaded petrol. Is the use of leaded petrol more fuel-efficient than unleaded petrol?

Motor cycle	Leaded (km/l)	Unleaded (km/l)
A	24	26
B	30	30
C	32	37
D	28	26
E	33	33
F	32	33
G	35	33
H	27	29
I	25	28
J	28	31

15.8. A company wants to evaluate the effectiveness of an advertising campaign on the sales of a particular product sold in 100 shops situated in various regional districts. The monthly sales (£) of a random sample of seven shops, before and after the advertising campaign, are given in the table below.

(a) Is the company justified in concluding that there has been a significant increase in sales?

(b) If the shops make a 10 per cent profit on sales and the advertising campaign cost £100, has it made an addition to profit?

Shop	1	2	3	4	5	6	7
Sales before advertising (£)	102	160	81	130	115	160	123
Sales after advertising (£)	100	185	106	151	115	187	139

15.9. In a time and motion study of workers on a production line, the time taken by a group of 40 workers to complete a particular operation was recorded. Certain changes to the production line were then made and the time taken by the same 40 workers to complete the same operation was recorded again. This provides a large matched sample of 40 paired observations. Denoting the difference between each pair of observations as D (i.e. $x_1 - x_2$) the results were

$$\overline{D} = -2.3 \text{ minutes} \qquad s_D = 3.48$$

Does this indicate a significant improvement in performance on the adapted production line at the 5 per cent level of significance?

Chapter 16

Estimation and hypothesis testing: proportions

In Chapters 12 to 15 we concentrated on estimation and the testing of hypotheses concerning sample means drawn from one or two populations. In this chapter we now focus our attention on situations involving estimation and the testing of hypotheses concerning *proportions* – i.e. situations in which we are interested in the *proportion* of a population that has a certain attribute. Attributes may relate to personal or physical characteristics such as the proportion of people intending to vote for a particular political party or the proportion of defective components in a production batch. It will be recognized that such situations involve the *binomial distribution* (considered in Chapter 7) because we are again concerned with only two possible outcomes – in this case, the proportion 'with' or 'without' the attribute in question.

The chapter is organized as follows:

- point and interval estimation of a population proportion based on sample information
- hypothesis tests concerning a single population proportion
- hypothesis tests concerning the difference between two proportions

- understand what is meant by a population proportion
- grasp the meaning of the sampling distribution of the population proportion
- understand the terminology associated with estimation and hypothesis testing for population proportions
- comprehend the principles of point and interval estimation for a population proportion
- comprehend the principles of statistical inference in the context of hypotheses about population proportions
- carry out hypothesis tests of a single population proportion
- conduct hypothesis tests of the difference between two population proportions.

Principles

We need to distinguish between proportions possessing a specified attribute in the *population* as against the proportion with the attribute in a *sample*. For the purpose of exposition we employ the following notation:

π *is used to denote the popoulation proportion, i.e. the* population parameter (*note that this should not be confused with the mathematical constant π*).

p *is used to denote a sample proportion, i.e. the* sample statistic (*this should not be confused with the p value used in the previous chapters to denote probability values in hypothesis testing*).

A proportion is, of course, simply the number x of items in a sample or population possessing the particular attribute divided by the total number in the sample (i.e. n) or population (i.e. N). Thus,

Sample proportion

$$p = \frac{x}{n}$$

The sampling distribution of p

Different random samples generate a variety of values for p (the sample proportion of successes), i.e. a sampling distribution of p. For the purposes of estimation and hypothesis testing, we need to know the properties of this sampling distribution.

It will be recalled from our discussion of the binomial distribution in Chapter 7 that the probability of observing x 'successes' (the number of items with the particular attribute) is given by the following expression:

$$P(x) = {}^nC_x(p^x)(q^{n-x})$$

where x is the number of 'successes' in the sample (i.e. $x = 0, 1, 2, \ldots, n$), n is the sample size, p is the proportion of 'successes', q is the proportion of 'failures' (i.e. proportion of non-successes, $1 - p$) and ${}^nC_x$ indicates the number of combinations of x from n given by

$$ {}^nC_x = \frac{n!}{x!(n-x)!} $$

For the binomial distribution, we also showed in Chapter 7 that the summary statistics, mean and variance, are as follows:

mean number of successes $\mu_x = np$

standard deviation $\sigma_x = \sqrt{npq} = \sqrt{np(1-p)}$

In terms of the *proportion* of successes – our concern in this chapter – division of these summary statistics by the sample size n provides equivalent measures for proportions. Thus,

mean proportion $\mu_p = \dfrac{np}{n} = p$

standard deviation (standard error) of the proportions

$$\sigma_p = \frac{\sqrt{np(1-p)}}{n} = \sqrt{\frac{p(1-p)}{n}}$$

Note that the expression for the standard deviation refers to the whole sampling distribution of p and is thus the standard error (as indicated).

The Standard Error of p

In terms of the population proportion, therefore, the population standard error of the proportion is:

$$\sigma_p = \sqrt{\frac{\pi(1-\pi)}{n}}$$

When π is not known, p substitutes for π and thus an estimate of σ_p (denoted s_p) is made as:

$$s_p = \sqrt{\frac{p(1-p)}{n}}$$

Naturally, s_p is used in making confidence interval estimates of π. In hypothesis testing procedures, however, a better candidate for estimating σ_p is available in the hypothesized value of π (denoted π_0) because under the null hypothesis in such tests it is assumed that π_0 is true. Thus the standard error is then taken as:

$$\sigma_p = \sqrt{\frac{\pi_0(1-\pi_0)}{n}}$$

Finite population correction factor

As with the distribution of sample means, $\bar{x}$, the standard deviation of p depends on whether the population is finite or infinite. The correction factor is the same as before (see pp. 151–2) so that for a finite population the standard error of p is given by the following expression:

$$\sigma_p = \sqrt{\frac{p(1-p)}{n}}\sqrt{\left(\frac{N-n}{N-1}\right)}$$

where n is sample size and N is population size.

In practice, again as before, use of the finite population correction factor $(N-n/N-1)$ follows the rule of thumb that it is recommended for use only when samples are large relative to the population, in particular when $n/N > 0.05$.

Normal approximation to the binomial distribution

We noted in Chapter 9 that under certain conditions the normal distribution provides a good approximation to the binomial distribution. The rule of thumb that defines these conditions is when $np \geq 5$ and $nq \geq 5$. The normal approximation may then be applied. A closer approximation of the two distributions is obtained by making a 'continuity correction' to allow for the fact that the normal distribution involves a *continuous* variable whereas the binomial distribution deals with a *discrete* variable (for further details see Chapter 9). But for reasonably large samples the correction factor is small and we ignore it in this chapter.

We are now in a position to set out the principles of point and interval estimation and hypothesis testing for proportions.

Point estimation of a population proportion

Just as we showed earlier (Chapter 11) that the mean of a sampling distribution of means $(\mu_{\bar{x}})$ is equal to the population mean μ, so in the case of proportions the mean of the sampling distribution of all possible proportions (μ_p) is equal to the population proportion π. Thus, just as $\bar{x}$ provides an unbiased estimator of μ, p provides an unbiased estimator of π. A sample proportion p, therefore, provides the required point estimate of the unknown population proportion π.

Interval estimation for a population proportion

We confine attention here to situations where the normal approximation to the binomial distribution applies. There is then a direct parallel with the procedure for making interval estimates of the population mean considered in Chapter 12. When the conditions for the normal approximation do not apply, the calculations involved require considerable repetition using the binomial formula and are therefore tedious if carried out manually. Exact calculations using the binomial distribution, however, may be obtained using MINITAB and its use is illustrated later in the Applications and Worked Examples section. Note that the t distribution does not apply when making interval estimates of a population proportion.

Using the normal approximation, a confidence interval for the population proportion π is computed as follows:

Confidence interval

$$\bar{\bar{\pi}} = p \pm Z_{\alpha/2}(s_p)$$

where

$$s_p = \sqrt{\frac{p(1-p)}{n}}$$

(assuming that the finite population correction factor does not apply).

Note that s_p is used here instead of σ_p since σ_p is unknown. Thus, for $\alpha = 0.05$, $Z_{\alpha/2} = \pm 1.96$ (from tables) and hence a 95 per cent confidence interval for π is

$$\bar{\bar{\pi}} = p \pm 1.96 s_p$$

Hypothesis testing for a single population proportion

Again, the procedure for conducting hypothesis tests of a population proportion, given that the normal approximation applies, is the same as that for tests of a population mean (see Chapter 14). Thus, the procedure for two-tailed and for one-tailed tests may be summarized as shown in Table 16.1 overleaf.

Hypothesis testing for two population proportions

Suppose that a sample survey of electors in one constituency shows that 52 per cent intended to vote in favour of the Labour Party while in another constituency only 48 per cent intended to vote Labour. We may be interested in testing the hypothesis that there is a significant difference between the voting intentions of the two constituencies. This provides an example of a situation requiring a test of the difference between two population proportions based on two independent samples. Once again, there is a direct analogy between the test procedure required here with that involving a test of the difference between two population means considered earlier in Chapter 15.

The procedural steps for two-tailed and for one-tailed tests are summarized in Table 16.2 below. Note that, as before, the test statistic Z is calculated as

$$Z = \frac{\text{sample value} - \text{hypothesized value}}{\text{standard error of statistic}}$$

In this case, however, note that under the null hypothesis we put $\pi_1 = \pi_2$ (i.e. the difference between the two proportions is zero) and we can obtain a better estimate of the hypothesized *common* population proportion by combining the information from the two samples. Thus this estimate, which we denote as $\hat{\pi}$ (pronounced 'pi hat'), may be derived as the sum of the combined observations divided by the combined sample sizes or, equivalently, as a weighted average of the two sample proportions. Thus we have

$$\hat{\pi} = \frac{x_1 + x_2}{n_1 + n_2} = \frac{n_1 p_1 + n_2 p_2}{n_1 + n_2}$$

Table 16.1 Summary of hypothesis tests of a single population proportion

	Two-tailed test	One-tailed test	
		Left-hand tail	Right-hand tail
Hypotheses	$H_0: \pi = \pi_0$ $H_1: \pi \neq \pi_0$	$H_0: \pi = \pi_0$ $H_1: \pi < \pi_0$	$H_0: \pi = \pi_0$ $H_1: \pi > \pi_0$
Significance level	α	α	
Standard error of p	$\sigma_p = \sqrt{\dfrac{\pi_0(1 - \pi_0)}{n}}$		$\sigma_p = \sqrt{\dfrac{\pi_0(1 - \pi_0)}{n}}$
Test statistic			
Actual value	$Z = \dfrac{p - \pi_0}{\sigma_p}$		$Z = \dfrac{p - \pi_0}{\sigma_p}$
Critical value	$\pm Z_{\alpha/2}$	$-Z_\alpha$	$+Z_\alpha$
Decision	Reject H_0 if $|Z| > |Z_{\alpha/2}|$ Accept H_0 otherwise	Reject H_0 if $Z < -Z_\alpha$ Accept H_0 otherwise	Reject H_0 if $Z > +Z_\alpha$ Accept H_0 otherwise
Conclusion	Acceptance of H_0 indicates that the population proportion (π) cannot be regarded as being different from the hypothesized (π_0) value at the α level of significance	Acceptance of H_0 indicates that the population proportion cannot be regarded, at the α level of significance, as being: less than π_0	greater than π_0

Table 16.2 Summary of hypothesis tests of two-population proportions

	Two-tailed test	One-tailed test	
		Left-hand tail	Right-hand tail
Hypotheses	$H_0: \pi_1 - \pi_2 = 0$ $H_1: \pi_1 - \pi_2 \neq 0$	$H_0: \pi_1 - \pi_2 = 0$ $H_1: \pi_1 < \pi_2$	$H_1: \pi_1 > \pi_2$
Significance level	α	α	
Standard error of $p_1 - p_2$	$s_{p_1-p_2} = \sqrt{\hat{\pi}(1-\hat{\pi})\left(\dfrac{1}{n_1}+\dfrac{1}{n_2}\right)}$ where $\hat{\pi} = \dfrac{x_1+x_2}{n_1+n_2} = \dfrac{n_1 p_1 + n_2 p_2}{n_1+n_2}$	$s_{p_1-p_2} = \sqrt{\hat{\pi}(1-\hat{\pi})\left(\dfrac{1}{n_1}+\dfrac{1}{n_2}\right)}$ where $\hat{\pi} = \dfrac{x_1+x_2}{n_1+n_2} = \dfrac{n_1 p_1 + n_2 p_2}{n_1+n_2}$	
Test statistic Actual value	$Z = \dfrac{(p_1-p_2)-(\pi_1-\pi_2)}{s_{p_1-p_2}}$	$Z = \dfrac{(p_1-p_2)-(\pi_1-\pi_2)}{s_{p_1-p_2}}$	
Critical value	$\pm Z_{\alpha/2}$	$-Z_\alpha$	$+Z_\alpha$
Decision	Reject H_0 if $|Z| > |Z_{\alpha/2}|$ Accept H_0 otherwise	Reject H_0 if $Z < -Z_\alpha$ Accept H_0 otherwise	Reject H_0 if $Z >> +Z_\alpha$ Accept H_0 otherwise
Conclusion	Acceptance of H_0 indicates that there is no difference between the two population proportions at the α level of significance	Acceptance of H_0 indicates that the population proportion, π_1, at the α level of significance is: less than π_2	greater than π_2

Using $\hat{\pi}$, the standard error of the difference between the two proportions, $s_{p_1-p_2}$, is estimated as follows:

Standard error of difference between two proportions

$$s_{p_1-p_2} = \sqrt{\frac{\hat{\pi}(1-\hat{\pi})}{n_1} + \frac{\hat{\pi}(1-\hat{\pi})}{n_2}} = \sqrt{\hat{\pi}(1-\hat{\pi})\left(\frac{1}{n_1}+\frac{1}{n_2}\right)}$$

Applications and Worked Examples — **Worked example 16.1 Point estimation of a population proportion**

General descriptions of the MINITAB and Excel computer programs and their application are given in Appendices B and C respectively.

A manufacturer supplies boxes of paper with 1000 sheets per box. A random sample of 25 boxes shows that the number with less than 1000 sheets was eight. Estimate the proportion π in the population of boxes that has less than 1000 sheets per box.

Solution

Point estimate of $\pi = p = x/n = 8/25 = 0.32$; i.e. 32 per cent.

Worked Example 16.2 Interval estimation of a population proportion

Given the data in Worked example 16.1 above, estimate a 95 per cent confidence interval for π, the proportion of boxes in the population with less than 1000 sheets in each.

Worked solution

The normal approximation to the binomial distribution applies here because

$np = 25(0.32) = 8 > 5$

and

$nq = n(1-p) = 25(0.68) = 17 > 5$

Thus, a 95 per cent confidence interval for π is given by

$$\overline{\pi} = p \pm Z_{\alpha/2}(s_p)$$

$$= 0.32 \pm 1.96\sqrt{\frac{0.32(0.68)}{25}}$$

$$= 0.14 \text{ to } 0.50$$

We can therefore be 95 per cent confident that the proportion of boxes in the population with less than 1000 sheets in each is in the range between 14 and 50 per cent.

Computer solution using MINITAB　Ⓜ

(a) Using the normal approximation to the binomial distribution

Choose **Stat → Basic Statistics → 1 Proportion**

Check **Summarized data** button
Enter **Number of trials** (25 in this case)
Enter **Number of successes** (8 in this case)
Select **Options**
Enter **Confidence level** (95% in this case)

Check **Use test and interval based on normal distribution** box
Click **OK**

> *Authors' comment:* Note that checking this box ensures the use of the normal distribution. Note also that MINITAB conducts a significance test and calculates a confidence interval at the same time (we consider significance testing in Worked example 16.3 below).

The output is as follows:

Test and Confidence Interval for One Proportion

Test of p = 0.5 vs p < 0.5

Sample	X	N	Sample p	95.0 % CI	Z-Value	P-Value
1	8	25	0.320000	(0.137145, 0.502855)	−1.80	0.036

95% confidence limits based on normal distribution

(b) Exact estimates (not using the normal approximation to the binomial)

The instructions remain the same as above except that the final step in which the use of the normal distribution is selected is omitted.
   The output is as follows:

Test and Confidence Interval for One Proportion

Test of p = 0.5 vs p < 0.5

Sample	X	N	Sample p	95.0 % CI	Exact P-Value
1	8	25	0.320000	(0.149495, 0.535001)	0.054

95% confidence limits based on binomial distribution

Computer solution using Excel

There is no pre-defined tool in Excel for solving this problem.

Worked example 16.3   Hypothesis test of a single sample proportion

It is claimed by a telephone company that, on any single day, of all public telephones, the proportion out of order is under 6 per cent. A random sample of 300 telephones shows that, on a particular day, 25 were out of order. Does this evidence show that the telephone company's claim is false (at the 1 per cent level of significance)?

Worked example 16.3 Hypothesis test of a single sample proportion *(continued)*

Worked solution

Given:	$n = 300; x = 25$
	$\pi_0 = 0.06$ (i.e. hypothesized population proportion)
Hypotheses:	$H_0: \pi_0 = 0.06$
	$H_1: \pi_0 < 0.06$
Significance level:	$\alpha = 0.01$
Standard error of p:	$\sigma_p = \sqrt{\dfrac{\pi_0(1-\pi_0)}{n}} = \sqrt{\dfrac{0.06(1-0.06)}{300}} = 0.0137$
	$p = \dfrac{x}{n} = \dfrac{25}{30} = 0.0833$
Test statistic:	Critical: $Z_\alpha = -Z_{0.01} = -2.33$
	Actual: $Z = \dfrac{p - \pi_0}{\sigma_p} = \dfrac{0.0833 - 0.06}{0.0137} = 1.70$

As $Z > -Z_{0.01}$, i.e. $1.70 > -2.33$, we cannot reject H_0, at the 1 per cent level of significance.

Conclusion: We conclude, therefore, that the company's claim that the proportion of telephones out of order on any one day is under 6 per cent is not justified by the evidence from the sample survey.

Computer solution using MINITAB

(a) Using the normal approximation to the binomial distribution

The command procedure is the same as in Worked example 16.2 part (a). The output is as follows:

Test and Confidence Interval for One Proportion

Test of p = 0.06 vs p < 0.06

Sample	X	N	Sample p	99.0 % CI	Z-Value	P-Value
1	25	300	0.083333	(0.042231, 0.124436)	1.70	0.956

Authors' comment and interpretation: The actual Z value is shown as 1.70 (as in the manual calculation above). As $Z > -Z_\alpha$, i.e. $1.70 > -2.33$, we cannot reject H_0 at the 1% level of significance. Thus we conclude that the company's claim is not justified. The same conclusion is reached using the P value as $0.956 > \alpha(= 0.01)$.

(b) Exact calculation (not using the normal approximation to the binomial)

The command procedure is the same as in Worked example 16.2 part (b). The output is as follows:

Test and Confidence Interval for One Proportion

Test of p = 0.06 vs p < 0.06

Sample	X	N	Sample p	99.0 % CI	Exact P-Value
1	25	300	0.083333	(0.047456, 0.133014)	0.960

> *Authors' comment and interpretation:* The test here is based solely on the P-value (as the normal distribution is not being used, the Z statistic does not apply). The exact P-value is shown as 0.960 (compare with the P-value in part (a) above which is 0.956). As $0.96 > \alpha(= 0.01)$, we cannot reject H_0 and the same conclusion follows, as before.

Computer solution using Excel

There is no pre-defined tool in Excel for solving this problem.

Worked example 16.4   Hypothesis test of the difference between two sample proportions

An opinion poll is conducted to determine whether a political leader is less popular with female voters than with male voters. The results were as shown below. Do these results show that the politician is more popular among male voters than female voters?

	Male voters	*Female voters*
Sample size	$n_1 = 400$	$n_2 = 400$
Number supporting politician	216	180
Proportion	$p_1 = 216/400$	$p_2 = 180/400$
	$= 0.54$	$= 0.45$

Worked example 16.4   Hypothesis test of the difference between two sample proportions *(continued)*

Worked solution

Hypotheses: $H_0: \pi_1 - \pi_2 = 0$
             $H_1: \pi_1 - \pi_2 > 0$

Significance level: $\alpha = 0.01$

Standard error: $s_{p_1-p_2} = \sqrt{\hat{\pi}(1-\hat{\pi})\left(\dfrac{1}{n_1}+\dfrac{1}{n_2}\right)}$

where

$$\hat{\pi} = \frac{216 + 180}{400 + 400} = 0.4950$$

Thus

$$s_{p_1-p_2} = \sqrt{0.4950(1-0.4950)\left(\frac{1}{400}+\frac{1}{400}\right)} = 0.03535$$

Test Statistic:

$$\text{Actual } Z = \frac{(p_1 - p_2) - (\pi_1 - \pi_2)}{s_{p_1-p_2}} = \frac{(0.54 - 0.45) - 0}{0.03535}$$

$$= 2.55$$

$$\text{Critical } Z_\alpha = +Z_{0.01} = +2.33$$

As $Z > + Z_\alpha$, i.e. $2.55 > + 2.33$, we must reject H_0 at the 1 per cent level of significance.

Conclusion: We conclude, therefore, that the politician is more popular with male voters than with female voters.

Computer solution using MINITAB

Choose **Stat** → **Basic statistics** →
   2 Proportions
Check **Summarized data** button
Enter number of **Trials** and number of
   Successes for **First sample** (400 and 216
   respectively in this case)
Enter number of **Trials** and number of
   Successes for **Second sample** (400 and
   180 respectively in this case)

Select **Options**
Enter **Confidence level** (99% in this case)
Enter **Test difference** (0.0 in this case)
Select **Alternative**, i.e. alternative hypothesis
   ('greater than' in this case)
Check **Use pooled estimate of p for test** box
Click **OK** twice

The output is as follows:

Test and Confidence Interval for Two Proportions

Sample	X	N	Sample p
1	216	400	0.540000
2	180	400	0.450000

Estimate for p(1)−p(2): 0.09
99% CI for p(1)−p(2): (−0.000695166, 0.180695)
Test for p(1)−p(2) = 0 (vs > 0): Z = 2.55   P-Value = 0.005

Authors' comment and interpretation: As $Z > + Z_\alpha$, i.e. 2.55 > +2.33, we must reject H_0 at the 1 per cent level of significance. Thus we conclude that the politician is more popular among male voters than among female voters. The same conclusion is reached using the P-value as P(= 0.005) < α(= 0.01).

Computer solution using Excel   **E**

There is no pre-defined tool in Excel for solving this problem.

Key terms and concepts

Hypothesis testing for two population proportions Procedure to test the difference between two population proportions based on two independent samples.

Hypothesis testing for a single population proportion Procedure using sample data to test hypotheses about a population proportion.

Interval estimation for a population proportion A confidence interval for the population proportion π computed as:

$$\overline{\overline{\pi}} = p \pm Z_{\alpha/2}(s_\mathrm{p})$$

where

$$s_\mathrm{p} = \sqrt{\frac{p(1-p)}{n}}$$

and p is the sample proportion (assuming that the finite population correction factor does not apply).

Point estimation for a population proportion A single numerical value used as an estimate of a population proportion.

Sample proportion The proportion in a set of sample data possessing a certain attribute.

Sampling distribution of a sample proportion A probability distribution of all possible values of a sample proportion.

Standard error of a sample proportion The standard deviation of the sampling distribution of a sample proportion, p:

$$\sigma_p = \sqrt{\frac{p(1-p)}{n}}$$

Standard error of difference between two proportions The standard deviation of the sampling distribution of the difference between two proportions, p_1 and p_2:

$$s_{p_1-p_2} = \sqrt{\frac{\hat{\pi}(1-\hat{\pi})}{n_1} + \frac{\hat{\pi}(1-\hat{\pi})}{n_2}} = \sqrt{\hat{\pi}(1-\hat{\pi})\left(\frac{1}{n_1} + \frac{1}{n_2}\right)}$$

where

$$\hat{\pi} = \frac{x_1 + x_2}{n_1 + n_2} = \frac{n_1 p_1 + n_2 p_2}{n_1 + n_2}$$

Self-study exercises

16.1. An enquiry into the employment and marital status of women in an area gave the following results:

	Married	Single
Employed full-time	23	37
Employed part-time	22	32
Not employed	25	11
Total	70	80

(a) Make a point estimate of the proportion of married women who are employed full-time.

(b) Is the proportion of part-time workers among those employed significantly different for married and single women?

16.2. In a survey to determine car parking requirements, a random survey of workers showed that 78 out of 250 travelled to work by car. Compute a 99 per cent confidence interval for the proportion who travel to work by car.

16.3. There are 250 public telephones in a town. For 20 per cent of the time each telephone is out of order. What is the probability of having the following numbers of telephones out of order at any one time:

(a) more than 60?

(b) between 60 and 65?

(c) less than 32?

16.4. The manager of a supermarket claims that three-quarters of people making purchases over £100 use a credit card. In a random sample of 50 people who had spent over £100, 31 used a credit card. Test the manager's claim at the 5 per cent and 1 per cent levels of significance.

16.5. **(a)** It is required to calculate the proportion of undergraduates who have cars on a university campus. Determine the upper bound for the sample size necessary to give 95 per cent confidence that our estimate differs from the true proportion by not more than 0.05.

(b) A random sample of 70 undergraduates shows that 21 have cars on campus. Use a 95 per cent confidence interval to estimate the proportion of undergraduates who have cars on campus.

(c) A random sample of 30 postgraduate students showed that 15 had cars on campus. Did this indicate that a significantly greater proportion of postgraduate students have cars on campus compared with undergraduates?

16.6. **(a)** Two opinion polls, in which random samples of 1600 and 2000 electors respectively were asked about voting intentions in the next general election, showed 48 per cent and 51 per cent respectively in favour of the ruling political party. Is this a significant difference?

(b) Compute the combined percentage from the two opinion polls above and say whether or not it represents a significant increase over the vote of 44 per cent in the previous general election for the ruling party at both the 5 per cent and 1 per cent levels of significance.

(c) In the run-up to the next general election, it is planned to take repeated weekly samples to signal trends in voting intentions. It is desired to measure the proportion intending to vote for the ruling party with a maximum margin of error of ± 1 per cent with 99 per cent confidence. How large a sample should be taken?

Chapter 17

Analysis of variance (ANOVA)

This chapter provides an introduction to the analysis of variance (ANOVA) technique. It is a very important technique in applied statistical analysis because of its usefulness in tackling a common type of problem across a wide range of fields including not only practical applications in industry and management but also the whole spectrum of research in natural and social sciences. A full treatment of the topic is beyond the scope of this book but its importance is such that it merits a short introduction to the principal elements of the technique and its application.

The essence of the technique is that where a particular subject of interest is influenced by several different factors, it permits statistical data on the subject to be broken down and the influence of each factor assessed. The technique does so by measuring the variation (variance) in a set of data and then partitioning, or subdividing, the variance into separate parts, each of which measures the contribution of a particular factor to the overall variation. The name of the technique – 'the analysis of variance' – is therefore self-explanatory. For example, we may have data on the performance of a group of people on a certain task (the time taken to assemble a component, to run a hundred metres etc.). The variation in performance may then be sub-divided according to factors thought likely to influence performance, such as sex, age, health, training received, years of experience etc. and the significance of these factors tested.

In the context of this part of the book, in which we have been concerned with comparisons involving two populations, the technique permits the analysis of significance testing to be extended to cases involving more than two populations. In Chapter 15 we considered how to compare the means of two populations using the t test. However, in practice, we may often wish to compare means from more than two populations and in this situation the *t* test is not efficient or appropriate. For example, with six sets of data there would be 15 pair-wise tests and as the potential for error in each test (Type I error) combine together, the level of significance of the test overall is reduced.

In this chapter we consider the principles and application of ANOVA to two types of situation:

■ Single-factor analysis of variance

■ Two-factor analysis of variance

Learning Outcomes After working through this chapter you will be able to:

■ test hypotheses about the difference between more than two population means

■ appreciate what is meant by the term analysis of variance (ANOVA)

■ understand the difference between single-factor and two-factor ANOVA

■ appreciate the difference between *variance between samples* and *variance within samples*

■ understand how to decompose overall variation in a set of observations into component parts

■ construct an ANOVA table

■ compute the *F* statistic

■ understand the *F* distribution

■ apply the *F* distribution to test for the statistical significance of the influence of different factors in contributing to the overall variation in a data set.

Principles

The basic principle behind the ANOVA technique is its use of the additive property of the 'variance' measure of variation in a set of data. This property means that if a set of data can be broken down into subsets of data for factors which contribute to the overall variation, then the overall variance is equal to the sum of the individual component variances (assuming the factors act independently). It is important to note that this additive property applies to the variance measure of variation not to the standard deviation.

At the outset it should also be noted that the application of the technique depends on certain assumptions about the data. These are that the samples of data taken from the populations under study are independent of one another, that they are random, and that the populations are normally distributed and have equal variance. But in practice the normality assumption is regarded as not being too important and the equal-variances assumption is also regarded as not being important (provided that the number of observations in each group is about the same); independence and random sampling, however, are essential.

Single-factor analysis of variance

Single-factor analysis of variance – also referred to as *one-way* analysis of variance – applies when the data are like those used in the test of the equality of two population means considered in Chapter 15: the only factor is the difference in the populations from which the sample observations are obtained. In Chapter 15 the *t* test is applied, on the assumption (as in ANOVA) that the populations have equal variance. Thus in this case, the ANOVA technique is an alternative to the *t* test for the equality of two population means (assuming equal population variances) and both give exactly the same (two-tail) results. The advantage of the ANOVA technique, however, is that it is not restricted to two populations.

In summary, therefore, we can say that the single-factor analysis of variance applies when we want to compare the means of several populations. The question to be answered is: 'Do all of the populations have the same mean'? Assuming we have 3 populations, then the null hypothesis to be tested is:

H_0: $\mu_1 = \mu_2 = \mu_3$

The alternative hypothesis is: H_1: all three population means are not equal.

The test procedure may be explained by analogy with the *t* test for the difference between two means. The first step in that test is to consider the variation in the two sample means by calculating the difference between them: $\bar{x}_1 - \bar{x}_2$. When more than two populations are involved, however, a single overall measure of difference is obtained by considering the differences between each sample mean and the mean of the samples *taken as a whole* (sometimes referred to as the *grand mean*). A summary measure is then obtained, not by taking the simple differences but by summing the *squared differences* taking account of the size of the samples in each case. Thus the summary measure of variation in the sample means is a weighted average of the squared deviations between each sample mean and the overall mean. Denoting the overall mean as $\bar{x}$ and the number of samples as k (in this case $k = 3$), this may be expressed as follows:

$$\frac{n_1(\bar{x}_1 - \bar{x})^2 + n_2(\bar{x}_2 - \bar{x})^2 + n_3(\bar{x}_3 - \bar{x})^2}{k - 1}$$

This produces, of course, a variance measure of the difference between the means and is part of the reason for the use of the term 'analysis of variance'. The numerator in the expression above – the sum of squares from the overall mean – is called the *Between the Samples Sum of Squares* (SSB). The denominator is the degrees of freedom and is always one less than the number of groups (thus in this case it is $3 - 1 = 2$). The expression as a whole is called the *Mean Square Between Samples* (MSB).

We next need a standard of comparison. This must be to compare the overall variation *between* the samples (SSB) with the variation that exists *within* the samples: intuitively we may consider that if the variance *between* the sample means is large relative to the variation *within* each sample, then we should be led towards the conclusion that the population means (μ_i) are different and thus to reject H_0. The variation within each sample is therefore measured by summing the squared deviations of the sample observations from their sample means. Thus for each sample we calculate $\sum(x - \bar{x})^2$ and sum over all samples. This measure is called the *Within Sample Sum of Squares* (SSW). It is often called *Sum of Squares Error* (SSE), the reason being that, as it measures variation *within* the samples, it cannot be attributed to a particular underlying cause. The corresponding *Mean Square Within Samples* (MSW) measure is obtained by dividing by the appropriate degrees of

Table 17.1 ANOVA table – single-factor ANOVA

Source of variation	Sum of squares	Degrees of freedom	Mean square
Between	SSB	$k - 1$	MSB $=$ SSB$/(k - 1)$
Within	SSW	$N - k$	MSW $=$ SSW$/(N - k)$
Total	SST	$N - 1$	

freedom which, in this case, is the total number of sample observations less one degree of freedom for each sample. Thus, given three samples (as above), MSW may be expressed as:

$$\frac{\sum(x_{i1} - \bar{x}_1)^2 + \sum(x_{i2} - \bar{x}_2)^2 + \sum(x_{i3} - \bar{x}_3)^2}{n_1 + n_2 + n_3 - 3}$$

It is helpful to summarize this information in an ANOVA table – shown in Table 17.1 above. We then proceed to consider testing the statistical significance of the difference between means.

Note that the total variation (SST) measures the variation of all the sample observations combined and is equivalent to the sum of the two component variances SSB and SSW.

Analysis of variance tests of statistical significance are made using the F distribution (described below). The value of the test statistic F is given by the ratio of the two component variances – the variance between samples (MSB) and the variance within samples (MSW). Thus the F *statistic* is given by the following expression:

F statistic

$$F = \frac{\text{MSB}}{\text{MSW}} = \frac{\text{SSB}/(k - 1)}{\text{SSW}/(N - k)}$$

where $(k - 1)$ and $(N - k)$ are the number of degrees of freedom for the numerator and the denominator respectively in the F distribution.

F distribution

The form of the F distribution is illustrated in Figure 17.1 below.

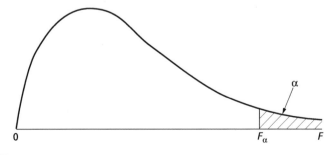

Figure 17.1 F distribution

Note that the distribution is non-symmetrical and the F values cannot be negative. The actual shape of the distribution depends upon its corresponding numerator and denominator degrees of freedom. Tables of the F statistic for probability values of 0.05 and 0.01 with degrees of freedom denoted v_1 and v_2, for the numerator and denominator respectively, ranging from 1 to ∞ are given in Appendix A Table A6. Thus, for example, it will be seen that, with $v_1 = 9$ and $v_2 = 12$, the critical F statistic equals 2.80 at the 5 per cent level of significance and 4.39 at the 1 per cent level. That is, there is only a 5 per cent chance of obtaining an F value greater than 2.80 and a 1 per cent chance of it exceeding 4.39. Note that the Appendix A Table A6 is in two parts, the first corresponding to $\alpha = 0.05$ and the second to $\alpha = 0.01$.

So, returning to the conduct of the F test, if the computed F statistic exceeds the critical F value, then we reject the null hypothesis and conclude that there is a statistically significant difference between the population means.

Summary of test procedure for single-factor analysis of variance

We conclude this section by summarizing the test procedure. Examples of its application are given in the Applications and Worked Examples section including examples of the use of the MINITAB and Excel packages, both of which contain inbuilt programs for conducting the test directly. Given that the assumptions necessary for conducting the test do in fact hold, the procedure is as follows:

1. Compute the sums of squares for the between-samples (SSB) and within-samples (SSW) variances and note the degrees of freedom which apply to each.
2. Compute the corresponding mean square values (MSB and MSW).
3. Compute the ratio between MSB and MSW to give the F statistic.
4. State the null and alternative hypotheses and the significance level for the test.
5. Compare the F statistic with its critical value given by the F tables and state the appropriate conclusion regarding statistical significance and level of probability.

Two-factor analysis of variance

Two-factor analysis of variance (also called *two-way* analysis of variance) is a direct extension of single-factor ANOVA in which, as the name implies, each data item is classified in two ways. Thus, in addition to comparing two or more different populations, two-factor ANOVA also controls for an additional factor which may influence the observations. An illustrative example, cited in the MINITAB handbook, is drawn from an experiment to study the driving abilities of different drivers. Data are collected to measure driving ability and then cross-classified according to two factors: driver experience (experienced and inexperienced) on the one hand and, on the other hand, according to road type (first class, second class or dirt). The data are summarized in a cross-classification table with one factor classified to the columns and the other factor classified to the rows. These two factors give two sources of variation which can be measured in the same way as in the one-factor case. Here we now have to compute the respective sums of squares between the columns (SSC) *and* between the rows (SSR). Total variation for the whole data set consists of the sum of the variation for the two component factors plus the residual variation which is not explained by the two measured factors (the 'sum of squares error' – SSE). The results, including the corresponding mean squares values (MSC, MSR

Table 17.2 ANOVA table – two-factor ANOVA

Source of variation	Sum of squares	Degrees of freedom	Mean square
Between columns	SSC	$k - 1$	$MSC = SSC/(k - 1)$
Between rows	SSR	$r - 1$	$MSR = SSR/(r - 1)$
Residual	SSE	$N - (k - 1) - (r - 1) - 1$	$MSE =$ $SSE/[N - (k - 1) - (r - 1) - 1]$
Total	SST	$N - 1$	

and MSE) may be set out in an ANOVA table as before and this is shown in Table 17.2 above. The degrees of freedom depend on the number of columns (k) and the number of rows (r) as shown. In practice, SSE and the degrees of freedom for it are determined by subtraction. Thus

$$SSE = SST - SSC - SSR$$

and

$$MSE = SSE/[N - (k - 1) - (r - 1) - 1]$$

In two-way ANOVA we simultaneously measure the differential effects of two factors. Two F tests may therefore be performed. These are:

(a) a test of variation between columns by comparing the 'between-columns' and 'residual' estimates of variance (MSC/MSE).

(b) a test of variation between rows by comparing the 'between-rows' and 'residual' estimates of variance (MSR/MSE).

The test assumes that the row and column effects, if any, are additive. It is possible, of course, that the measured factors may interact. This gives rise to an *interaction effect* which may itself be amenable to measurement but this takes us beyond the scope of the introduction to the ANOVA technique given here.

Applications and Worked Examples Worked example 17.1 Single-factor ANOVA

General desciptions of the MINITAB and Excel computer programs and their application are given in Appendices B and C respectively.

Three machines are used to perform a packaging process but, for various reasons, the hourly output of each machine varies. It is desired to test whether or not the machines are different. Six observations of hourly output for each machine are collected at random and are shown below. Is there a significant difference between the three machines?

Machine 1:	31	23	29	30	34	21
Machine 2:	30	31	28	20	21	26
Machine 3:	28	34	38	35	29	37

Worked Solution

The main steps are:

1. Calculate sample means and the overall (grand) mean.
2. Calculate the *SSB* (Between Samples Sum of Squares) and note the degrees of freedom.

3. Calculate the *SSW* (Within Samples Sum of Squares) and note the degrees of freedom.
4. Compile an ANOVA table (this is not essential but it is helpful and, in particular, helps to avoid error).
5. Calculate the F statistic, conduct a significance test and state the conclusion.

Step 1   Calculation of means

	Machine 1	*Machine 2*	*Machine 3*
	31	30	28
	23	31	34
	29	28	38
	30	20	35
	34	21	26
	21	26	37
Means:	$\bar{x}_1 = 28$	$\bar{x}_2 = 26$	$\bar{x}_3 = 33$

Overall (grand) mean $(\bar{x}) = [6(28) + 6(26) + 6(33)]/18 = 29$

Step 2   Calculation of variation between machines and SSB

	Machine 1	*Machine 2*	*Machine 3*
Deviations $(\bar{x}_i - \bar{x})$	$28 - 29 = -1$	$26 - 29 = -3$	$33 - 29 = 4$
(Deviations)^2 $(\bar{x}_i - \bar{x})^2$	$(-1)^2 = 1$	$(-3)^2 = 9$	$(4)^2 = 16$

Thus the Between Samples Sum of Squares
$$(\text{SSB}) = n_1(\bar{x}_1 - \bar{x})^2 + n_2(\bar{x}_2 - \bar{x})^2 + n_3(\bar{x}_3 - \bar{x})^2 = 6(1) + 6(9) + 6(16) = 156$$

Step 3   Calculation of variation within machines and SSW

For machine 1: $\sum(x_i - \bar{x}_1)^2 = (31 - 28)^2 + (23 - 28)^2 + \cdots = 124$
For machine 2: $\sum(x_i - \bar{x}_2)^2 = (30 - 26)^2 + (31 - 26)^2 + \cdots = 106$
For machine 3: $\sum(x_i - \bar{x}_3)^2 = (28 - 33)^2 + (34 - 33)^2 + \cdots = 120$

Thus the Within Samples Sum of Squares (SSW) $= 124 + 106 + 120 = 350$

Step 4   ANOVA table

Source of variation	Sum of squares	Degrees of freedom	Mean square
Between	SSB $= 156$	$k - 1 = 2$	MSB $= $ SSB$/(k - 1) = 156/2 = 78$
Within	SSW $= 350$	$N - k = 15$	MSW $= $ SSW$/(N - k) = 350/15 = 23.33$
Total	SST $= 506$	$N - 1 = 17$	

Worked example 17.1 Single-factor ANOVA *(continued)*

Step 5 Calculation of *F* statistic and conduct of test

$F = \text{MSB}/\text{MSW} = 78/23.33 = 3.34$

Hypotheses:
$$H_0: \mu_1 = \mu_2 = \mu_3$$
$$H_1: \text{all three population means are not equal}$$

Significance level $\qquad\qquad \alpha = 0.05$

Critical value of test statistic F_α (with $v_1 = 2$ and $v_2 = 15\,\text{df}$) $= 3.68$ (from tables)

$F < F_\alpha$, therefore accept H_0.

Conclusion: null hypothesis of no difference between the three population means cannot be rejected.

Computer solution using MINITAB

With the data for the three machines in columns C1, C2 and C3 of a MINITAB worksheet, the procedure is as follows:

Select **Stat** on the menu bar
Choose **ANOVA – One-way (unstacked)**
Select Columns 1–3 by highlighting them in the dialogue box itself and clicking on **Select**
Click **OK**

The resulting output is shown below.

One-way Analysis of Variance

Analysis of Variance

Source	DF	SS	MS	F	P
Factor	2	156.0	78.0	3.34	0.063
Error	15	350.0	23.3		
Total	17	506.0			

```
                                     Individual 95% CIs For Mean
                                     Based on Pooled StDev
Level     N     Mean    StDev    --------+---------+---------+---------
C1        6    28.000   4.980          (--------*--------)
C2        6    26.000   4.604       (-------*--------)
C3        6    33.000   4.899                     (-------*---------)
                                     --------+---------+---------+------
Pooled StDev =    4.830             25.0      30.0      35.0
```

Authors' comment and interpretation: Note that the output is in two parts: an ANOVA table together with descriptive and other statistics. The ANOVA table includes the usual information together with the value of the *F* statistic and its probability ('P'). Unlike the Excel output (described below), the critical value of *F* is not given but, as the probability of *F* is shown, the significance of the results can be judged by comparing the P-value with specified levels of significance (α). As before, the *F* value of 3.34 has a probability of 0.063 and, as this is greater than $\alpha(= 0.05)$, the results are not statistically significant (the null hypothesis of equality between the three sample means cannot be rejected). The descriptive and other statistics presented provide supplementary information and are not directly relevant to the conduct of the test.

Computer solution using Excel

Enter the data for the three machines in columns A, B and C respectively of an Excel worksheet.

Select **Tools** on the toolbar
Choose **Data Analysis**
Select **Anova: Single Factor**
Complete **Input range** by highlighting the data cells in columns A–C (if the columns have been
    named click on **Labels in First Row** box
Input desired value for **Alpha** (significance level α, 0.05 is shown by default)
Click **OK**

The resulting output is shown below:

Anova: Single Factor

SUMMARY

Groups	Count	Sum	Average	Variance
Machine 1	6	168	28	24.8
Machine 2	6	156	26	21.2
Machine 3	6	198	33	24

ANOVA

Source of Variation	SS	df	MS	F	P-value	F crit
Between Groups	156	2	78	3.342857	0.063005	3.682316
Within Groups	350	15	23.33333			
Total	506	17				

> *Authors' comment and interpretation*: Note that, like MINITAB, the output gives a summary
> of the data together with the ANOVA table. But note that, unlike MINITAB, the *critical* as well
> as the actual, values of F are shown, together with the *P-value* (i.e. probability) of
> $F(= 0.063005)$. The latter gives the significance level directly, thus as it is greater than
> 0.05 the null hypothesis cannot be rejected – the results are not significant at the 5% level of
> significance.

Worked example 17.2   Two-factor ANOVA

The data in Worked example 17.1 related to the performance of three packaging machines. A single-
factor ANOVA test showed that there was no significant difference between the three machines. On
further consideration, however, it was thought that differences in performance may be related to
the person operating the machines. The sample data below now show performance cross-classified
by both machine and operator. Are differences in performance significantly affected by the two
factors?

Worked example 17.2 Two-factor ANOVA *(continued)*

Operator	Machine 1	Machine 2	Machine 3
A	31	31	28
B	23	30	34
C	30	31	38
D	34	23	33
E	21	39	24
F	29	26	35

Worked Solution

The main steps are:

1. Calculate sample means and the overall (grand) mean.

2. Calculate the *SSC* (Between Columns Sum of Squares) and note the degrees of freedom.

3. Calculate the *SSR* (Between Rows Sum of Squares) and note the degrees of freedom.

4. Calculate the *SST* (Total Sum of Squares) over the sample data as a whole.

5. Compile an ANOVA table (this is not essential but it is useful particularly in helping to avoid error).

6. Calculate the *F* statistic, conduct a significance test and state the conclusion.

Step 1 Calculation of means

Operator	Machine 1	Machine 2	Machine 3	Row means
A	31	31	28	30
B	23	30	34	29
C	30	31	38	33
D	34	23	33	30
E	21	39	24	28
F	29	26	35	30
Column means:	28	30	32	30 = grand mean

Step 2 Calculation of the Between Columns Sum of Squares (SSC)

	Machine 1	Machine 2	Machine 3
Column deviations $= (\bar{x}_i - \bar{x})$	$28 - 30 = -2$	$30 - 30 = 0$	$32 - 30 = 2$
(Deviations)$^2 = (\bar{x}_i - \bar{x})^2$	$(-2)^2 = 4$	$(0)^2 = 0$	$(2)^2 = 4$

Thus the Between Columns Sum of Squares:
$$SSC = n_1(\bar{x}_1 - \bar{x})^2 + n_2(\bar{x}_2 - \bar{x})^2 + n_3(\bar{x}_3 - \bar{x})^2 = 6(4) + 6(0) + 6(4) = 48$$

Step 3 Calculation of the Between Rows Sum of Squares (SSR)

	Operator					
	A	B	C	D	E	F
Row deviations $(\bar{x}_i - \bar{x})$	$30 - 30 = 0$	$29 - 30 = -1$	$33 - 30 = 3$	$30 - 30 = 0$	$28 - 30 = -2$	$30 - 30 = 0$
(Deviations)^2 $(\bar{x}_i - \bar{x})^2$	$(0)^2 = 0$	$(-1)^2 = 1$	$(3)^2 = 9$	$(0)^2 = 0$	$(-2)^2 = 4$	$(0)^2 = 0$

Thus the Between Rows Sum of Squares
$$SSR = n_1(\bar{x}_1 - \bar{x})^2 + n_2(\bar{x}_2 - \bar{x})^2 + n_3(\bar{x}_3 - \bar{x})^2 + n_4(\bar{x}_4 - \bar{x})^2 + n_5(\bar{x}_5 - \bar{x})^2 + n_6(\bar{x}_6 - \bar{x})^2$$
$$= 3(0) + 3(1) + 3(9) + 3(0) + 3(4) + 3(0) = 42$$

Step 4 Calculation of Total Sum of Squares (SST)

$$SST = \sum(x_i - \bar{x})^2 = (31 - 30)^2 + (23 - 30)^2 + \cdots + (35 - 30)^2 = 450$$

Step 5 ANOVA table

Source of variation	Sum of squares	Degrees of freedom	Mean square
Between columns	$SSC = 48$	$k - 1 = 2$	$MSC = 48/2 = 24$
Between rows	$SSR = 42$	$r - 1 = 5$	$MSR = 42/5 = 8.4$
Residual (Error)	$SSE = 360$	$N - (k - 1) - (r - 1) - 1 = 10$	$MSE = 360/10 = 36$
Total	$SST = 450$	$N - 1 = 17$	

Step 6 Calculation of F statistics and conduct of test

The values of F for differences between columns and between rows respectively are:

Actual values   *Critical values for $\alpha = 0.05$*

$F = MSC/MSE = 24/36 = 0.667$ F with 2 and 10 df $= 4.10$ (from tables)
$F = MSR/MSE = 8.4/36 = 0.233$ F with 5 and 10 df $= 3.33$ (from tables)

Hypotheses: H_0: no difference between population means for the machines and for operators.
           H_1: there is a difference between population means for the machines and for operators.
Significance level: $\alpha = 0.05$
Test: in each case $F < F_\alpha$, therefore accept H_0

Conclusion: We cannot reject the null hypothesis that there is no difference between the population means for machines and operators.

Worked example 17.2   Two-factor ANOVA *(continued)*

Computer solution using MINITAB

The data need to be entered into the MINITAB worksheet in columns as shown below (this is different from the Excel worksheet format).

Then choose **Stat** → **ANOVA** → **Two-way**
In **Response** enter C1 Output
In **Row factor** enter C2 Operator
In **Column factor** enter C3 Machine
Click **OK**

The resulting output is shown below, after the worksheet.

Worksheet

C1 Output	C2 Operator	C3 Machine
31	1	1
23	2	1
30	3	1
34	4	1
21	5	1
29	6	1
31	1	2
30	2	2
31	3	2
23	4	2
39	5	2
26	6	2
28	1	3
34	2	3
38	3	3
33	4	3
24	5	3
35	6	3

Two-way Analysis of Variance

Analysis of Variance for Output

Source	DF	SS	MS	F	P
Operator	5	42.0	8.4	0.23	0.939
Machine	2	48.0	24.0	0.67	0.535
Error	10	360.0	36.0		
Total	17	450.0			

Authors' comment and interpretation: As with single-factor ANOVA, the output gives a summary of the data together with the ANOVA table showing the actual values of F along with the *P-values* (i.e. probability) of the calculated values of F for the two factors ($= 0.67$ and 0.23 respectively). The *P-values* give the significance levels for the two tests directly and, as they are both greater than 0.05, the results are not significant at the 5% level of significance – i.e. one cannot reject the null hypotheses of no difference between the respective population means. The summary statistics provide supplementary information which is not needed for the significance test.

Computer solution using Excel

Enter the data for the three machines and six operators in columns A–C and rows 1–6 respectively of an Excel worksheet – the same format as shown above.

Choose **Tools → Data Analysis → Anova: Two-Factor Without Replication***
Complete **Input range** by highlighting the data cells in columns A–C
Input desired value for **Alpha** (significance level $\alpha = 0.05$ is shown by default)
Click **OK**

The resulting output is shown below.

Anova: Two-Factor
Without Replication

SUMMARY	Count	Sum	Average	Variance
Row 1	3	90	30	3
Row 2	3	87	29	31
Row 3	3	99	33	19
Row 4	3	90	30	37
Row 5	3	84	28	93
Row 6	3	90	30	21
Column A	6	168	28	24.8
Column B	6	180	30	29.6
Column C	6	192	32	26

Worked example 17.2   Two-factor ANOVA *(continued)*

ANOVA

Source of Variation	SS	df	MS	F	P-value	F crit
Rows	42	5	8.4	0.233333	0.939089	3.325837
Columns	48	2	24	0.666666	0.534824	4.102815
Error	360	10	36			
Total	450	17				

Authors' comment and interpretation: The ANOVA table is, of course, identical to that given in the MINITAB example above, except that the critical values for the F statistic are given. The same comments and interpretation apply.

* *Note*: The alternative choice of 'Two-Factor with Replication' applies when there is more than one sampling per group – further details about this option are given in the Excel manual and in the Help facility in the program.

Key terms and concepts

Analysis of variance (ANOVA) A statistical technique for testing whether or not the means of two or more populations are equal.

Analysis of variance table A tabular summary of the results of an analysis of variance procedure with entries showing the sources of variation, the degrees of freedom, the sum of squares and the mean squares (variances).

Between samples sum of squares (SSB) The sum of squared deviations between each sample mean and the overall (grand) mean.

Error sum of squares (SSE) The variation which remains after subtracting from the total variation (SST) any variation arising from measured factors (also referred to as *residual sum of squares* or *sum of squares error*).

***F* distribution** A family of probability distributions based on the ratio between two sample variances.

***F* statistic** The ratio between two estimates of variance in random samples from normal distributions: if both samples are drawn from normal distributions with the *same* variance then the ratio of the two estimates follows the F distribution.

Factor The variable of interest in an ANOVA procedure.

Grand mean In the context of ANOVA, the mean of all the sample observations.

Mean square between columns (MSC) The sum of squares between columns (SSC) divided by degrees of freedom – a measure of the variation among means of samples (arranged in columns) taken from different populations.

Mean square between rows (MSR) The sum of squares between rows (SSR) divided by degrees of freedom – a measure of the variation among means of samples (arranged in rows) taken from different populations.

Mean square between samples (MSB) Between samples sum of squares (SSB) divided by its degrees of freedom – a measure of the variation among means of samples taken from different populations.

Mean square error (MSE) Error sum of squares divided by its degrees of freedom.

Mean square within samples (MSW) Within sample sum of squares (SSW) divided by its degrees of freedom – a measure of the variation within samples of all samples taken from different populations.

MSB Mean square between samples.

MSC Mean square between columns.

MSE Mean square error.

MSR Mean square between rows.

MSW Mean square within samples.

One-way analysis of variance The application of the ANOVA technique to test the equality of population means when classification is by one variable (also called *single-factor analysis of variance*).

Residual sum of squares (SSE) The variation which remains after subtracting from the total variation any variation arising from measured factors (also referred to as *error sum of squares or sum of squares error*).

Single-factor analysis of variance The application of the ANOVA technique to test the equality of population means when classification is by one variable (also called *one-way analysis of variance*).

SSB Between samples sum of squares.

SSC Sum of squares between columns.

SSE Error sum of squares.

SSR Sum of squares between rows.

SST Total sum of squares.

SSW Within sample sum of squares.

Sum of squares between columns (SSC) In the context of two-factor ANOVA where the observations for the samples relating to one factor are arranged in columns and the corresponding observations for the other factor are arranged in rows, the sum of squared deviations between the observations in a column and the column mean, summed over all columns.

Sum of squares between rows (SSR) In the context of two-factor ANOVA where the observations for the samples relating to one factor are arranged in columns and the corresponding observations for the other factor are arranged in rows, the sum of squared deviations between the observations in a row and the row mean, summed over all rows.

Sum of squares error (SSE) The variation which remains after subtracting from total variation (SST) any variation arising from measured factors (also referred to as *error sum of squares* or *sum of squares error*).

Total sum of squares (SST) Sum of squared deviations between the observations in a data set and the mean of the data set.

Two-factor analysis of variance The application of the ANOVA technique to test the equality of population means when classification is by two variables or factors (also called *two-way analysis of variance*).

Two-way analysis of variance The application of the ANOVA technique to test the equality of population means when classification is by two variables or factors (also called *two-factor analysis of variance*).

Within sample sum of squares (SSW) The sum of squared deviations between the observations in a sample and the mean of the sample, summed over all samples.

Self-study exercises

17.1. What assumptions are required in using the ANOVA procedure?

17.2. A single-factor ANOVA table is shown in the following table with certain elements deleted.

Source of variation	Sum of squares	Degrees of freedom	Mean square
Between	SSB = 208	☐	MSB = ☐
Within	SSW = 232	18	MSW = 23.33
Total	SST = ☐	20	

(a) What is the total variation (SST) within the sample as a whole?

(b) What is the total sample size?

(c) What percentage of the total variation is represented by variation between samples?

(d) How many samples are being tested?

(e) How many degrees of freedom are there for the between-sample variation?

(f) What is the value of MSB?

(g) What is the value of the F statistic?

(h) State the degrees of freedom to be used in determining the critical value of the F statistic.

(i) What is the critical value of the F statistic at the 5 per cent level of significance?

(j) Should the null hypothesis of no difference between the population means be rejected at the 5 per cent level of significance?

17.3. A multi-national company suspects that the absentee rate of its employees in three countries is different. The table below shows the days of absence recorded in random samples of 6 employees in each country last year.

Country 1	Country 2	Country 3
3	6	9
8	5	7
5	2	15
12	12	9
9	8	8
5	15	6

At the 5 per cent level of significance test the null hypothesis that the mean number of days absence per employee is the same in each country.

17.4. A two-factor ANOVA table is shown below with two elements deleted – shown by blank boxes.

Source of variation	Sum of squares	Degrees of freedom	Mean square
Between columns	SSC $= 102$	2	MSC $= 51$
Between rows	SSR $= 188$	☐	MSR $=$ ☐
Residual (error)	SSE $= 110$	10	MSE $= 11$
Total	SST $= 400$	16	

(a) How many samples are displayed by column?

(b) How many samples are displayed by row?

(c) What are the degrees of freedom for the variation between rows?

(d) What is the value of MSR?

(e) What hypotheses can be tested using these data?

(f) What are the degrees of freedom for the critical values of F for these tests at the 5 per cent level of significance?

(g) What are the actual values of F?

(h) State your conclusions regarding the significance tests.

17.5. The Managing Director of a chain of ladies' fashion boutiques wishes to test whether or not sales are affected by playing music in the shops and the type of music played. It is also thought that sales may depend on the towns in which the shops are located. An experiment is conducted in which shops play either classical or pop music or no music at all and daily sales (£000) recorded. A random sample of 21 shops gave the results for average daily sales as shown below.

Town	No music	Classical music	Pop music
A	15	29	41
B	27	51	12
C	8	11	123
D	65	62	98
E	75	23	55
F	18	112	21
G	87	84	61

Is there a statistically significant difference in sales at the 5 per cent level of significance according to the playing of music and town location?

Part IV

Relationships between variables

Never make forecasts, especially about the future.

SAMUEL GOLDWYN

The following diagram provides an overview of the whole of this part of the book.

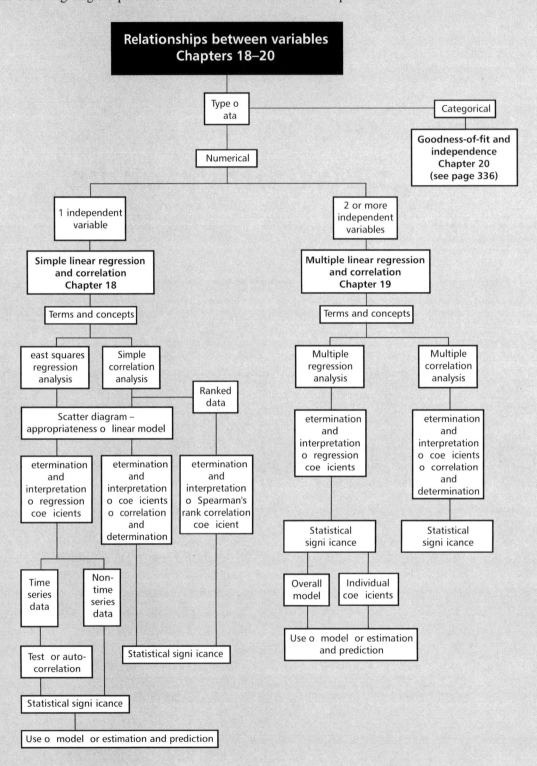

Chapter 18

Simple linear regression and correlation

In Part III of the book we were concerned with making statistical inferences from sample data, either by making point and interval estimates of a population parameter or by testing hypotheses about the parameter. We now turn in this part of the book to the analysis of relationships between variables.

This chapter introduces simple linear regression and correlation analysis. *Regression analysis* deals with the *nature* of the relationship between variables and *correlation analysis* is concerned with measuring the *strength* of the relationship between variables. Thus, for example, we may be interested in how sales of a product are related to its sales promotion expenditure, or how wheat yield depends on the amount of fertilizer applied.

The essence of regression and correlation analysis is conveyed by Figures 18.1(a) – 18.1(e). Each of the diagrams (known as *scatter diagrams*) consists of a set of points of paired observations on two variables x and y. For example, the x variable could be various values of advertising expenditure and the y variable the corresponding level of sales achieved. Figures 18.1(a) and 18.1(c) show that y increases as x increases and therefore show different strengths of *positive* correlation, while 18.1(b) and 18.1(d) show that y decreases as x increases with different strengths of *negative* correlation. Diagram 18.1(e) shows no correlation between y and x at all.

It will be also be seen that in Figures 18.1(a) – 18.1(d) a line has been drawn through the scatter of points in each case. The objective of regression analysis is to arrive at an expression ('model') defining the line which 'best fits' each set of plotted points: such lines are referred to as *regression lines*. No such line is drawn in Figure 18.1(e) because the scatter of points is such that there is no unique best-fitting regression line.

As noted in Figure 18.1, the diagrams illustrate different degrees of correlation. The objective of correlation analysis is to arrive at a single, summary, measure of the degree of correlation (i.e. association) between any two variables, such as x and y.

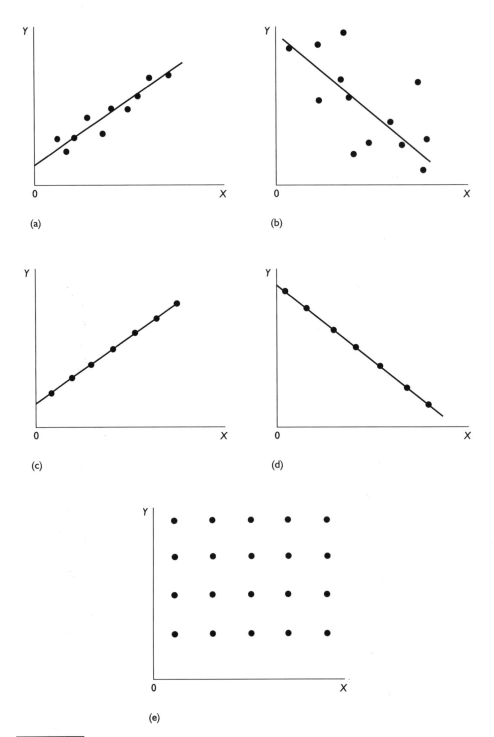

Figure 18.1 Scatter diagrams: (a) high degree of positive correlation; (b) low degree of negative correlation; (c) perfect positive correlation; (d) perfect negative correlation; (e) no correlation

In this chapter we set out the general principles underlying these techniques in the context of cases involving two variables only – referred to as *simple* linear regression and correlation analysis. In the next chapter these principles are extended to cases involving more than two variables – referred to as *multiple* linear regression and correlation analysis.

The chapter is structured as follows:

- regression analysis
- testing the statistical significance of the regression model
- estimation and prediction using the regression model
- dangers of extrapolation and forecasting
- correlation analysis
- testing the statistical significance of the correlation coefficient r
- Spearman's rank correlation coefficient r_s

Learning Outcomes   After working through this chapter you will be able to:

Regression

- appreciate the usefulness of measuring relationships between variables
- analyze the nature of the relationship between two variables using simple linear regression and correlation
- create scatter diagrams
- understand the criteria for defining the 'line of best fit' between data for two variables and the *principle of least squares*
- comprehend the form of the equation for a straight line
- understand the meaning of the terms *intercept* and *slope coefficients*
- grasp the concepts of dependent and independent variables
- calculate the intercept and regression (slope) coefficients
- test the statistical significance of the calculated coefficients
- use the regression results for estimation and prediction and create confidence and prediction intervals
- recognize the dangers of extrapolation and the pitfalls in forecasting
- appreciate the potential problem of autocorrelation (serial correlation) and how to detect its presence
- test the statistical significance of autocorrelation using the Durbin-Watson test statistic

Correlation

In addition, you will be able to:

- appreciate the link between regression analysis and correlation
- understand the meaning of positive and negative correlation
- measure the degree of association between two variables using correlation analysis
- calculate and interpret the coefficients of correlation and determination
- decompose the total variation in a dependent variable into explained and unexplained variations
- calculate and test the statistical significance of the correlation coefficient, r
- recognise the pitfalls in interpreting statistical correlation
- grasp the distinction between correlation and causation
- understand the distinction between ordinary and rank correlation
- calculate and test the statistical significance of the rank correlation coefficient, r_s

Principles

Regression analysis

Regression analysis is concerned with the derivation of an equation which defines a best-fitting regression line. In Figure 18.1 above, the best-fitting regression lines in diagrams (a) – (d) are drawn as straight lines. The principles of regression analysis are explained here in the context of straight lines involving only two variables. But the principles are readily extended to cases involving non-linear relationships and to cases involving more than two variables.

The linear regression lines shown in Figure 18.1 can be described, in general, by the expression for the equation of a straight line:

Equation of a straight line

$$y = a + bx$$

It is conventional to assume that y varies as x varies rather than the other way round, i.e. that y depends on x. Hence, y is known as the *dependent* variable and x as the *independent* (or *explanatory*) variable. The interpretation of this equation is that, as x changes, y changes by b times the change in x – and therefore b measures the *slope* or *gradient* of the line. It follows that when x takes the value of zero, $y = a$ (a is referred to as the *intercept*). Given the equation for a straight line, it is possible to 'predict' values of y for any given values of x. Examples of different straight-line equations are shown in Figures 18.2(a) and

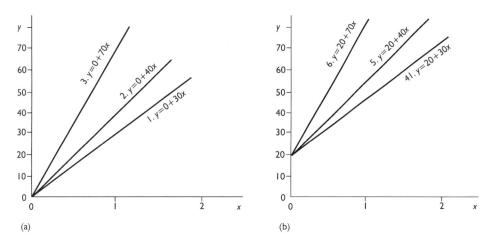

Figure 18.2 Equations of straight lines: (a) with zero intercept; (b) with positive intercept

18.2(b). Figure 18.2(a) shows three lines with a common intercept $a = 0$ (i.e. the lines 1, 2 and 3 all pass through the *origin* but with different slopes b, equal to 30, 40 and 70 respectively). Thus, for example, line 1 shows that when $x = 1$, $y = 30$ (i.e. $a + bx = 0 + 30 \times 1$), when $x = 2$, $y = 60$, etc. The lines 4, 5 and 6 in Figure 18.2(b) have the same slopes as in 18.2(a) but a common positive intercept equal to 20. Thus, for line 4, for $x = 1$, $y = 20 + 30 \times 1 = 50$.

The relationships shown in Figure 18.2 demonstrate precise relationships between an independent variable (x) and the dependent variable (y). In practice, of course, such relationships are rarely precise. We are generally faced with situations in which the relationships are imprecise as shown by the scatter diagrams in Figures 18.1(a) and 18.1(b). The problem is therefore to determine the equation that best describes the relationship, i.e. the *best-fitting regression line*.

Deriving a best-fitting regression line

The best-fitting regression line is determined using the criterion of *least squares*. Consider Figure 18.3. This shows a scatter of points indicating the expenditure y of eight households in relation to their incomes x and a regression line fitted to these data denoted by:

$$y_c = a + bx$$

where y_c stands for the value of variable y *computed* from the relationship for a given value of the variable x (y_c is therefore distinguished from y which represents *actual*, observed values). The line $y_c = a + bx$ expresses the average relationship between the two variables. It is called the *linear regression of y on x* and the slope coefficient b is referred to as the *regression coefficient*. The problem remaining is to determine the values of a and b. This will then give us the equation for the best-fitting line.

The values of a and b must be calculated in such a way that the fitted line is as close as possible to the plotted points corresponding to the observed values of x and y (i.e. x_i, y_i).

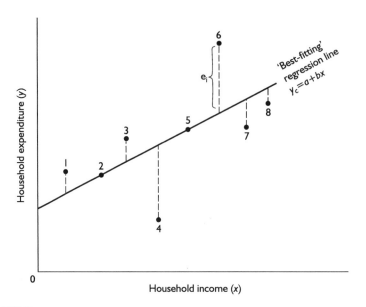

Household expenditure (y)

Household income (x)

Figure 18.3 Deriving the best-fitting regression line

Consider, for example, point 6 in Figure 18.3. This point corresponds to a specific pair of observed values of x and y. The computed y value (y_c) corresponding to the given x_i value can be determined from the relationship $y_c = a + bx_i$. There will usually be deviations (or differences) between the actual values of y and their computed values, depending on the closeness of the relationship. These deviations between the actual and computed y values, referred to as 'residuals' or 'errors', are measured by:

$$e_i = y_i - y_c$$

These deviations e_i are also illustrated in Figure 18.3. Naturally they can take positive or negative values for different points, i.e. actual points may be above (e_i positive) or below (e_i negative) the fitted line. The relevance of this analysis of deviations is that the best-fitting line can be obtained by ensuring that the sum of the *squared* deviations ($\sum e_i^2$) is as small as possible. The deviations are squared in order to give weight to both positive and negative values (otherwise they could cancel each other out). This is the basic principle of the least squares method.

Principle of the least squares method

minimize $\sum (y_i - y_c)^2 \equiv$ minimize $\sum e_i^2$

where y_i is the observed value of the dependent variable for the ith observation and y_c is the computed value of the dependent variable for the ith observation.

The procedure for finding the values of a and b from first principles, using the method of least squares, involves differential calculus. Fortunately, its demonstration is not

Chapter 18 ▪ Simple linear regression and correlation

essential to understanding the principles of regression analysis and hence it is not included here. It leads to the following expressions for determining the values of a and b.

Formulae for regression coefficients

$$b = \frac{\sum x_i y_i - n\bar{x}\bar{y}}{\sum x_i^2 - n\bar{x}^2}$$

$$a = \bar{y} - b\bar{x}$$

Once a and b have been determined, we can then express the equation for the best-fitting line, i.e. $y_c = a + bx$. Note that the coefficients a and b can have negative as well as positive values. A negative value for b indicates an inverse relationship between x and y, so that y decreases as x increases and vice versa. A negative value for a indicates, of course, a negative intercept on the y axis. The application of these formulae is simple, especially for small data sets, merely involving the summation of x_i^2 and $x_i y_i$ and the calculation of $\bar{x}$ and $\bar{y}$. Worked examples are given in the Applications and Worked Examples section showing both worked solutions and computer solutions.

A note of warning

It should be appreciated that calculation of a regression equation for a bivariate set of data x and y, in which x is taken as an independent variable and y as a dependent variable, should not be taken to imply causation, i.e. that changes in x necessarily *cause* changes in y. Inferences about cause and effect relationships must depend upon the judgement of the analyst.

Testing the statistical significance of the regression model

The calculation of the regression coefficients a and b from a sample of data naturally provides only *estimates* of the corresponding intercept and slope coefficients for the population from which the sample has been drawn (we shall denote these population coefficients by A and B respectively). As with the calculation of other statistics from samples, these are generally only of interest as estimates of the corresponding population parameters. In the case of regression, the population coefficients are, of course, unknown and the sample coefficients represent only one result amongst a number of possible results that could have been obtained depending on the sample. In other words, the sample coefficients are subject to sampling error. Thus, the fact that a sample regression coefficient is calculated as having a positive or a negative value does not necessarily indicate that the corresponding population parameter is also positive or negative. A significance test is required to test the probability of obtaining such positive or negative (i.e. non-zero) sample coefficients even though the population parameters are really zero. Most commonly, we are concerned with testing the null hypothesis H_0: $B = 0$, i.e. that the slope of the population regression line is zero. Acceptance of this hypothesis means that the population regression line is a horizontal line so that the value of y does *not* vary as x varies. In this case, information about x would be of no value in helping us to predict values of y. On the other hand, if we accept one of the possible alternative hypotheses that we may wish to test ($B \neq 0, B > 0$ or $B < 0$), then the population regression line must slope upwards (or downwards) so that information about x will help us to predict values of y.

To test these hypotheses we must turn to the only information available, namely the sample regression equation, and test H_0 against the sample result. It is possible to test both a and b, but we focus only on b here as this is normally of most interest. The test is therefore concerned with comparing the sample coefficient b with the hypothesized value of the population parameter B.

The principles of the test procedure are identical to those set out in Chapter 14 relating to the test of a single sample mean. In other words, the difference between a sample statistic and its corresponding population parameter is compared with the size of the standard error of the statistic and assessed in terms of its probability of occurrence. Thus, as before, the general form of the test statistic is computed as:

$$\frac{\text{sample statistic} - \text{hypothesized population parameter}}{\text{standard error of the sample statistic}}$$

In the case of regression, therefore, we next need to consider the derivation of the standard error of the sample statistic b, denoted s_b. This is given by the expression:

Standard error of b

$$s_b = \frac{\text{SEE}}{\sqrt{\sum x_i^2 - n\bar{x}^2}}$$

where SEE denotes the standard error of estimate, computed as

$$\text{SEE} = \sqrt{\frac{\sum e_i^2}{n-2}} = \sqrt{\frac{\sum y_i^2 - a\sum y_i - b\sum x_i y_i}{n-2}}$$

The test statistic t is used in regression analysis because appropriate population information is not available and hence a Z test cannot be used. The test statistic t is therefore given by:

Test statistic

$$t = \frac{b - B}{s_b}$$

This has a t distribution with $(n-2)$ degrees of freedom (where n is the number of paired x, y observations). This is then compared with the critical value of t from the t tables (Appendix A Table A4) for the required level of significance, α. Note that since under H_0, $B = 0$ is being tested, this is equivalent to testing $t = b/s_b$. As with the other tests of significance, if the calculated value of t falls outside the critical limits t_α (one-tailed test) or $t_{\alpha/2}$ (two-tailed test), we accept the alternative hypothesis and conclude that $B > 0$ or $B < 0$ (according to the one-tailed test), $B \neq 0$ (two-tailed test).

Testing for autocorrelation

It is common for regression studies in business and economics to use data collected over time (time series data). It often arises that values of the dependent variable, y, are related to the values of y in previous time periods. This gives rise to a problem in regression analysis which is referred to as *autocorrelation* (also called *serial correlation*). If the value of the variable y in time period t is related to its value in $t-1$, *first-order autocorrelation* is present. If the value of y in time period t is related to its value in $t-2$, *second-order autocorrelation* is present, and so on. In such cases, the standard error (s_b) associated with the estimated regression coefficient (b) will be understated, leading to less precise confidence intervals for the estimated coefficient than would otherwise be the case and thus unreliable tests of significance. This will result in the null hypothesis (H_0: $B = 0$) being rejected too frequently.

Autocorrelation is a matter of degree – the problem will always be present to some extent in time series data. The question that needs to be asked is how serious is the problem. To answer this it is important to be able to detect the degree of autocorrelation. A common approach is to examine the relationship (correlation) between residuals (e) in successive time periods. So in the case of testing for first-order autocorrelation, we would test for correlation between residuals in time periods t and $t-1$. If a positive residual in one time period is followed by a positive residual in the next time period, or if a negative residual is followed by another negative residual, then positive autocorrelation is said to exist. Conversely, if a positive residual is followed by a negative residual and so on, negative autocorrelation is said to exist.

A common test for first-order autocorrelation is the *Durbin-Watson test*. This is based on the *Durbin-Watson statistic* (denoted d) computed as follows:

Durbin-Watson statistic

$$d = \frac{\sum\limits_{t=2}^{n}(e_t - e_{t-1})^2}{\sum\limits_{t=1}^{n}e_t^2}$$

A small value of d indicates that successive values of the residuals are close together (positive autocorrelation), while a large value is indicative of negative autocorrelation. The Durbin-Watson statistic ranges in value from zero to four, thus the closer d is to zero the greater the evidence of positive autocorrelation, while the closer d is to 4 the greater the evidence of negative autocorrelation. As a rule of thumb, if d is found to be 2, one may assume that there is no first-order autocorrelation, either positive or negative.

The exact probability distribution of d is difficult to derive but Durbin and Watson have developed tables which represent the critical values of d (see Appendix A Table A8 which gives values for three different levels of significance: $\alpha = 0.05, 0.025, 0.01$). It will be seen from this table that there are two values of $d(d_L$ and $d_U)$ for each combination of sample size, n, and the number of independent variables in the regression model, k. When $k = 1$, we are dealing with a simple regression model and when $k > 1$, we are dealing with a multiple regression model (see Chapter 19).

We now turn to the application of the Durbin-Watson statistic to test for autocorrelation. To take an example, it will be seen from the tables in Appendix A Table A8 that with $\alpha = .05, n = 15$ and $k = 1, d_L$ and d_U are given as 1.08 and 1.36 respectively. The interpretation of these two values is illustrated in Figure 18.4 below.

Panel (a) shows that if the actual value of d is less than d_L (1.08 in this case) then positive autocorrelation is present. The null hypothesis to be tested is always that there is no autocorrelation and thus, in this case, the null hypothesis would be rejected. Values of d between d_L and d_U are inconclusive, while values greater than d_U (1.36 in this example) mean that there is no evidence of positive autocorrelation.

Panel (b) shows the corresponding test for negative autocorrelation. Note that the corresponding critical values of d in this case are 2.64 and 2.92. These values are not read directly from the tables of d_L and d_U but are obtained by symmetry around the mid-value of 2. Thus 2.64 is obtained as either $2 + (2 - 1.36)$ or, alternatively, as $4 - 1.36$ and 2.92 is obtained as either $2 + (2 - 1.08)$ or as $4 - 1.08$.

The interpretation of the diagram then follows as before: values of d greater than 2.92 indicate negative autocorrelation, values between 2.64 and 2.92 are inconclusive while values less than 2.64 show that there is no evidence of negative autocorrelation.

Note that the Durbin-Watson tables list the smallest sample size as 15. The reason is that the test is generally inconclusive for sample sizes smaller than this. On the whole, sample sizes of at least 50 are preferred in order to give reliability to the results.

If autocorrelation is found to be present in time series data, a number of steps may be taken in an attempt to correct for it – see under 'Autocorrelation' in Chapter 19 which deals with multiple linear regression. It is also important to note that the Durbin-Watson test should not be used to test for autocorrelation in regression models containing lagged values of the dependent variable as explanatory variable(s). If the test is applied mistakenly in such cases, there is a built-in bias against discovering the presence of autocorrelation – i.e. we may mistakenly conclude that autocorrelation is not present when in fact it is! In such situations a special test statistic must be used known as Durbin's h statistic. Treatment of this may be found in more specialized textbooks.

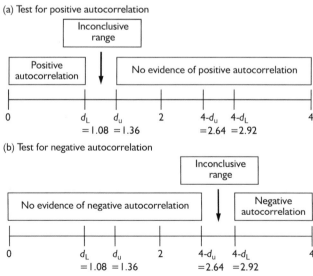

Figure 18.4 Application of the Durbin-Watson statistic – critical values of d for $n = 15$, $k = 1$, $\alpha = .05$

Estimation and prediction using the regression model

The regression equation (model) can be used, of course, to predict values of the dependent variable y for any given values of x. It will be appreciated, however, that such single-point estimates are subject to error and, as with estimates of population means (see Chapter 12), it is possible to construct an interval around such estimates to provide a given degree of confidence for the prediction, e.g. 95 per cent confidence that the true population value would lie within the specified interval.

There are two situations to consider:

1. interval estimation of the *average* value of y for a given value of x and

2. interval estimation of a *single* value of y for a given value of x.

Thus in our earlier example (Figure 18.3) the value of y may refer to the mean expenditure of a group of households or the expenditure of an individual household. In both cases the point estimate of y (i.e. y_c, where $y_c = a + bx$) would be the same but an interval estimate of y_c would differ in each case. This is because the variation for a mean value is less than that for individual values within the group. These intervals are referred to respectively as:

- confidence intervals
- prediction intervals.

In summary, we can use the regression model to estimate *average* values or to predict *individual* values of the dependent variable y with specified levels of confidence.

Confidence interval

To construct a $100(1 - \alpha)$ per cent confidence interval for y_c for a given value of x (denoted x_0), we use the following formula:

Confidence interval

$$y_c \pm t_{\alpha/2} \left\{ \text{SEE} \sqrt{\frac{1}{n} + \frac{(x_0 - \bar{x})^2}{\sum (x_i - \bar{x})^2}} \right\}$$

where $y_c = a + bx_0$ and $t_{\alpha/2}$ has $n - 2$ degrees of freedom.

Prediction interval

To construct a $100(1 - \alpha)$ per cent prediction interval for the estimated individual value of y (i.e. y_c) for a given value of x (i.e. x_0), the formula is the same as that above except that '1' is added to the expression within the square root sign:

Prediction interval

$$y_c \pm t_{\alpha/2} \left\{ \text{SEE} \sqrt{1 + \frac{1}{n} + \frac{(x_0 - \bar{x})^2}{\sum (x_i - \bar{x})^2}} \right\}$$

where $y_c = a + bx_0$ and $t_{\alpha/2}$ has $n - 2$ degrees of freedom.

The addition of '1' to the expression naturally widens the resulting interval. We may note, in passing, that the use of computer packages such as MINITAB and Excel take the burden out of the calculation of confidence and prediction intervals, as well as regression itself – see the Applications and Worked Examples section later in this chapter.

Dangers of extrapolation and forecasting

Naturally, the regression equation may be used to estimate y_c for any given value of x. However, there are great dangers in using the equation to forecast the values of y_c *outside* the range of the observed data set from which the regression equation itself was derived. Computing values of the dependent variable (y_c) *within* the range of recorded observations on the independent variable x is referred to as *interpolation*. In contrast, projecting values (backwards as well as forwards) outside this range, giving rise to y^f values, is referred as *extrapolation*. This distinction is illustrated in Figure 18.5.

Thus, computed values of y within the range AB in Figure 18.5 give rise to interpolation while any projection outside this range gives rise to extrapolation. It must be stressed that extrapolation may, in some circumstances, give rise to very inaccurate or dubious forecasts. This is because a *linear* relationship which may be a true or reasonably good approximation for a particular range of observations (such as between AB) may not be true for observations outside this range, as illustrated in Figure 18.5 by the curves CC and DD. Forecasts are only estimates, and so a *forecast error* is likely to arise in almost every case. It is for this reason that such forecasts are often referred to as *projections* rather than *predictions*. True predictions would attempt to take account of other factors not reflected in the data.

A further warning is also in order. Consider Figure 18.6. This shows the upper and lower limits for various confidence and prediction intervals around a regression line. It will be seen that the width of the intervals increases markedly as the value of x in which we are

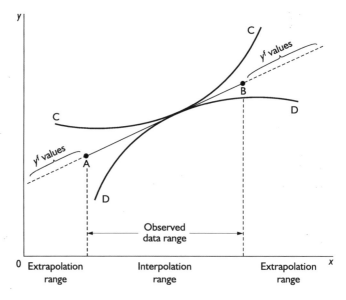

Figure 18.5 Extrapolation and interpolation

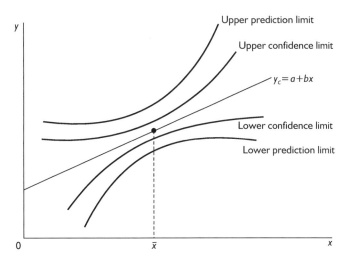

Figure 18.6 Confidence and prediction limits

interested (x_0) moves further and further away from the mean value of x (i.e. $\bar{x}$). Forecasts within such areas, therefore, are inevitably subject to wider margins of error. It will also be seen from this diagram that the prediction interval is wider than the confidence interval, as noted earlier.

Correlation analysis

Correlation analysis, as we explained in the introduction to this chapter, is a means of measuring the strength of 'closeness' of the relationship between two variables. It should be clear that the concept of correlation is very closely linked to regression analysis. If all the paired points (x_i, y_i) lie on a straight line then the correlation between the variables x and y is perfect. Whether the correlation is perfectly positive or negative depends, of course, on whether the straight line through the points has a positive or negative slope. Figure 18.1(c), discussed earlier, illustrates perfect positive correlation while 18.1(d) illustrates perfect negative correlation. The more scattered and random are the points (x_i, y_i) then the less the degree of correlation – Figures 18.1(a), 18.1(b) and 18.1(e) refer.

Correlation analysis provides a numerical summary measure of the degree of correlation between two variables x and y – a *correlation coefficient* denoted by r. This is defined so that its value must be within the range from -1 to $+1$ so that:

- $r = +1$ denotes perfect positive correlation, as in Figure 18.1(c)
- $r = 0$ denotes no correlation, as in Figure 18.1(e)
- $r = -1$ denotes perfect negative correlation, as in Figure 18.1(d)

Correlation and causation

When discussing regression we sounded a warning note about the danger of assuming that calculating a regression relationship between two variables cannot be taken to mean that there is a cause and effect relationship. It is important to recall this warning here. The same point applies: *correlation does not imply causation*. There must be some theoretical justification for believing that there is a meaningful relationship between the two variables. Otherwise, an apparently significant degree of correlation may simply represent *spurious correlation*.

The *correlation coefficient, r*, is defined and calculated as follows:

Definition of correlation coefficient, *r*

$$r = \frac{s_{xy}}{s_x s_y}$$

where s_{xy} is the sample covariance, given by

$$\frac{\sum(x_i - \bar{x})(y_i - \bar{y})}{n - 1}$$

and s_x is the sample standard deviation of x values and s_y is the sample standard deviation of y values.

Computing the value of r using this expression is tedious. However, it reduces to a quicker, convenient formula, expressed as follows:

Computational formula for correlation coefficient, *r*

$$r = \frac{n \sum x_i y_i - \sum x_i \sum y_i}{\left[\sqrt{n \sum x_i^2 - (\sum x_i)^2}\right]\left[\sqrt{n \sum y_i^2 - (\sum y_i)^2}\right]}$$

This coefficient is often referred to as the *product-moment correlation coefficient* or *Pearson's correlation coefficient*.

Interpretation of correlation and the coefficient of determination

The coefficient of determination is simply the squared value of the correlation coefficient above, i.e. r^2. It is particularly useful in describing the closeness of the relationship between x and y; i.e. how closely do the actual points cluster around the regression line – how 'good' is the fit? In the estimated regression of y on x ($y_c = a + bx$), we are trying to explain the total variation in the observations of y by variations in x. Rarely, however, is *all* of the variation in y explained *totally* by the variation in x. As will be seen in Figure 18.7, for each observation of x, such as x_0, we may divide the deviation of the corresponding y_0 from the mean $\bar{y}$ into two parts: a part 'explained' by the regression and an 'unexplained' part. Thus:

total variation in y = explained variation + unexplained variation, i.e:

$$\sum(y_i - \bar{y})^2 = \sum(y_c - \bar{y})^2 + \sum(y_i - y_c)^2$$

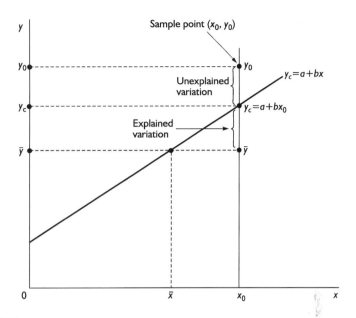

Figure 18.7 Explained and unexplained variations

where $\sum(y_i - \bar{y})^2$ is referred to as the *total sum of squares*, denoted SST; $\sum(y_c - \bar{y})^2$ is referred to as the *regression sum of squares*, denoted SSR; and $\sum(y_i - y_c)^2$ is referred to as the *error sum of squares*, denoted SSE. Thus:

$$SST = SSR + SSE$$

The error (i.e. unexplained) sum of squares, SSE, is the sum of the residual 'error' terms, $\sum e_i^2$, referred to earlier in this chapter.

It follows from the definition of the correlation coefficient r that the coefficient of determination, r^2, is equivalent to:

Coefficient of determination, r^2

$$r^2 = \frac{\text{explained variation in } y}{\text{total variation in } y} = \frac{\sum(y_c - \bar{y})^2}{\sum(y_i - \bar{y})^2} = \frac{SSR}{SST}$$

Thus, r^2 measures the proportion of total variation in y that is explained by the regression equation. So, for example, a value of $r = 0.60$ gives a value of $r^2 = 0.36$ and indicates that 36 per cent of the variation in the variable y is explained by the regression equation, leaving 64 per cent unexplained. Naturally, it is easier to measure r^2 by simply taking the squared value of r. By definition, the unexplained proportion is measured by $1 - r^2$.

The reader may very well ask whether a value of $r = 0.60$ (or, equivalently, $r^2 = 0.36$) indicates a 'strong' or 'weak' correlation between y and x. This is a question that cannot be answered simply by noting the size of r (or r^2). It requires a test of statistical significance.

Testing the significance of the correlation coefficient r

It will be appreciated, as before, that the calculated value of r is a sample statistic and is thus only an estimate of the corresponding population parameter value for the correlation coefficient, commonly denoted by the Greek letter ρ (pronounced 'rho'). We must therefore test the significance of r.

The test involves the difference between the value of r calculated from the sample data and the hypothesized population correlation coefficient ρ. As in the case of the regression coefficient, b, we need to determine the probability of this difference occurring, and thus whether the difference is statistically significant. Hence, we calculate the test statistic t as:

$$t = \frac{r - \rho}{\text{standard error of } r}$$

In correlation analysis, we are usually concerned with testing whether the value of r could have come from a population in which there was no linear correlation between x and y (i.e. $\rho = 0$). Hence, the above expression reduces to:

$$t = \frac{r}{\text{standard error of } r}$$

where the standard error of r (denoted s_r) is given by:

$$s_r = \sqrt{\frac{1 - r^2}{n - 2}}$$

Thus, the test statistic for the null hypothesis that $\rho = 0$ reduces to

Test statistic for r

$$t = \frac{r}{\sqrt{(1 - r^2)/(n - 2)}}$$

which will have a t distribution with $n - 2$ degrees of freedom.

One- and two-tailed tests may then be conducted as appropriate. Application of the significance test for r is shown below in the Applications and Worked Examples section.

Spearman's rank correlation coefficient r_s

In some situations the values of the variables y and x are expressed in rank order form, e.g. the rank order in which international contestants in a race are placed. One might wish to test, for example, whether or not there is any correlation between the rank order of the contestants in relation to the population size of the countries which they represent.

The measure of correlation which deals with this type of situation is *Spearman's rank correlation coefficient* r_s defined as follows:

Formula for Spearman's rank correlation coefficient, r_s

$$r_s = 1 - \frac{6 \sum D^2}{n(n^2 - 1)}$$

where D is the difference between each pair of x and y ranks and n is the number of paired values of x and y.

In terms of our example above, the x ranks would represent the rank order of the countries in terms of population size in ascending or descending order and the y ranks would represent the order in which the contestants finished the race, again in ascending or descending order. Note that we must rank *both* x and y either in ascending order or in descending order. As with r, the value of r_s varies within the range -1 to $+1$.

What happens if there are ties *within* the x or y ranks? In this case, the ranks are then obtained by averaging. Thus, for example, if there is a dead heat for the first place between two people, then the ranks 1 and 2 are averaged, each being given the rank value of 1.5, and the person placed next is given a rank of 3 and so on. The formula for r_s then has to be modified as follows:

Correction for tied ranks

add $\frac{1}{12}(t^3 - t)$ to $\sum D^2$

for *each* tied set, where t is the number of tied observations.

An example using this correction term is given in the Applications and Worked Examples section below.

Testing the statistical significance of r_s

As with r, we can test the statistical significance of r_s by using the sample results to make an inference about the value of the corresponding population parameter ρ_s. The sampling distribution of r_s is symmetrical around the value 0 and is approximately normal, provided that the number of paired ranks n is greater than or equal to 10, and a t test may then be applied. Under the null hypothesis H_0: $\rho_s = 0$, the test statistic is the same as for r, given by:

Test statistic for r_s

$$t = \frac{r_s}{\sqrt{(1 - r_s^2)/(n - 2)}}$$

which will have a t distribution with $n - 2$ degrees of freedom.

Note that for situations in which there are fewer than ten paired ranks (i.e. $n < 10$), special tables have been constructed for determining the significance of the Spearman rank correlation coefficient r_s – see Appendix A Table A5.

Applications and Worked Examples	Worked example 18.1 Deriving and using the regression model

General descriptions of the MINITAB and Excel computer programs and their application are given in Appendices B and C respectively.

The following data record the sales of heating fuel (thousands of litres) by a distributor over ten winter weeks along with the average temperature (°C) for each week.

Week:	1	2	3	4	5	6	7	8	9	10
Fuel oil sales (000s litres):	26	17	7	12	30	40	20	15	10	5
Average temperature (°C):	4	10	14	12	4	5	8	11	13	15

(a) Compute the best-fitting regression equation between sales (y) and temperature (x), assuming that fuel oil sales are dependent on temperature.

(b) What would the regression model estimate for fuel sales be when the temperature is 5°C? How does this compare with the observed data?

(c) What fuel oil sales would be expected if the temperature falls to 3°C?

(d) What fuel oil sales would be expected if the temperature rises to 25°C?

Worked solution

The computations involved to calculate the regression coefficients, a and b, are shown below (note that y_i^2 values are not needed here but are included for use in analysing correlation later).

x_i	y_i	x_i^2	y_i^2	$x_i y_i$
4	26	16	676	104
10	17	100	389	170
14	7	196	49	98
12	12	144	144	144
4	30	16	900	120
5	40	25	1600	200
8	20	64	400	160
11	15	121	225	165
13	10	169	100	130
15	5	225	25	75
$\sum x_i = 96$	$\sum y_i = 182$	$\sum x_i^2 = 1076$	$\sum y_i^2 = 4408$	$\sum x_i y_i = 1366$
$\bar{x} = 9.6$	$\bar{y} = 18.2$			

(a)

$$b = \frac{n\sum x_i y_i - \sum x_i \sum y_i}{n\sum x_i^2 - (\sum x_i)^2}$$

$$= \frac{10(1366) - 96(182)}{10(1076) - (96)^2}$$

$$= -2.4689$$

$$a = \bar{y} - b\bar{x}$$

$$= 18.2 - (-2.4689)(9.6)$$

$$= 41.901$$

Thus the equation for the estimated regression model $y_c = a + bx$ is equal to:

$$y_c = 41.901 - 2.4689x$$

It will be seen, not surprisingly, that fuel sales are found to be negatively related to temperature.

(b) For a temperature of 5°C, an estimate of fuel sales is given by the regression model:

$$y_c = 41.901 - 2.4689(5)$$

$$= 29.5565; \text{ i.e. } 29\,556.5 \text{ litres.}$$

The observed fuel sales, at a temperature of 5°C, were 40 000 litres.

(c) For a temperature of 3°C, estimated sales are given by:

$$y_c = 41.901 - 2.4689(3)$$

$$= 34.4943; \text{ i.e. } 34\,494.3 \text{ litres.}$$

Note that since a temperature of 3°C falls *outside* the range of observed temperature values given above, we are *extrapolating* fuel sales y_c in this example.

(d) For a temperature of 25°C, estimated fuel sales are given by:

$$y_c = 41.901 - 2.4689(25)$$

$$= -19.8215; \text{ i.e. } -19\,821.5 \text{ litres.}$$

The meaningless answer of -19.8215×1000 litres highlights the danger of *extrapolation*, i.e. forecasting values of the dependent variable y far beyond the range of the previously observed values.

Computer solutions

The computer solutions are given on pp. 291–5. These cover all the computer solutions relating to Worked examples 18.1 – 18.4.

Worked example 18.2 Testing the statistical significance of the regression model

Test the significance of the regression coefficient b at the 5 per cent level of significance for the estimated regression model in Worked example 18.1 above.

Worked solution

Hypotheses: $H_0: B = 0$

$H_1: B < 0$

(Note that the alternative hypothesis is $B < 0$ since, *a priori*, we would expect sales to rise as temperature falls.)

Significance level: $\alpha = 0.05$

Standard error: $s_b = \dfrac{\text{SEE}}{\sqrt{\sum x_i^2 - n\bar{x}^2}}$

where

$$\text{SEE} = \sqrt{\frac{\sum y_i^2 - a\sum y_i - b\sum x_i y_i}{n-2}}$$

$$= 4.394$$

Thus

$$s_b = \frac{4.394}{1076 - 10(9.6)^2}$$

$$= 0.3556$$

Test statistic: critical $t = -t_\alpha$ with $n - 2\,\text{df} = -1.86$

$$\text{actual } t = \frac{b - B}{s_b}$$

$$= \frac{-2.4689 - 0}{0.3536}$$

$$= -6.98$$

Conclusion: $t < -t_\alpha$, i.e. $-6.98 < -1.86$. Therefore we reject H_0 and conclude that $B < 0$, i.e. that there is a significant negative relationship between fuel sales and temperature at the 5 per cent level of significance.

Computer solutions

The computer solutions are given on pp. 291–5. These cover all the computer solutions relating to Worked examples 18.1–18.4.

Worked example 18.3 Constructing confidence intervals and prediction intervals

Using the data in Worked example 18.1,

(a) construct 95 per cent confidence intervals for an individual value of y when
$x = 3$ and $x = 25$,

(b) construct 95 percent prediction intervals for an individual value of y when
$x = 3$ and $x = 25$.

Worked solution

(a) A 95 per cent confidence interval when $x = 3$ is given by

$$(a + bx_0) \pm t_{\alpha/2} \left\{ \text{SEE} \sqrt{\frac{1}{n} + \frac{(x_0 - \bar{x})^2}{\sum(x_i - \bar{x})^2}} \right\}$$

(Recall that $\sum(x_i - \bar{x})^2$ is equivalent to $\sum x_i^2 - n\bar{x}^2$ ($= 154.4$) which can be readily calculated from the data given in the Worked solution to Worked example 18.1.)

Thus, using the information given in Worked examples 18.1 and 18.2 above, we have

$$[41.901 - 2.4689(3)] \pm 2.306 \left\{ 4.394 \sqrt{\frac{1}{10} + \frac{(3 - 9.6)^2}{154.4}} \right\}$$

$$= 34.4943 \pm 6.2636$$
$$= 28.23 \text{ to } 40.76$$

A 95 per cent confidence interval when $x = 25$ is given by

$$[41.901 - 2.4689(25)] \pm 2.306 \left\{ 4.394 \sqrt{\frac{1}{10} + \frac{(25 - 9.6)^2}{154.4}} \right\}$$

$$= -19.8215 \pm 12.9602$$
$$= -32.78 \text{ to } -6.86$$

(b) A 95 per cent prediction interval is given by

$$(a + bx_0) \pm t_{\alpha/2} \left\{ \text{SEE} \sqrt{1 + \frac{1}{n} + \frac{(x_0 - \bar{x})^2}{\sum(x_i - \bar{x})^2}} \right\}$$

For $x = 3$:

$$34.4943 \pm 2.306 \left\{ 4.394 \sqrt{1 + \frac{1}{10} + \frac{(3 - 9.6)^2}{154.4}} \right\}$$

$$= 34.4943 \pm 11.9122$$
$$= 22.58 \text{ to } 46.41$$

Worked example 18.3   Constructing confidence intervals and prediction intervals
                                          (continued)

For $x = 25$:

$$-19.8215 \pm 2.306 \left\{ 4.394 \sqrt{1 + \frac{1}{10} + \frac{(25 - 9.6)^2}{154.4}} \right\}$$

$$= -19.8215 \pm 16.4510$$

$$= -36.28 \text{ to } -3.37$$

Note that $x = 25$ is an extreme value, well outside the observed range of the data, producing negative values for the confidence and prediction intervals. This illustrates again the dangers of extrapolation.

Computer solutions

The computer solutions are given on pp. 291–5. These cover all the computer solutions relating to Worked examples 18.1–18.4.

Worked example 18.4   Computation and significance of the correlation coefficient *r*

Using the information given in Worked example 18.1 above

(a) what is the degree of correlation r between fuel sales and temperature?

(b) test the statistical significance of this value of r at the 5 per cent level of significance;

(c) what proportion of the variation in fuel oil sales is explained by variations in temperature?

Worked solution

(a) The appropriate summations needed to compute r are given in the Worked solution to Worked example 18.1. Thus:

$$r = \frac{n \sum x_i y_i - \sum x_i \sum y_i}{\left[\sqrt{n \sum x_i^2 - \left(\sum x_i\right)^2} \right]\left[\sqrt{n \sum y_i^2 - \left(\sum y_i\right)^2} \right]}$$

$$= \frac{10(1366) - 96(182)}{\left[\sqrt{10(1076) - (96)^2} \right]\left[\sqrt{10(4408) - (182)^2} \right]}$$

$$= -0.927$$

(b)

Hypotheses: $H_0: \rho = 0$

                          $H_1: \rho < 0$

Significance level: $\alpha = 0.05$

Standard error: $s_r = \sqrt{\dfrac{1 - r^2}{n - 2}} = \sqrt{\dfrac{1 - (-0.927)^2}{10 - 2}}$

$$= 0.1326$$

Test statistic: critical $t = -t_\alpha$ with $n - 2$ df $= -1.86$

actual $t = \dfrac{r - \rho}{s_r} = \dfrac{-0.927 - 0}{0.1326}$

$$= -6.99$$

Conclusion: $t < -t_\alpha$, i.e. $-6.99 < -1.86$. Therefore, we reject H_0 and conclude that $\rho < 0$, i.e. that there is a significant negative correlation between fuel sales and temperature. (Note that, in a simple regression involving only one dependent variable y and one explanatory variable x, a significance test on r is equivalent to a significance test on the regression coefficient b.)

(c) The proportion of the variation in fuel sales explained by variations in temperature is given by the coefficient of determination r^2:

$$r^2 = (-0.927)^2$$
$$= 0.859$$

That is, 85.9 per cent of the variation in fuel sales is explained by variations in temperature.

Computer solutions

The computer solutions which follow cover Worked examples 18.1–18.4. The solutions to the Worked examples are listed below the printout and are also labelled as indicated on the printout itself.

Computer solution using MINITAB

(Solutions to Worked examples 18.1–18.4)

Worksheet

Temp (x)	Fuel (y)	Values of x for which predictions are required
4	26	3
10	17	25
14	7	
12	12	

Computer solutions   Worked examples (18.1–18.4) *(continued)*

4	30
5	40
8	20
11	15
13	10
15	5

Choose **Stat → Regression → Regression**

Select **Response** variable (*y*) and **Predictor** variable (*x*)

Choose **Results**

Check an appropriate button under **Control the display of results** (four options are available; the fourth option is chosen here in order to display a full set of results)

Click **OK**

The output is as follows:

Choose **Options**

In the **Prediction intervals for new observations** box enter the location (either column or constant – K number – reference) of the values of *x* for which predicted *y* values and confidence and prediction intervals are required (in this case the *x* values 3 and 25 were entered in C3 of the worksheet)

Click **OK** twice

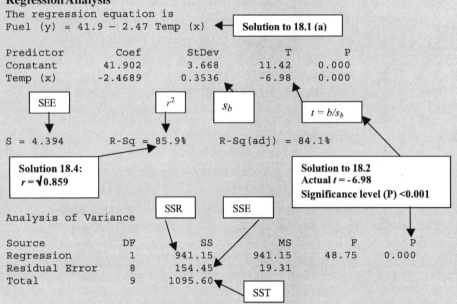

Regression Analysis

```
The regression equation is
Fuel (y) = 41.9 - 2.47 Temp (x)      ◄── Solution to 18.1 (a)

Predictor       Coef        StDev            T        P
Constant       41.902       3.668        11.42    0.000
Temp (x)       -2.4689      0.3536       -6.98    0.000
```

SEE → r^2 s_b $t = b/s_b$

```
S = 4.394      R-Sq = 85.9%      R-Sq(adj) = 84.1%
```

Solution 18.4: $r = \sqrt{0.859}$

Solution to 18.2
Actual $t = -6.98$
Significance level (P) <0.001

SSR   SSE

```
Analysis of Variance

Source          DF          SS           MS        F        P
Regression       1        941.15       941.15    48.75    0.000
Residual Error   8        154.45        19.31
Total            9       1095.60
```

SST

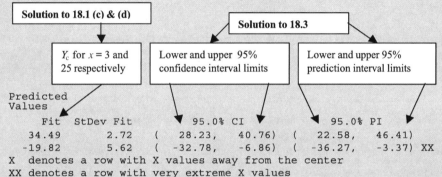

```
Obs   Temp (x)   Fuel (y)      Fit   StDev Fit    Residual    St Resid
  1        4.0      26.00     32.03       2.42       -6.03       -1.64
  2       10.0      17.00     17.21       1.40       -0.21       -0.05
  3       14.0       7.00      7.34       2.09       -0.34       -0.09
  4       12.0      12.00     12.27       1.63       -0.27       -0.07
  5        4.0      30.00     32.03       2.42       -2.03       -0.55
  6        5.0      40.00     29.56       2.14       10.44        2.72R
  7        8.0      20.00     22.15       1.50       -2.15       -0.52
  8       11.0      15.00     14.74       1.48        0.26        0.06
  9       13.0      10.00      9.81       1.84        0.19        0.05
 10       15.0       5.00      4.87       2.36        0.13        0.04
```

| Solution to 18.1 (b) | y_c | $e = y - y_c$ |

```
R denotes an observation with a large standardized residual
```

| Solution to 18.1 (c) & (d) | | Solution to 18.3 |

| Y_c for $x = 3$ and 25 respectively | Lower and upper 95% confidence interval limits | Lower and upper 95% prediction interval limits |

```
Predicted
Values
     Fit   StDev Fit        95.0% CI            95.0% PI
   34.49       2.72    (   28.23,   40.76)  (   22.58,   46.41)
  -19.82       5.62    (  -32.78,   -6.86)  (  -36.27,   -3.37) XX
X   denotes a row with X values away from the center
XX denotes a row with very extreme X values
```

Computation of *r* using MINITAB

In addition to the MINITAB output shown above which includes the correlation coefficient along with other measures, MINITAB can also give the value of *r* alone using the **Correlation** tool as follows:

Choose **Stat → Basic statistics → Correlation**

Select the **Variables** it is desired to correlate (C1 and C2 in this case)

Click **OK**

The output is as follows:

Authors' comment: Note that in addition to giving the value of the coefficient of correlation, the P-value (probability) of the result is also shown. The interpretation of the P-value is the same as in the regression example. The coefficient is called the Pearson coefficient.

Correlations (Pearson)

Correlation of Temp (*x*) and Fuel (*y*) $= -0.927$, P–Value $= 0.000$

Computer solution using Excel **Ⓔ**

Solutions to Worked examples 18.1, 18.2 and 18.4 (there is no pre-defined tool in Excel for solving Worked example 18.3).

Use the **REGRESSION** analysis tool With the data on temperature and fuel entered in cells A2:A11 and B2:B11 respectively of the worksheet:

Computer solutions   Worked examples (18.1–18.4) *(continued)*

Choose **Tools** → **Data Analysis** ... →
   Regression
In **Input Y Range** enter *y* values by high-
   lighting cells B2:B11 in the worksheet
In **Input X Range** enter *x* by highlighting
   cells A2:A11 in the worksheet

Check the **Confidence Level** box and enter
the required level of confidence (95% is
shown by default)
Click **OK**

The output is as follows:

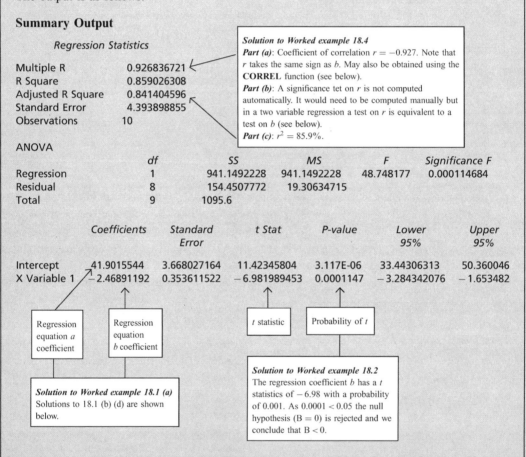

Summary Output

Regression Statistics

Multiple R	0.926836721
R Square	0.859026308
Adjusted R Square	0.841404596
Standard Error	4.393898855
Observations	10

> *Solution to Worked example 18.4*
> *Part (a)*: Coefficient of correlation $r = -0.927$. Note that
> r takes the same sign as b. May also be obtained using the
> **CORREL** function (see below).
> *Part (b)*: A significance tet on r is not computed
> automatically. It would need to be computed manually but
> in a two variable regression a test on r is equivalent to a
> test on b (see below).
> *Part (c)*: $r^2 = 85.9\%$.

ANOVA

	df	SS	MS	F	Significance F
Regression	1	941.1492228	941.1492228	48.748177	0.000114684
Residual	8	154.4507772	19.30634715		
Total	9	1095.6			

	Coefficients	Standard Error	t Stat	P-value	Lower 95%	Upper 95%
Intercept	41.9015544	3.668027164	11.42345804	3.117E-06	33.44306313	50.360046
X Variable 1	−2.46891192	0.353611522	−6.981989453	0.0001147	−3.284342076	−1.653482

> Regression equation *a* coefficient

> Regression equation *b* coefficient

> *t* statistic

> Probability of *t*

> *Solution to Worked example 18.1 (a)*
> Solutions to 18.1 (b) (d) are shown below.

> *Solution to Worked example 18.2*
> The regression coefficient *b* has a *t*
> statistics of −6.98 with a probability
> of 0.001. As 0.0001 < 0.05 the null
> hypothesis (B = 0) is rejected and we
> conclude that B < 0.

Computer solution using Excel to Worked example 18.1 Parts (b) – (d)

Use the **FORECAST** function
With the data entered in columns A and B of
the worksheet, as above:

Click on = in the formula bar and select
   FORECAST

In *x* enter the value of *x* for which you want to
   predict the value of *y*
In **Known_y's** enter the data array for *y* and in
   Known_x's enter the data array for *x*
Click **OK**
   The output is shown opposite:

Solution to Part (b)

With x entered as 5 this procedure returns: 29.55699 (thousand litres) to a pre-selected cell in the worksheet (say D1) as shown below

Solution to Part (c)

With x entered as 3 this returns: 34.49482 (thousand litres) to a pre-selected cell in the worksheet (say C1) as shown below

Solution to Part (d)

With x entered as 25 this returns: -19.8212 (thousand litres) to a pre-selected cell in the worksheet (say D2) as shown opposite:

Excel Output for Worked example 18.1 parts (b) – (d)

Input		Output	
A	B	C	D
x	y	34.49482	29.55699
4	26		-19.8212
10	17		
14	7		
12	12		
4	30		
5	40		
8	20		
11	15		
13	10		
15	5		

Worked example 18.5   Testing for autocorrelation

The data below show the value of Gross National Product (GNP) and Consumers' Expenditure (£bn measured at constant prices) over a period of twenty years. The regression of Consumers' Expenditure (y) on GNP (x) gives the equation: $y = 4.5649 + 0.56512x$ and is statistically significant at the 5 per cent level. Test whether or not statistically significant autocorrelation is present and, if so, whether it is positive or negative.

Year	y	x	Year	y	x
1	46.7	74.5	11	59.7	98.1
2	47.7	75.4	12	63.3	100.3
3	49.7	78.5	13	66.3	108.6
4	51.3	82.6	14	65.1	107.1
5	52.1	84.6	15	64.7	105.7
6	53.2	86.2	16	64.8	109.9
7	54.4	88.4	17	64.6	110.2
8	56.0	91.9	18	68.2	114.2
9	56.3	93.5	19	71.4	115.9
10	57.8	95.6	20	71.5	113.8

Worked solution

It is first necessary to compute the Durbin-Watson statistic. The calculations necessary to do this are set out below.

Worked example 18.5 Testing for autocorrelation *(continued)*

Actual y y	Predicted y y_c	Residuals $e_t = y - y_c$	Lagged Residuals e_{t-1}	e_t^2	$e_t - e_{t-1}$	$(e_t - e_{t-1})^2$
46.7	46.666	0.034		0.0011		
47.7	47.175	0.525	0.034	0.2759	0.491	0.2415
49.7	48.927	0.773	0.525	0.5981	0.248	0.0616
51.3	51.244	0.056	0.773	0.0032	−0.717	0.5141
52.1	52.374	−0.274	0.056	0.0750	−0.330	0.1091
53.2	53.278	−0.078	−0.274	0.0061	0.196	0.0383
54.4	54.521	−0.121	−0.078	0.0147	−0.043	0.0019
56.0	56.499	−0.499	−0.121	0.2492	−0.378	0.1428
56.3	57.403	−1.103	−0.499	1.2174	−0.604	0.3650
57.8	58.590	−0.790	−1.103	0.6243	0.313	0.0981
59.7	60.003	−0.303	−0.790	0.0918	0.487	0.2374
63.3	61.246	2.054	−0.303	4.2182	2.357	5.5542
66.3	65.937	0.363	2.054	0.1320	−1.690	2.8577
65.1	65.089	0.011	0.363	0.0001	−0.352	0.1241
64.7	64.298	0.402	0.011	0.1618	0.391	0.1530
64.8	66.671	−1.871	0.402	3.5017	−2.273	5.1688
64.6	66.841	−2.241	−1.871	5.0213	−0.370	0.1366
68.2	69.101	−0.901	−2.241	0.8123	1.340	1.7943
71.4	70.062	1.338	−0.901	1.7902	2.239	5.0145
71.5	68.875	2.625	1.338	6.8893	1.287	1.6557
Sum				**25.6839**		**24.2687**

The Durbin-Watson statistic $(d) = \dfrac{\sum\limits_{t=2}^{n}(e_t - e_{t-1})^2}{\sum\limits_{t=1}^{n}e_t^2} = \dfrac{24.2687}{25.6839} = 0.94$

From Appendix A Table A8 the critical values of d at the 5 per cent level of significance for $n = 20$ and $k = 1$ are: $d_L = 1.2$ and $d_U = 1.41$. As actual $d(= 0.94) < d_L(= 1.2)$, significant positive autocorrelation is present.

M

Computer solution using Minitab

With the data for y and x entered in columns C1 and C2 of a MINITAB worksheet:

Choose **Stat → Regression → Regression**

Select **Response** (y) variable and **Predictor** (x) variable by clicking on the variable names or columns numbers

Select **Results** and click on **Display nothing**

[*Note*: as we only wish to obtain the Durbin-Watson statistic, there is no need to display other regression output]

Select **Options**, click on **Durbin-Watson statistic**

Click **OK** twice

The resulting output is:

Durbin-Watson statistic = 0.94

Authors' comment and interpretation: The output simply gives the value of the Durbin-Watson statistic. Its statistical significance has to be determined by reference to the Tables of the Durbin-Watson statistic in Appendix A Table A8 – see the *Worked solution* above for details and conclusion.

Computer solution using Excel

Unlike MINITAB, the Excel program does not compute the Durbin-Watson statistic directly but does give some of the data needed to calculate it manually (as in the *Worked solution* table above).

Enter the data for x and y in two columns of the Excel spreadsheet

Choose **Tools → Data Analysis → Regression**

Complete the **Input y range** and the **Input x range** by selecting in each case the appropriate range of cells on the spreadsheet

Complete the output options as desired

Click on **Residuals** in order to obtain output of the residuals

Click **OK**

The minimum output obtained is shown below. Note that the final, 'RESIDUAL OUTPUT', part gives the predicted y values and the residuals. Using the residuals it is a simple matter to use the Excel spreadsheet to compute the Durbin-Watson statistic in the same way as in the manual *Worked solution* set out above.

Summary Output

Regression Statistics

Multiple R	0.98872437
R Square	0.97757589
Adjusted R Square	0.9763301
Standard Error	1.19452136
Observations	20

ANOVA

	df	SS	MS	F	Significance F
Regression	1	1119.684137	1119.684	784.7073	2.68621E-16
Residual	18	25.68386289	1.426881		
Total	19	1145.368			

Worked example 18.5   Testing for autocorrelation *(continued)*

	Coefficients	Standard Error	t Stat	P-value	Lower 95%	Upper 95%
Intercept	4.56488769	1.969994045	2.317209	0.032476	0.426080583	8.703694805
X Variable 1	0.56511744	0.020173668	28.01263	2.69E-16	0.522734104	0.607500776

RESIDUAL OUTPUT

Observation	Predicted Y	Residuals
1	46.6661370	0.033863037
2	47.1747427	0.525257341
3	48.9266067	0.773393277
4	51.2435882	0.056411774
5	52.3738231	−0.273823106
6	53.2780110	−0.078011010
7	54.5212694	−0.121269377
8	56.4991804	−0.499180417
9	57.4033683	−1.103368320
10	58.5901149	−0.790114944
11	60.0029085	−0.302908544
12	61.2461669	2.053833089
13	65.9366417	0.363358338
14	65.0889655	0.011034498
15	64.2978011	0.402198913
16	66.6712943	−1.871294334
17	66.8408296	−2.240829566
18	69.1012993	−0.901299325
19	70.0619990	1.338001027
20	68.8752523	2.624747650

Worked Example 18.6   Computation of Spearman's rank correlation coefficient r_s

In a survey, ten popular television programmes were ranked in order by groups of men and women as shown below. Is there a significant relationship between the ranking of programmes by men and women?

Television programme	Ranking by men	Ranking by women
1	1	5
2	5	10
3	8	6
4	7	4
5	2	7
6	3	2
7	10	9
8	4	8
9	6	1
10	9	3

Solution

$$r_s = 1 - \frac{6 \sum D^2}{n(n^2 - 1)}$$

We need first to calculate $\sum D^2$, i.e. the sum of the squared differences between the two ranks. This is given below:

D		D²
1 − 5 =	−4	16
5 − 10 =	−5	25
8 − 6 =	2	4
7 − 4 =	3	9
2 − 7 =	−5	25
3 − 2 =	1	1
10 − 9 =	1	1
4 − 8 =	−4	16
6 − 1 =	5	25
9 − 3 =	6	36
		$\sum D^2 = 158$

$$r_s = 1 - \frac{6(158)}{10(99)} = 0.0424$$

From Appendix A Table A5, the critical value of r_s for $n = 10$ and $\alpha = 0.05$ is 0.564.

Since the computed value of r_s (0.0424) is less than the critical value (0.564) we cannot reject the null hypothesis. We conclude, therefore, that the rankings of men and women are not significantly related at the 5 per cent level of significance.

Computer solution using MINITAB

Ⓜ

MINITAB and rank correlation MINITAB can be used to compute r_s as follows:

1. Use the program to assign rank scores to the values in each column, if the data are not already ranked (as in the example above), and store the results in separate columns: Choose **Manip → Rank**
2. Choose **Stat → Basic Statistics → Correlation**; select columns containing the ranked data; Click **OK** – see the correlation example at the end of Worked example 18.4 above

Note, however, that this procedure is inappropriate when tied ranks occur because, although the **Rank** command assigns appropriate rank scores to the tied values, the **Correlation** command does not make a tied ranks adjustment.

Computer solution using Excel

Ⓔ

There is no pre-defined tool in Excel for solving this problem, but, as in MINITAB, the **CORREL** function can be applied to calculate the correlation coefficient, if the series are ranked (as in the table above) and this is shown below. Note, however, that if the data are not ranked neither of the two ranking tools in Excel provides an appropriately

Worked example 18.6   Computation of Spearman's rank correlation coefficient r_s (continued)

ranked series directly. The **Data Analysis Tools** include a tool called **Rank and Percentile** but this tool ranks each data set separately and *rearranges* the series in rank order whereas what is required here is an allocation of ranks for both data sets jointly without rearranging – as in the table above. Note that there is also a **RANK** function in Excel but, unlike the **Rank and Percentile** tool, it returns the rank of a *single* number only in a list of numbers, not the whole array of numbers. Further, it should be noted that neither of these tools allocates an appropriate ranking to tied ranks – see Worked example 18.7 below.

Given the two sets of ranks entered in, say, cells A1:A10 and B1:B10 of an Excel worksheet, Spearman's rank correlation coefficient r_s may be calculated as follows:

Click on the $=$ sign in the formula bar and select function **CORREL**

In **Array 1** enter the array cells for the first set of ranks (A1:A10)

In **Array 2** enter the array cells for the second set of ranks (B1:B10)

Click **OK**

This returns the coefficient: 0.042424

A significance test is not conducted by Excel.

Worked example 18.7   Spearman's rank correlation – tied ranks

A company attempts to evaluate the potential for a new bonus plan by selecting a random sample of ten sales persons to use the bonus plan for a trial period. The weekly sales volumes before and after implementing the plan are shown below. Are the rankings of the individual sales persons significantly different between the two schemes?

Salesperson	Weekly sales before	Weekly sales after
1	30	36
2	24	28
3	36	38
4	30	36
5	32	36
6	45	37
7	28	42
8	37	39
9	34	34
10	26	30

Solution

First it is necessary to rank the data. The rank order of each salesperson, ranked from high to low, is given in the table below together with D (the difference between the ranks) and D^2. Note that there are two sets of tied ranks. In the first column salesperson number 1 and salesperson number 4, each with weekly sales of 30, are ranked equal sixth and are given the average of ranks 6 and 7, i.e. 6.5. In the second column there are three tied ranks – in positions 5, 6 and 7 (salespersons 1, 4 and 5) – and each is therefore ranked 6. Thus two adjustments for tied ranks need to be made in the formula for r_s with $t = 2$ in the first case and $t = 3$ in the second case, as shown below.

Sales-person	Ranks Before	After	D	D^2
1	6.5[a]	6.0[a]	0.5	0.25
2	10.0	10.0	0.0	0
3	3.0	3.0	0.0	0
4	6.5[a]	6.0[a]	0.5	0.25
5	5.0	6.0[a]	−1.0	1.00
6	1.0	4.0	−3.0	9.00
7	8.0	1.0	7.0	49.00
8	2.0	2.0	0.0	0.00
9	4.0	8.0	−4.0	16.00
10	9.0	9.0	0.0	0.00
				$\sum D^2 = 75.5$

[a]Tied ranks

$$r_s = 1 - 6\left[\frac{\sum D^2 + (t^3 - t)/12 + (t^3 - t)/12}{n(n^2 - 1)}\right]$$
$$= 1 - 6\left[\frac{75.5 + (8 - 2)/12 + (27 - 3)/12}{10 \times 99}\right]$$
$$= 1 - 6\left(\frac{75.5 + 0.5 + 2.0}{990}\right) = 0.527$$

Test:

$H_0: \rho_s = 0$
$H_1: \rho_s \neq 0$

From Appendix A Table A5 the critical value of r_s at the 5 per cent level of significance and $n = 10$ is 0.564.

Actual $r_s <$ critical r_s (0.527 < 0.564). Thus H_0 cannot be rejected and we must conclude that the rank correlation is not significantly different from zero at the 5 per cent level.

Computer solutions

There is no pre-defined tool in MINITAB for solving this problem – see the MINITAB solution to Worked example 18.6 above for comments. There is also no pre-defined tool in Excel for solving the problem. The same remarks apply as in Worked example 18.6 above but, in addition, the Excel ranking tools do not rank tied observations appropriately (each being assigned the same rank, not the average rank). Further, the function **CORREL** cannot be used because of the need to make an adjustment for tied ranks.

Key terms and concepts

Autocorrelation Situation where values of the dependent variable y are related to the values of y in previous time periods. Also referred to as *serial correlation*.

Coefficient of determination (r^2) A measure of the variation explained by the estimated regression equation. It is a measure of how well the estimated regression equation fits the data.

$$r^2 = \frac{\text{Explained variation in } y}{\text{Total variation in } y} = \frac{\sum(y_c - \bar{y})^2}{\sum(y_i - \bar{y})^2}$$

Confidence interval The interval estimate of the mean value of y for a given value of x:

$$y_c \pm t_{\alpha/2}\left\{SEE\sqrt{\frac{1}{n} + \frac{(x_0 - \bar{x})^2}{\sum(x_i - \bar{x})^2}}\right\}$$

Correction for tied ranks Correction required in ranking data (in the context of calculating Spearman's rank correlation coefficient) when there are tied ranks:

$$\text{add } \frac{1}{12}(t^3 - t) \text{ to } \sum D^2$$

Correlation analysis Concerned with providing a statistical measure of the strength of any relationship between variables.

Correlation coefficient (r) Provides a numerical summary measure of the degree of correlation between two variables:

$$r = \frac{n\sum x_i y_i - (\sum x_i)(\sum y_i)}{\left[\sqrt{n\sum x_i^2 - (\sum x_i)^2}\right]\left[\sqrt{n\sum y_i^2 - (\sum y_1)^2}\right]}$$

Dependent variable The variable that is being predicted or explained by the independent variable. It is denoted by y in the regression equation.

Durbin-Watson statistic Statistic used to test for the presence of first-order autocorrelation:

$$d = \frac{\sum\limits_{t-2}^{n}(e_t - e_{t-1})^2}{\sum\limits_{t-1}^{n} e_t^2}$$

Equation of a straight line:

$$y = a + bx$$

Error sum of squares In regression, the sum of the squared differences between the actual and predicted values of y – the part of the total sum of squares (SST) that is not explained by the regression model.

Estimated regression equation The estimate of the regression equation obtained by the method of least squares:

$$y_c = a + bx$$

Extrapolation Predicting values of y from a regression equation using values of x which lie outside the range of the recorded values of x from which the regression equation itself was derived.

First-order autocorrelation Situation where the value of y in time period t is related to its value in time period $t - 1$.

Formula for regression coefficients

$$b = \frac{\sum x_i y_i - n\bar{x}\bar{y}}{\sum x_i^2 - n\bar{x}^2}$$

$$a = \bar{y} - b\bar{x}$$

Independent (or explanatory) variable The variable used to predict the value of the dependent variable. It is denoted by x in the regression equation.

Intercept The point at which a regression line intersects the vertical axis on which the dependent variable is measured. It is the value of y when x is zero.

Interpolation Predicting values of y from a regression equation using values of x which lie inside the range of recorded values of x from which the regression equation itself was derived.

Method of least squares The approach used to develop the estimated regression equation which minimizes the sum of squares of the deviations between the actual values and computed (i.e. estimated) values of the dependent variable, i.e.: minimize $\sum e_i^2$ (also referred to as the *principle of least squares*).

Prediction interval The interval estimate of an individual value of y for a given value of x:

$$y_c \pm t_{\alpha/2}\left\{SEE\sqrt{1 + \frac{1}{n} + \frac{(x_0 - \bar{x})^2}{\sum(x_i - \bar{x})^2}}\right\}$$

Principle of least squares The approach used to develop the estimated regression equation which minimizes the sum of squares of the deviations between the actual values and computed (i.e. estimated) values of the dependent variable, i.e.: minimize $\sum e_i^2$ (also referred to as the *method of least squares*)

Rank correlation Correlation between variables which are measured in rank order.

Regression analysis Concerned with measuring the way in which one variable is related to another.

Regression coefficient The coefficient b in the regression equation. It measures the slope of the regression line.

Regression line Line of best fit between two related variables.

Regression sum of squares In regression, the part of the total sum of squares (SST) that is explained by the regression model.

Residual The difference between the actual value of the dependent variable and the value computed using the estimated regression equation.

$$e_i = y_i - y_c$$

Scatter diagram A diagram which consists of a set of points of paired observations on two variables.

SEE Standard error of estimate

$$\sqrt{\frac{\sum e_i^2}{n-2}} = \sqrt{\frac{\sum y_i^2 - a \sum y_i - b \sum x_i y_i}{n-2}}$$

Serial correlation Situation where values of the dependent variable, y are related to the values of y in previous time periods (also referred to as *autocorrelation*).

Spearman's rank correlation coefficient (r_s) Provides a numerical summary measure of the degree of correlation between two variables measured in rank order:

$$r_s = 1 - \frac{6 \sum D^2}{n(n^2 - 1)}$$

SSE Error sum of squares In regression, the sum of the squared differences between the actual and predicted values of y – the part of the total sum of squares (SST) that is not explained by the regression model.

SSR Regression sum of squares In regression, the part of the total sum of squares (SST) that is explained by the regression model.

SST Total sum of squares In regression, the sum of the squared differences between the actual y values and the mean value of y.

Standard error of b

$$s_b = \frac{SEE}{\sqrt{\sum x_i^2 - n\bar{x}^2}}, \; where \; SEE = \sqrt{\frac{\sum y_i^2 - a \sum y_i - b \sum x_i y_i}{n-2}}$$

Testing the significance of b

$$t = \frac{b - B}{s_b}$$

Testing the significance of r

$$t = \frac{r}{\sqrt{(1 - r^2)/(n - 2)}}$$

Testing the significance of r_s

$$t = \frac{r_s}{\sqrt{(1 - r_s^2)/(n - 2)}}$$

Total sum of squares In regression, the sum of the squared differences between the actual y values and the mean value of y.

Self-study exercises

18.1. It is thought that the number of industrial stoppages of work that occur each year is related to the level of unemployment. The data given below are collected over a ten-year period.

  (a) What would you regard as the dependent and independent variables?

  (b) Fit an appropriate regression line using the principle of least squares, test its significance and explain the meaning of your results.

  (c) Plot a scatter diagram and draw in the regression line.

Year		1	2	3	4	5	6	7	8	9	10
Stoppages (S) (000s)		2.5	2.9	2.9	2.3	2.0	2.7	2.5	2.1	1.3	1.3
Unemployment (U) (%)		3.8	2.7	2.6	4.0	5.5	5.8	5.7	5.3	6.8	10.5

$\sum S = 22.5$ $\sum U = 52.7$ $\sum S^2 = 53.69$ $\sum U^2 = 325.45$ $\sum SU = 108.6$

18.2. The following data show how a manufacturer's total costs of production $C(\pounds)$ vary with the number of items produced Q.

Q	10	14	17	20	22	25
C	70	100	120	136	150	162

  (a) Is the relationship a linear one?

  (b) Comment on the goodness-of-fit of the estimated regression line to the data.

  (c) Calculate 95 per cent confidence limits for the regression slope.

  (d) Give an estimate of the manufacturer's total fixed costs.

  (e) What is the marginal cost of an extra 10 units of production?

18.3. It is postulated that the amount people save depends on their income. The following data show the average weekly savings $y(\pounds)$ and the average weekly incomes $x(\pounds)$ of people in different income groups.

x	19	22	27	30	36	43	47	51	61	64
y	1.0	1.4	1.8	2.4	3.0	3.8	4.3	4.5	5.8	6.3

  (a) Given the following summations, calculate the linear regression of savings on income:

$$\sum x = 400 \quad \sum x^2 = 18\,246 \quad \sum y = 34.3$$
$$\sum y^2 = 147.47 \quad \sum xy = 1630.4$$

(b) Calculate the coefficient of correlation and test its significance.

(c) How much of the variation in savings is 'explained' by the variation in income?

(d) What savings would you expect an income group with an average income of £50 to make?

(e) Compute the 95 per cent confidence interval for savings for an income group with an average income of £50.

18.4. In an attempt to determine to what extent investment expenditure I is influenced by the rate of interest R, the data given below were collected over an eight-year period.

Year	1	2	3	4	5	6	7	8
Investment (I) (£m)	2.1	1.8	1.8	2.2	2.8	4.1	3.6	3.1
Average rate of interest (R) (%)		9.6	9.9	10.5	9.8	12.1	7.7	9.5	9.2

$\sum R = 78.3$ $\sum R^2 = 777.05$ $\sum I = 21.5$ $\sum I^2 = 62.95$ $\sum RI = 206.61$

(a) Is the expected sign of the regression slope coefficient positive or negative?

(b) By how much does investment change in response to a 1 percentage point decline in the rate of interest?

(c) Does the rate of interest have a significant influence on the level of investment?

(d) How much of the variation in investment is explained by variation in the rate of interest?

18.5. Suppose that the residuals for a set of data collected over 15 consecutive time periods are as follows:

Time period	Residual
1	+3
2	−1
3	−5
4	+8
5	+2
6	+6
7	−4
8	−1
9	+3
10	+3
11	−5
12	−2
13	−1
14	+6
15	+2

(a) Compute the Durbin-Watson statistic.

(b) What are the critical values of the Durbin-Watson statistic for this set of data ($k = 1$) at:

   (i) the 5 per cent level of significance?

   (ii) the 1 per cent level of significance?

(c) What conclusion can be reached about the presence of positive or negative autocorrelation at the 5 per cent or 1 per cent levels of significance?

18.6. The manager of a large telephone call-answering centre wishes to develop a statistical model to assess how long it takes to deal with calls for different clients. Part of the data collected over a period of 21 days shows the following:

Day	Calls answered	Time taken (minutes)
1	11	15
2	54	63
3	20	34
4	31	18
5	18	22
6	22	33
7	18	36
8	52	99
9	10	28
10	28	62
11	30	74
12	19	31
13	22	48
14	26	47
15	18	34
16	31	74
17	15	33
18	19	47
19	27	61
20	15	22
21	27	72

Is there evidence of significant autocorrelation in these data?

18.7. A manufacturer who is interested in determining the effect of changes in his price on sales varies his prices in different regions of the country. Sales are shown below.

Region	1	2	3	4	5	6	7	8	9	10
Price (P) (£)	0.7	1.0	0.8	0.9	1.5	0.6	0.7	1.4	1.5	0.9
Sales per 1,000 population (S)	140	80	120	80	75	135	130	70	70	100

$\sum P = 10.0 \qquad \sum P^2 = 11.06 \qquad \sum S = 1000 \qquad \sum S^2 = 107\,350 \qquad \sum PS = 923.5$

    (a) Using regression techniques give:

       (i) a point estimate and

       (ii) a prediction interval estimate

       of sales per 1000 population if the price is 50p.

    (b) Show that tests on whether the regression coefficient and the correlation coefficient differ significantly from zero are identical.

    (c) To what extent were sales affected by factors other than price? How could these factors be taken into account?

18.8. A personnel manager is investigating a selection procedure for salesmen in which applicants are given an aptitude test and have an interview. The sales records of ten recruits during their first year are noted and compared with their score in the aptitude test and rating in their interview.

Salesman	A	B	C	D	E	F	G	H	I	J
Interview ranking (Y)	1	2	3	4	5	6	7	8	9	10
Aptitude test score (A)	45	56	68	68	50	78	76	72	68	60
Sales performance (S) (000s)	27	28	32	45	17	55	18	66	46	36

$\sum Y = 55$ $\sum Y^2 = 385$ $\sum A = 641$ $\sum A^2 = 42\,177$ $\sum S = 370$ $\sum S^2 = 15\,968$

    (a) Using an appropriate statistical test discuss the usefulness of each selection procedure in placing salesmen in the correct order of their actual selling ability.

    (b) Examine the usefulness of:

       (i) the ranking at interview and

       (ii) score in the aptitude test

    as predictors of *actual* sales performance.

    (c) In the light of the answers to parts (a) and (b) what advice would you give the personnel manager about the recruitment of staff?

18.9. **(a)** What are the main differences between the product-moment correlation coefficient and a rank correlation coefficient for two characteristics each measured on a number of individuals?

    (b) Calculate a coefficient of rank correlation between cinema admissions and TV licences per 1000 of population for ten towns from the data given below.

    (c) Given that $r = -0.54$, comment on the nature and significance of the relationship in the light of your discussion in (a) and your results in (b).

Town	1	2	3	4	5	6	7	8	9	10
Cinema admission (000s)	12.8	10.9	13.7	12.4	8.8	13.7	10.1	11.6	10.6	9.4
TV licences per 1000 population	18	122	39	76	83	42	151	21	99	55

18.10. Daily expenditure per household on two commodity groups (food and housing) in the standard regions of Great Britain in a particular year was as shown below. Using an appropriate statistical measure, say whether there is a statistically significant tendency for expenditure on both food and housing to be high in the same regions and low in the same regions, or whether there is an opposite tendency. Explain your methodology clearly.

Region	Food (£)	Housing (£)
North	9.2	4.3
Yorks and Humberside	9.3	3.9
N. West	9.5	5.2
E. Midlands	8.9	5.1
W. Midlands	9.7	5.4
East Anglia	8.7	4.7
South East	10.1	6.9
South West	9.4	5.5
Wales	9.8	3.6
Scotland	9.7	4.2

18.11. MINITAB output for the regression of data on annual percentage increases in wage rates (denoted W) and the percentage rate of unemployment (denoted U) for a country over a period of 17 years is given below. The raw data are as follows:

Year	1	2	3	4	5	6	7	8	9	10	11	12	13	14	15	16	17
W	1.8	8.5	8.4	4.5	4.3	6.9	8.0	5.0	3.6	2.6	2.6	4.2	3.6	3.7	4.8	4.3	4.6
U	1.4	1.1	1.5	1.5	1.2	1.0	1.1	1.3	1.8	1.9	1.5	1.4	1.8	2.1	1.5	1.3	1.4

Regression Analysis

The regression equation is
$W = 10.3 - 3.81\ U$

Predictor	Coef	StDev	T	P
Constant	10.344	2.127	4.86	0.000
Income	−3.808	1.430	−2.66	0.018

$S = 1.717$ R–Sq = 32.1%   R–Sq (adj) = 27.6%

Analysis of Variance

Source	DF	SS	MS	F	P
Regression	1	20.900	20.900	7.09	0.018
Residual Error	15	44.198	2.947		
Total	16	65.098			

Durbin-Watson statistic = 1.60

(a) Comment on the results, explaining what the terms in the MINITAB results mean. (*Hint*: Among other things you should give an indication of goodness-of-fit, statistical significance and economic meaning of the regression, saying whether this is what you expected and why.)

(b) Are you satisfied with these results? Explain your answer giving an indication of how they might be improved.

18.12. With reference to the MINITAB regression results given below, where 'balance' is the average monthly balance (£) in an individual's current account in a bank and 'income' is the annual income (£) of the account holder, answer the following questions:

(a) Are the intercept and slope significantly different from zero?

(b) Is the slope significantly less than 0.06?

(c) What percentage of the variation in current account balances is explained by factors other than income?

(d) Give the point prediction of the size of balance for an individual earning

(i) £25,000 and

(ii) £5000.

Regression Analysis

The regression equation is
Balance = − 427 + 0.0584 Income

Predictor	Coef	StDev
Constant	− 426.56	13.65
Income	0.0583844	0.0009358

$S = 72.74$

Analysis of Variance

Source	DF	SS	MS	F	P
Regression	1	20593646	20593646	3892.50	0.000
Residual Error	98	518478	5291		
Total	99	21112124			

18.13. The following MINITAB output shows the results of a regression on two variables showing the heights (cms) and the corresponding weights (kgs) of ten people. Parts of the output have been deleted – these are shown by blank boxes – but the remaining output is sufficient to allow the boxes to be filled in. Fill in the empty boxes.

Worksheet		Output

C1	C2
Height	Weight

Height	Weight
164	64
185	82
179	74
154	60
159	78
187	88
192	90
172	70
177	82
185	94

Regression Analysis

The regression equation is

Height $= 101 + 0.947$ Weight

Predictor	Coef	StDev	T	P
Constant	101.37	17.87		0.000
Weight	0.9466	0.2265		0.003

S = ☐   R-Sq = ☐

Analysis of Variance

Source	DF	SS	MS	F	P
Regression	1	1014.0	1014.0	17.47	0.003
Residual Error	8	464.4	58.0		
Total	9	1478.4			

Chapter 19

Multiple linear regression and correlation

In Chapter 18 we considered simple linear regression and correlation involving a dependent variable (y) and only one independent variable (x). In this chapter we turn to consider the extension of the simple model to applications involving two or more independent variables as the basis for estimating the value of y. This extension gives rise to *multiple* linear regression and correlation.

The estimated multiple regression equation is expressed in the following general form, for the case of k independent variables:

Estimated multiple regression equation

$$y_c = a + b_1x_1 + b_2x_2 + b_3x_3 + \cdots + b_kx_k$$

---

where y_c is the computed (i.e. estimated) value of y, a is the value of the intercept term and $b_1, b_2, b_3, \ldots, b_k$ are the values of the estimated regression coefficients corresponding to each of the explanatory variables $x_1, x_2, x_3, \ldots, x_k$.

It will be appreciated that given the existence of several explanatory (independent) variables the regression line is a line in multi-dimensional space. The calculations required for determining the regression coefficients and their standard errors are tedious and complex and are not normally carried out manually. Nowadays, the use of computers and appropriate statistical packages, such as MINITAB and Excel, makes the estimation of the regression model a simple task.

In this chapter we focus on the key concepts involved in multiple linear regression and correlation analysis, the interpretation of the results and the application of MINITAB and Excel to various situations. In practice, use of multiple linear regression involves a number of technical issues which are beyond the scope of this book. We refer to these briefly below but do not develop them in detail.

Learning Outcomes   After working through this chapter you will be able to:

- analyse the nature of the relationship between a dependent variable and two or more explanatory variables using multiple linear regression and correlation
- interpret the estimated coefficients of a regression model
- test the statistical significance of individual regression coefficients
- understand the meaning of the overall explanatory power of an estimated regression equation
- measure the overall explanatory power using the multiple coefficient of determination, R^2
- derive the coefficient of multiple correlation, R
- understand the use of, and calculate, the adjusted multiple coefficient of determination, $\overline{R}^2$
- assess the statistical significance of the overall explanatory power of the regression model using the F test
- appreciate the pitfalls and limitations of multiple regression.

Principles

As noted above the multiple linear regression equation takes the following general form:

$$y_c = a + b_1x_1 + b_2x_2 + b_3x_3 + \cdots + b_kx_k$$

The coefficients are naturally estimates based on sample data of their corresponding population parameters which may be denoted, using capital letters, as

$$A, B_1, B_2, B_3, \ldots, B_k$$

The values of the b_k coefficients are derived using the principles of the method of least squares (described in the context of simple linear regression in Chapter 18). The b_k regression coefficients are defined such that the sum of the squares of the residuals (i.e. the differences between the actual values of y and the computed values, y_c) are as small as possible. Thus, the objective is to minimize $\sum(y - y_c)^2$ as before. As noted above, while it is possible to compute the values of the b_k coefficients by hand, based on this principle, it is not normally done nowadays given the availability of computer packages which carry out the necessary calculations automatically and at great speed. We therefore confine our attention here to the interpretation of results, their statistical significance and the pitfalls and limitations associated with multiple regression and correlation analysis.

As in the case of simple regression analysis, there are four aspects of the results to consider in the case of multiple regression:

- interpretation of the individual regression coefficients
- statistical significance of the regression coefficients
- overall explanatory power of the estimated equation
- statistical significance of the overall explanatory power

We discuss the principles of each of these in turn.

Interpretation of regression coefficients

In the multiple regression case, the intercept term a is the estimated value of y (i.e. y_c) when the values of *all* independent variables are zero. Thus, for the case of three independent variables,

$$y_c = a$$

when $x_1 = x_2 = x_3 = 0$.

The interpretation of any b_i coefficient in multiple regression analysis is as follows: b_i represents the change in y_c corresponding to a unit change in x_i when all other independent variables are held constant. For example, consider the case where the monthly sales of heating fuel (S) can be explained by three variables: price P, advertising expenditure E and mean temperature T for each month. Thus, the estimated relationship between sales and the explanatory variables can be expressed as

$$S_c = a + b_1 P + b_2 E + b_3 T$$

where S_c now represents the value of monthly sales as predicted by the equation. The coefficients a, b_1, b_2 and b_3 are derived from a data set providing past monthly observations of the value of sales and the three independent variables over a period of time (see Worked example 19.1 below).

The interpretation of the intercept a in this context need not detain us: conceptually it simply denotes the average value of sales when each of the three explanatory variables has a value of zero simultaneously. The interpretation of the other coefficients, however, requires more attention. The coefficient b_1 represents the average change in sales associated with a unit change in the price variable P when the other independent variables are held constant. Similarly, b_2 represents the average change in sales associated with a unit change in advertising expenditure E when the other independent variables are held constant (and so on for b_3). By this means of control, we are able to separate out the effect of each independent variable on sales, free of any distorting influences from the other independent variables. For this reason, the b_1, b_2 and b_3 values are referred to as *partial regression coefficients*.

Statistical significance of individual regression coefficients

The principles of testing the statistical significance of the individual regression coefficients are the same as for the simple linear regression case discussed in Chapter 18. For example, suppose we wish to know whether or not the partial regression coefficient b_1 is significantly different from zero. As in Chapter 18, we set up the null hypothesis which states that the variable associated with b_1, i.e. x_1, does not influence the dependent variable y, i.e.:

$$H_0: B_1 = 0$$

where B_1 represents the hypothesized population parameter. To say, as here, that $B_1 = 0$ means that it is being hypothesized that the variable x_1 has no effect on the dependent variable y. We are thus using b_1, which may have a non-zero value simply as a result of sampling error, to test the likelihood that the true population parameter B_1 is really zero. The alternative hypothesis, H_1, may be stated in either a one-tailed or a two-tailed form depending on the hypothesis of interest, i.e.:

$$H_1: B_1 \neq 0 \text{ (a two-tailed test)}$$

or

$$H_1: B_1 < 0 \text{ (a one-tailed test)}$$

or

$$H_1: B_1 > 0 \text{ (a one-tailed test)}$$

Similar hypotheses may be set up for the other regression coefficients.

These hypotheses may be tested using a t test (as before), where the value of the test statistic is defined as:

Test statistic

$$t = \frac{b_i - B_i}{s_{b_i}}$$

where, as before, s_{b_i} is the standard error of b_i. Since under the null hypothesis $H_0: B_i = 0$, this reduces to:

$$t = \frac{b_i}{s_{b_i}}$$

This calculated value of t is then compared with its critical value, given in the t table (see Appendix A Table A4). Note that the degrees of freedom for the t statistic in the multiple regression case, however, are:

Degrees of freedom

$n - k - 1$

where n is the number of observations on the dependent variable and k is the number of independent variables.

Overall explanatory power of the estimated regression equation

Multiple coefficient of determination R^2

In the last chapter we explained that the coefficient of determination r^2 provides a measure of the overall explanatory power of a simple regression model $y_c = a + bx$. Thus, an r^2 value of 0.65 would indicate that 65 per cent of the variation in the dependent variable y is explained by variation in the independent variable x. In the case of multiple regression analysis we compute a similar measure which shows how much of the variation in y is explained by the *joint* variation in all of the independent variables. This measure is known as the *multiple coefficient of determination*, denoted now as R^2 (rather than r^2). It is thus a measure of *overall* explanatory power (i.e. goodness-of-fit for the regression model as a whole). The value of R^2 is given by:

Multiple coefficient of determination

$$R^2 = \frac{\text{SSR}}{\text{SST}} = 1 - \frac{\text{SSE}}{\text{SST}}$$

---

where SSR is the sum of squares for the regression, $\sum(y_c - \bar{y})^2$; SSE is the sum of squares for the error, $\sum(y_i - y_c)^2$; and SST is the total sum of squares, $\sum(y_i - \bar{y})^2$, as explained in Chapter 18. The values of R^2, SSR, SSE and SST are all automatically given on the MINITAB and Excel computer outputs (see the Applications and Worked Examples section) and hence we do not need to calculate them directly. The range for R^2 is from 0 to 1. If $R^2 = 1$, then 100 per cent of the total variation in the dependent variable y has been explained by the model. Thus, in this case SSE $= \sum(y_i - y_c)^2 = 0$ and therefore $y_i = y_c$ for each observation in the sample, i.e. the computed model provides a *perfect predictor*. This does not occur in practice, but in general a large value of R^2 is desirable. The fit of the model is said to be 'better' the closer the value of R^2 is to 1. Naturally, judgements of overall explanatory power also require a test of the statistical significance of R^2 (see below).

Coefficient of multiple correlation

The *coefficient of multiple correlation*, denoted by R, is a measure of the degree of association between y and all the explanatory variables jointly. Conceptually, it is thus akin to the coefficient of correlation r in the simple regression case but, whereas r can be positive or negative, R is always taken to be positive. Note, however, that in practice R is of little importance and can be ignored. Most attention is paid to the multiple coefficient of determination, R^2.

Adjusted multiple coefficient of determination $\overline{R}^2$

It can be shown that adding additional independent variables to a multiple regression model will never lead to a decrease in the value of R^2 – in general, R^2 will increase and, at worst, it will remain unchanged. In view of this, in comparing two regression models with the *same dependent variable* but a different number of independent x variables, one must be very wary of choosing that model which has the highest R^2 value. To compare two R^2

values, it is necessary to take into account the number of x variables included in each model. This is done by computing an *adjusted R^2* value, denoted $\bar{R}^2$ (pronounced 'R-bar squared'), in order to avoid over-estimating the impact of any additional independent variable on the amount of explained variation. This is measured as:

Adjusted multiple coefficient of determination

$$\bar{R}^2 = 1 - (1 - R^2)\frac{n - 1}{n - k - 1}$$

where n is the sample size and k is the number of independent variables.

This adjustment is especially important when the sample size n is small. Note that if the sample size n is very large (relative to the number of independent variables k) then $n - 1$ and $n - k - 1$ in the above expression will be nearly identical and hence $\bar{R}^2 = 1 - (1 - R^2)1 = R^2$. Thus, $\bar{R}^2$ can only be significantly smaller than R^2 if the sample size n is small relative to k. As a rule of thumb it is normally recommended that the sample size should be at least four times the number of independent variables.

Statistical significance of overall explanatory power

F statistic

We showed earlier how the statistical significance of each of the partial regression coefficients b_k may be tested using a t test. Similarly, we can test the statistical significance of the regression model as a whole, i.e. its *overall* explanatory power, given by R^2. This requires a different procedure involving an F test.

Whereas the test of the statistical significance of an individual coefficient (b_k) involves a test of the null hypothesis, H_0: $B_k = 0$, a test of *overall* explanatory power is a test of whether or not *all* the Bs are equal to zero. The *F statistic* is defined as the ratio of the explained to the unexplained *variance*. It will be recalled that the total variation in the dependent variable $\sum(y_i - \bar{y})^2$, denoted SST, may be decomposed into a part explained by the regression, $\sum(y_c - \bar{y})^2$, denoted SSR, and a residual, unexplained, part $\sum(y_i - y_c)^2$, denoted SSE. Thus,

$$SST = SSR + SSE$$

This procedure for decomposing total variation into its components is called *analysis of variance*, abbreviated to ANOVA (discussed in detail in Chapter 17). The relevant expressions for explained variance and unexplained variance are simply SSR and SSE divided by their respective degrees of freedom, giving corresponding *mean square* expressions MSR and MSE respectively. The degrees of freedom are as follows:

Sum of squares	Degrees of freedom (df)
SSR	k
SSE	$n - k - 1$
SST	$n - 1$

where k is the number of independent variables and n is the number of observations of the dependent variable. Thus the *F statistic* is given by the following expression:

F statistic

$$F = \frac{\text{MSR}}{\text{MSE}} = \frac{\text{SSR}/k}{\text{SSE}/(n - k - 1)}$$

Before turning to the conduct of the F test, it is useful to provide a tabular summary of the key elements of the analysis of variance in an ANOVA table as shown in Table 19.1 below. The conduct of the F test follows the usual procedures, first specifying the test hypotheses and then comparing the calculated F ratio with a critical value obtained from tables of the F distribution (see Appendix A Table A6) with a given level of probability.

The hypotheses may be stated as:

H_0: $B_1 = B_2 = \cdots = B_k = 0$

H_1: one or more of the B coefficients is not equal to zero.

If we reject H_0 we can conclude that there is a significant relationship between the dependent variable and at least one of the independent variables and that the regression equation *as a whole* is significant. Note that, even if the regression as a whole is significant, some of the individual regression coefficients may not be statistically significant. It will be appreciated that in the case of simple regression analysis with only one explanatory variable, however, an F test of overall explanatory power is necessarily equivalent to a t test of the single regression coefficient B. In this case, it may be shown that $F = t^2$.

So, let us return now to the conduct of the F test. Tables of the F statistic for probability values of 0.05 and 0.01 with degrees of freedom denoted v_1 and v_2, for the numerator and denominator respectively, ranging from 1 to ∞ are given in Appendix A Table A6. The degrees of freedom in this case are $v_1 = k$ and $v_2 = n - k - 1$. If the computed F statistic exceeds the critical F value, then we reject the null hypothesis and conclude that the overall regression is significant.

Pitfalls and limitations of multiple linear regression

A number of pitfalls await the unwary in the use of multiple regression. A detailed discussion of these and the limitations which they pose is beyond the scope of this book

Table 19.1 ANOVA table – regression

Source of variation	Sum of squares	Degrees of freedom	Mean square
Regression (explained)	SSR	k	$\text{MSR} = \text{SSR}/k$
Error (unexplained)	SSE	$n - k - 1$	$\text{MSE} = \text{SSE}/(n - k - 1)$
Total	SST	$n - 1$	

and hence we only comment on them briefly in order to sound a note of caution. Detailed discussions will be found in specialized books on regression analysis.

The main problems are associated with the following aspects:

- the choice of an inappropriate *functional form* for the estimated regression equation (i.e. linear versus non-linear relationships), referred to as *functional form misspecification*;

- the extent to which two or more of the independent variables are correlated with one another, referred to as a problem of *multicollinearity*;

- the possibility that observations on the dependent variable *y* are themselves correlated over time, referred to as a problem of *autocorrelation*;

- the possibility that the prediction errors may not be constant and instead may be correlated with the size of the independent variables, a problem referred to as *heteroscedasticity*;

- the possibility that the independent variables in the regression model include measurement errors, a problem referred to as *errors in variables*.

We discuss each of these problems briefly and highlight the limitations they impose on multiple regression analysis.

Functional form misspecification

We have only considered in this and the last chapter situations involving *linear* relationships in which the dependent variable *y* is linearly related to one or more independent variables. In principle, many other functional forms are possible which are *non-linear* in nature, any one of which may provide a better statistical fit than the linear case. Unfortunately, there is often no theoretical guidance available to the analyst as to which form – linear or non-linear – is the most appropriate to particular models. The solution to this problem therefore reduces to an empirical one. Special statistical tests exist to assist in the choice of functional form, the basis of selection still being that of 'best fit' as explained in Chapter 18.

Multicollinearity

Multicollinearity refers to the situation in which some or all of the explanatory variables are very highly correlated with one another and are therefore not independently distributed. A high degree of multicollinearity can cause several problems with respect to the following aspects of regression analysis:

- estimated regression coefficients may not be uniquely determined;

- estimates of the coefficients may fluctuate markedly from sample to sample;

- less reliability may be placed on the relative importance of individual explanatory variables as indicated by the regression coefficients b_k.

It should be appreciated, however, that multicollinearity is inevitably present to some degree in most multiple regression analyses. It can rarely be eliminated completely and the aim of the researcher therefore is to minimize its influence as much as possible. The classic way of dealing with the problem is to discard variables from the regression analysis. If two variables are very highly correlated, use of either one in the regression (rather than both) can capture the effect of both. Alternatively, the problem may be reduced by collecting more data, by utilizing *a priori* information or by transforming the functional form of the relationship.

Autocorrelation

As noted in the previous chapter, the problem of autocorrelation refers to those situations where observations on the dependent variables *y* are correlated with each other. This is particularly common in the case of time series data, in which case the value of the dependent variable in one time period is almost invariably related to values in adjoining time periods. As noted earlier, the problem is also referred to as *serial correlation*. For example, over time, incomes generally rise year by year so that the value of income this year will be closely correlated with the value of income last year and, very likely, with the value of income next year. In such a case, the standard error associated with each estimated regression coefficient will be understated, leading to less precise confidence intervals for estimated coefficients than would otherwise be the case. This will result in the null hypothesis (H_0: $B = 0$) being rejected too frequently. The presence of autocorrelation may be tested using the *Durbin–Watson Statistic*, as explained in Chapter 18.

One way to correct for autocorrelation is to transform both the dependent and/or independent variables by expressing them in first-difference form (i.e. current value minus previous value). In addition, if significant autocorrelation is identified, it is important to investigate whether or not one or more key independent variables that have time-ordered effects on the dependent variable have been omitted. If no such variables can be identified, the inclusion of an additional variable that measures the time of the observation will sometimes eliminate or reduce the degree of autocorrelation. For instance, a time variable, denoted *T*, could take the values 1, 2, 3, etc. for first, second and third observations respectively, and so on.

Heteroscedasticity

One assumption underlying regression analysis is that the error term e_i given by $y_i - y_c$ has a constant variance, i.e. it should stay the same regardless of the value of the independent variable. For example, take the case where the amount of money saved by individuals is related to income levels. If the variation in the amount saved stays the same across *all* income groups then the assumption is satisfied and the error term is said to be *homoscedastic*. However, if the variance of savings is not constant across all levels of income (e.g. if individuals at high income levels save more than those at low income levels) then we have the problem of *heteroscedasticity*. The consequence is that misleading conclusions are likely to be drawn about the significance of the estimated regression coefficients using the *t* and *F* tests. This problem is more commonly encountered in cross-sectional data, i.e. observations across a range of variables recorded at a point in time as opposed to observations recorded over a period of time. The potential solutions to the problem involve, in principle, transformations of one or more of the explanatory variables. The principles involved are considered in more detail in more advanced textbooks.

Errors in variables

The problem of errors in variables refers to the case in which the variables in the regression model include measurement errors. It is important, however, to distinguish between the dependent and independent variables. Measurement errors in the dependent variable are incorporated into the error term and do not create a problem. But errors in the explanatory variables lead to biased and inconsistent parameter estimates. One solution to the problem is to replace the explanatory variable subject to measurement errors with another variable – called an *instrumental variable* – that is highly correlated with the original explanatory variable but is independent of the error term. In practice, the simplest instrumental variable to use is often a lagged version of the explanatory variable in question.

Summary

The purpose of this brief overview of the potential pitfalls in, and limitations of, regression analysis has been to warn the reader of the dangers of taking the results of regression analysis at face-value without considering the possible shortcomings that may be inherent in them. This is a highly technical subject which, as we have indicated, is dealt with fully in more advanced textbooks. As a final note of caution, the warning given in Chapter 18 dealing with simple linear regression should be recalled, namely that *correlation does not necessarily imply causation*. There is always the danger that the unwary user of statistics mistakenly interprets significant statistical results as implying the existence of a causal relationship with no theoretical or justifiable foundation.

In the next section we turn to applications, two involving time series data (including prediction) and the other involving cross-sectional data. Because of the complexity of carrying out the required calculations by hand, we consider only the computer-based solutions using MINITAB and Excel.

| Applications and Worked Examples | Worked example 19.1   Multiple regression analysis using time series data |

General descriptions of MINITAB and Excel computer programs and their application are given in Appendices B and C respectively

The data below show the monthly sales of heating fuel (in thousands of litres) for a firm over the past 12 months together with the average price (£) charged per litre in each month, the advertising expenditure (£000) per month and the mean daily temperature (°F) recorded during each month. Using these data, compute the regression equation which can be used to estimate the influence of the three explanatory variables (price, advertising expenditure and temperature) on heating fuel sales. Comment on the results.

Month	Sales (000s litres)	Price (£ per litre)	Advertising expenditure (£000s)	Mean daily temperature (°F)
January	450	0.6	25	42.5
February	380	1.2	17	40.3
March	298	1.8	14	44.1
April	350	1.5	18	44.0
May	201	3.0	10	55.5
June	215	2.7	11	56.3
July	220	2.7	12	65.7
August	240	2.1	11	62.9
September	192	3.0	7	56.7
October	201	2.4	7	51.7
November	202	2.4	8	46.3
December	235	2.3	10	38.6

The computer printouts below shows the estimated regression equation for the linear model:

$$\text{sales} = a + b_1(\text{price}) + b_2(\text{advertising expenditure}) + b_3(\text{temperature})$$

Computer solution using MINITAB

With the data for the four variables entered in columns C1–C4 respectively of the MINITAB worksheet, the procedure is the same as for simple regression considered in Chapter 18:

Choose **Stat** → **Regression** → **Regression**
Select **Response** variable (C1) and **Predictor** variables (C2, C3, C4)

Choose **Results**
Check an appropriate button under **Control the display of results** (the third option is chosen here). No predictions or other results are required in this example so:
Click **OK**

The output is as follows:

Regression Analysis

The regression equation is
Sales = 290 − 47.2 Price + 9.03 Advertising − 0.731 Temp

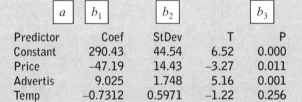

Predictor	Coef	StDev	T	P
Constant	290.43	44.54	6.52	0.000
Price	−47.19	14.43	−3.27	0.011
Advertis	9.025	1.748	5.16	0.001
Temp	−0.7312	0.5971	−1.22	0.256

S = 12.95   R−Sq = 98.3%   R−Sq(adj) = 97.7%

P-values

Degrees of freedom   Sums of squares   Mean squares   F statistic

Analysis of Variance

Source	DF	SS	MS	F	P
Regression	3	77962	25987	155.04	0.000
Residual Error	8	1341	168		
Total	11	79303			

Source	DF	Seq SS
Price	1	73428
Advertis	1	4282
Temp	1	251

Authors' comment: This gives the sequential sum of squares, i.e. the contribution of each independent variable to the regression sum of squares (SSR)

Unusual Observations

Obs	Price	Sales	Fit	StDev Fit	Residual	St Resid
2	1.20	380.00	357.76	6.71	22.24	2.01R

R denotes an observation with a large standardized residual

Worked example 19.1   Multiple regression analysis using time series data *(continued)*

Computer solution using Excel

The procedure using Excel is the same as for simple regression (see Worked example 18.1) except that in **Input x range** the range of cells containing *all* the independent variables is entered – these cells must be in contiguous columns (e.g. in this case with the variables for price, advertising and temperature in cells B2:B13, C2:C13 and D2:D13 respectively, enter the full range from B2:D13). If you want the variable names to appear in the output, enter the names in cells B1, C1 and D1 and check the **Labels** box

Click **OK**

The output is as follows:

> *Authors' comment*: compare these results with those from MINITAB above.

SUMMARY OUTPUT

Regression Statistics

Multiple R	0.991509333
R Square	0.983090757
Adjusted R Square	0.97674979
Standard Error	12.94675683
Observations	12

ANOVA

	df	SS	MS	F	Significance F
Regression	3	77961.71857	25987.24	155.038	1.9982E-07
Residual	8	1340.9481	167.6185		
Total	11	79302.66667			

	Coefficients	Standard Error	t Stat	P-value
Intercept	290.4252844	44.54476861	6.519852	0.000184
Price	−47.19210825	14.42800182	−3.27087	0.011341
Advertise	9.025346681	1.748464919	5.161869	0.000862
Temp	−0.731174715	0.597069827	−1.22461	0.255558

Interpretation

To interpret the results we pay attention to the following factors:

▪ the size and sign of the estimated regression coefficients (b_1, b_2, b_3)

- the statistical significance of the regression coefficients
- the statistical significance of the regression as a whole
- the overall explanatory power of the regression model
- the presence of unusual observations (i.e. so-called *outliers*)

Size and sign of coefficients

A priori, one would expect sales to be inversely related to price (i.e. sales increase as price falls), for sales to be positively related to advertising expenditure and, in the case of heating fuel, for sales to rise as temperature falls (another inverse relationship). In summary, therefore, *a priori* expectations are that the regression coefficients for price (b_1) and temperature (b_3) should be negative and that for advertising expenditure (b_2) should be positive. It will be seen from the computer output that this is in fact the case. Interpretation of the size of the coefficients, bearing in mind the units of measurement of the original data, is that sales fall by 47.2(000) litres for a unit increase (£1) in price, rise by 9.03(000) litres for a unit increase (£l000) in advertising expenditure and fall by 0.731(000) litres for a unit rise (1°F) in temperature.

Statistical significance of the coefficients

It will be seen that the *t* ratio for price is −3.27 with a probability value *p* of 0.011 indicating that the coefficient b_1 is different from zero at the 1.1 per cent level of significance (thus, if the required level of significance α is 0.05, the coefficient on price is statistically significant). Similarly, the coefficient on advertising expenditure, b_2, is also statistically significant – the *t* ratio of 5.16 is shown to have a probability value of 0.000 to 3 decimal places, indicating that the probability of this result being obtained by chance is very small and thus it may be said to

be highly significant. Finally, it will be seen that the third coefficient – on temperature – b_3, with a *t* ratio of −1.22 and a probability value of 0.256, is not significant at the 5 per cent level.

The non-significance of the temperature coefficient may seem surprising but may perhaps be explained in terms of consumers building up stock in advance of need and thus buying fuel in the summer months when temperatures are higher. The implication of this finding for further modelling of the determinants of fuel sales should involve consideration of a number of options such as lagging the temperature variable, or even dropping it altogether, collecting a longer run of data, including additional variables and, possibly, investigating a non-linear relationship. Further investigation would also need to take account of the pitfalls and limitations noted earlier.

Statistical significance of the regression has a whole

It will be seen that the *F* statistic (155.04) has a probability value of 0.000 indicating that the regression as a whole is very highly significant.

Overall explanatory power

The R^2 value (denoted R–sq in the computer printout) indicates that 98.3 per cent of the variation in sales is explained by the regression as a whole (i.e. by the joint variation in the three independent variables). The adjusted R^2 value equals 97.7 per cent.

Unusual observations

MINITAB highlights those observations which have especially large residuals (observed minus predicted values) for the dependent variable. (Note that the letter R used in MINITAB to denote such observations should not be confused with the coefficient of multiple correlation). It will be seen from the output that the second observation is relatively far from the estimated regression line and hence may be regarded as an outlier. The purpose of identifying such an unusual observation is to direct attention to the possibility that it may be mis-recorded or simply an unusual event (which should perhaps be ignored in further modelling).

Worked example 19.2   Use of multiple regression for prediction

Using the estimated regression model for heating fuel sales given in Worked example 19.1 above, predict the volume of sales for a month when price is £2.5 per litre, advertising expenditure (ADVTG) amounts to £20,000 in the month and the mean daily temperature (TEMP) for the month is 50°F.

Worked solution based on MINITAB

The estimated regression equation shown in the MINITAB printout above is:

$$\text{SALES} = 290.43 - 47.19\,\text{PRICE} + 9.025\,\text{ADVTG} - 0.7312\,\text{TEMP}$$

Substituting the values specified for the three explanatory variables gives:

$$\begin{aligned}
\text{predicted sales} &= 290.43 - 47.19(2.5) + 9.025(20) - 0.732(50) \\
&= 290.43 - 117.975 + 180.500 - 36.560 \\
&= 316.395 \; (000 \text{ litres})
\end{aligned}$$

Computer solution using MINITAB

With the data for the four variables entered in columns C1–C4 respectively of the MINITAB worksheet, as in Worked example 19.1, the procedure is the same as in that example (except that **Control the display of results** may be set at the second level to shorten the output) plus the specification of the required prediction values (as shown below):

Choose **Stat → Regression → Regression**
Select **Response** variable (C1) and **Predictor** variables (C2, C3, C4)

Choose **Results** and check an appropriate button under **Control the display of results**
Click **OK**
Choose **Options** and in the **Prediction intervals for new observations** box enter the location of the values of the independent variables for which the predicted y_c value is required (in this case the values for the three independent variables were entered in columns C5, C6 and C7 respectively).
Click **OK** twice.

The output is as follows:

Regression Analysis

The regression equation is
Sales = 290 − 47.2 Price + 9.03 Advertising − 0.731 Temp

Predictor	Coef	StDev	T	P
Constant	290.43	44.54	6.52	0.000
Price	−47.19	14.43	−3.27	0.011
Advert	9.025	1.748	5.16	0.001
Temp	−0.7312	0.5971	−1.22	0.256

S = 12.95   R-Sq = 98.3%   R-Sq(adj) = 97.7%

Analysis of Variance

Source	DF	SS	MS	F	P
Regression	3	77962	25987	155.04	0.000
Residual Error	8	1341	168		
Total	11	79303			

Predicted values

Authors' comment: required prediction

Fit	StDev Fit	95.0% CI	95.0% PI
316.39	18.33	(274.13, 358.65)	(264.65, 368.14) XX

X   denotes a row with X values away from the centre
XX denotes a row with very extreme X values

Computer solution using Excel

E

To **obtain** predicted y values from the estimated multiple regression equation, it is necessary to use the **TREND** function. The procedure is as follows:

Click = in the formula bar and select **TREND**
In **Known_Ys** enter the range of cells containing the y values
In **Known_Xs** enter the range of cells containing *all* the x values (as when using the multiple regression tool, the x variables must be in contiguous columns)

In **New_Xs** enter the values of the x variables for which predicted y values are required
In **Const** enter **True** if a non-zero intercept is to be used in the equation; otherwise enter **False**
Click **OK**

This returns the value: 316.3932

Authors' comment: Compare this result with the MINITAB output above.

Worked example 19.3   Multiple regression analysis using cross-sectional data

The table below gives the value of investment expenditures y, sales x_1 and a productivity index x_2 for 30 firms in an industry. It is expected that y will be directly related to x_1 and x_2 Regress y on x_1 and x_2 and interpret the results.

Worked example 19.3   Multiple regression analysis using cross-sectional data
(continued)

Firm	Investment expenditure y (£000s)	Sales x_1 (£000s)	Productivity index x_2
1	31.5	150	10.7
2	31.9	165	10.8
3	38.1	250	14.1
4	47.9	350	18.3
5	40.2	260	14.9
6	35.1	200	11.5
7	31.2	140	10.5
8	32.8	180	10.9
9	30.9	135	10.3
10	43.2	300	16.4
11	32.3	170	10.9
12	40.8	275	15.5
13	30.2	100	9.9
14	45.2	320	17.3
15	44.9	315	17.1
16	30.5	110	10.0
17	36.3	215	12.5
18	39.1	250	14.6
19	30.7	120	10.1
20	31.2	145	10.6
21	31.5	150	10.8
22	42.1	280	15.6
23	45.9	330	17.5
24	49.3	355	18.7
25	37.8	245	13.9
26	33.4	195	11.3
27	37.4	230	13.1
28	33.4	190	11.0
29	43.2	310	16.7
30	42.9	285	15.9

Computer solution using MINITAB

The commands required are the same as in Worked example 19.1

The output is as follows:

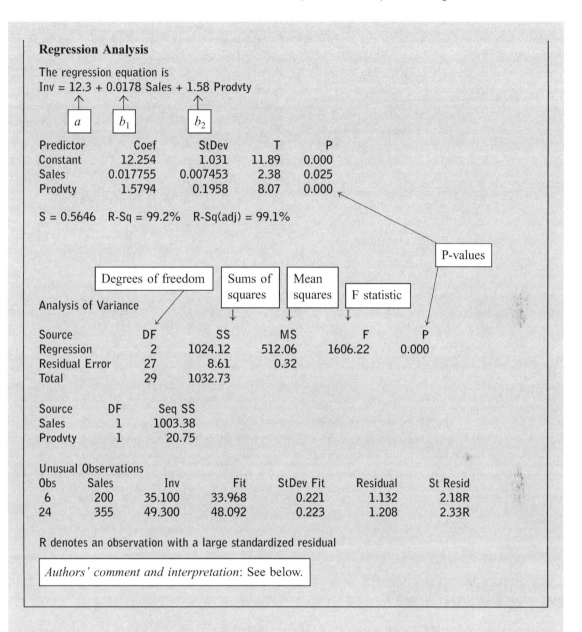

Regression Analysis

The regression equation is
Inv = 12.3 + 0.0178 Sales + 1.58 Prodvty

a b_1 b_2

Predictor	Coef	StDev	T	P
Constant	12.254	1.031	11.89	0.000
Sales	0.017755	0.007453	2.38	0.025
Prodvty	1.5794	0.1958	8.07	0.000

S = 0.5646   R-Sq = 99.2%   R-Sq(adj) = 99.1%

P-values

Degrees of freedom   Sums of squares   Mean squares   F statistic

Analysis of Variance

Source	DF	SS	MS	F	P
Regression	2	1024.12	512.06	1606.22	0.000
Residual Error	27	8.61	0.32		
Total	29	1032.73			

Source	DF	Seq SS
Sales	1	1003.38
Prodvty	1	20.75

Unusual Observations

Obs	Sales	Inv	Fit	StDev Fit	Residual	St Resid
6	200	35.100	33.968	0.221	1.132	2.18R
24	355	49.300	48.092	0.223	1.208	2.33R

R denotes an observation with a large standardized residual

Authors' comment and interpretation: See below.

Computer solution using Excel

E

The procedure is the same as shown in Worked example 19.1 and interpretation is the same as for MINITAB given below.

Worked example 19.3   Multiple regression analysis using cross-sectional data
(continued)

Interpretation

The interpretation given here follows the same format as that set out in Worked example 19.1. A full account of the computer printout and its interpretation is given there and thus, for brevity, we focus only on the key points of the results here and do not repeat the discussion given in Worked example 19.1.

Size, sign and significance of coefficients

The *a priori* expectation is a positive sign for b_1 (the coefficient on sales) and this is confirmed by the results showing that investment expenditure increases by roughly 0.0178 (£000s) for a unit increase (£1000) in sales. The coefficient on sales is significantly different from zero (probability value $P = 0.025$). With regard to the coefficient on the productivity index variable, expectations are less clear-cut: on the one hand, one could anticipate a negative sign if higher productivity led to a need for less capital investment, on the other hand, a positive sign might be expected if higher productivity allows price reductions which boost demand and hence, in turn, increase the need for more capital investment

to increase supply. The results show a positive relationship between productivity and investment expenditure, with a coefficient of roughly 1.59, indicating that investment increases by 1.59 (£000s) for each unit increase in the productivity index. The coefficient on the productivity variable is shown to be highly significant (probability value $P = 0.000$).

Significance and explanatory power of the regression as a whole

The F statistic (1606.22) with a probability value P of 0.000 indicates that the regression as a whole is highly significant. The R^2 value shows that 99.2 per cent of the variation in investment expenditure is explained by the regression as a whole.

Unusual observations

Two extreme observations – numbered 6 and 24 – are indicated on the MINITAB printout. As before, the possibility that these are errors or unusual events which should perhaps be ignored in modelling should be investigated.

Key terms and concepts

Adjusted multiple coefficient of determination ($\overline{R}^2$) Allows comparison of the overall explanatory power of regression models which have the same dependent variable but a different number of explanatory variables. The value of R^2 is adjusted as follows:

$$\overline{R}^2 = 1 - (1 - R^2)\frac{n - 1}{n - k - 1}$$

Coefficient of multiple correlation (R) A statistical measure of the degree of association between the dependent variable and all the explanatory variables jointly.

$$R = \sqrt{R^2}$$

Errors in variables Situation in which variables in a regression model include measurement errors.

F test In the context of multiple regression, a test of the statistical significance of the regression model as a whole, i.e. its overall explanatory power, given by R^2. It is based on the F statistic which follows the F distribution.

Functional form Form of the relationship between variables.

Heteroscedasticity Term used to describe the situation where the error terms (e_i) do not have a constant variance.

Homoscedasticity Term used to describe the situation where the error terms (e_i) have a constant variance.

Multicollinearity Situation in which some of the independent variables are correlated with one another.

Multiple coefficient of determination (R^2) Provides a statistical measure of the overall goodness-of-fit for the regression model as a whole:

$$R^2 = \frac{\text{Explained variation in y}}{\text{Total variation in y}} = \frac{\sum(y_c - \bar{y})^2}{\sum(y_i - \bar{y})^2}$$

Multiple correlation analysis Refers to the measuring and testing of the overall explanatory power of a multiple regression model.

Multiple regression model A regression equation in which more than one independent variable is used to explain the dependent variable. The general form of the equation is:

$$y_c = a + b_1 x_1 + b_2 x_2 + b_3 x_3 + \cdots + b_k x_k$$

Partial regression coefficients The b coefficients in a multiple regression model.

Statistical significance of the regression coefficient Use of a t test to test whether or not regression coefficients are significantly different from, greater than, or less than zero.

Test statistic (t) In the context of regression:

$$t = \frac{b_i - B_i}{s_{b_i}} = \frac{b_i}{s_{b_i}} \text{ under the null hypothesis } H_0: B_i = 0$$

Self-study exercises

19.1. The following MINITAB output shows the results of running a multiple regression on a set of data relating to the prices at which houses were sold (variable called PRICE, measured in £000s) and four variables relating to the characteristics of the houses themselves as follows:

   (i) number of bedrooms, called BEDS

   (ii) age of the house in years, called AGE

   (iii) number of bathrooms, called BATHS

   (iv) number of garages or car spaces, called CARS

The regression equation is
PRICE = 19.5 + 7.47 BEDS – 0.308 AGE + 13.4 BATHS + 2.61 CARS

43 cases used. 8 cases contain missing values

Predictor	Coef	StDev	T	P
Constant	19.46	11.25	1.73	0.092
BEDS	7.470	3.541	2.11	0.042
AGE	–0.30759	0.09794	–3.14	0.003
BATHS	13.439	5.956	2.26	0.030
CARS	2.609	2.915	0.89	0.376

S = 14.19   R-Sq = 55.8%   R-Sq(adj) = 51.1%

Analysis of Variance

SOURCE	DF	SS	MS	F	P
Regression	4	9648.0	2412.0	11.98	0.000
Error	38	7651.3	201.4		
Total	42	17299.4			

SOURCE	DF	Seq SS
BEDS	1	2215.6
AGE	1	5710.3
BATHS	1	1560.8
CARS	1	161.3

Unusual Observations

Obs.	BEDS	PRICE	FIT	StDev Fit	Residual	St Resid
43	4.00	125.00	73.45	6.05	51.55	4.02R
46	6.00	95.00	103.10	9.73	–8.10	–0.78X
50	3.00	16.00	47.62	4.48	–31.62	–2.35R

R denotes an obs. with a large st. resid.
X denotes an obs. whose X value gives it large influence.

(a) What was the total sample size and how many observations were used in the regression?

(b) What is the independent variable?

(c) How many independent variables were used in the regression?

(d) Interpret the coefficient on AGE.

(e) Test whether or not there is a significant relationship between the price of a house and the group of independent variables at the 5 per cent level of significance. Be sure to state the null and alternative hypotheses.

(f) Test the significance of the BEDS variable at the 5 per cent level. Be sure to state the null and alternative hypotheses.

(g) Interpret the value of R^2.

(h) Estimate the price of a house that has five bedrooms, is 30 years old, has two bathrooms and has a double garage.

19.2. It is argued that the amount of direct foreign investment (INV) in a country is positively related to the size of its gross national product (GNP) and its trade balance (BAL) but inversely related to its level of political risk (POL).

Data were collected for 24 countries. INV is measured in £million, GNP in £billion, and BAL in £million and POL is an index which runs from 0 to 10 where the higher the number the greater the level of political risk.

The following abbreviated MINITAB results were obtained:

```
The regression equation is
INV = 290 + 0.860 GNP + 0.174 BAL – 33.0 POL

Predictor        Coef         StDev
Constant        289.68        25.04
GNP             0.85952       0.05725
BAL             0.1743        0.1000
POL            –33.001        3.978

S = 46.93   R-Sq = 96.6%   R-Sq(adj) = 96.1%

Analysis of Variance

Source          DF           SS
Regression       3        1251632
Error           20          44052
Total           23        1295684

Source          DF         Seq SS
GNP              1        1054149
BAL              1          45890
POL              1         151593
```

(a) Do the parameter estimates take the expected signs?

(b) Test the null hypothesis that the explanatory variables, as a set, do not significantly affect direct foreign investment at the 1 per cent level of significance.

(c) Taken individually, do the variables have a significant influence on direct foreign investment at the 1 per cent level of significance?

(d) What is the expected effect on direct foreign investment of a worsening of the balance of payments by £10 million?

(e) Give the point prediction for direct foreign investment for a country that has a GNP of £200 billion, a balance of payment current account *deficit* of £100 million and a political risk factor of 5.

(f) Explain why predictions should be made within confidence limits and discuss the factors that would affect the size of these limits.

19.3. The production manager of a factory in charge of an assembly line is interested in the amount of time in minutes that it takes workers to perform a certain task relative to their score on an aptitude test and their age. Twelve of the workers were selected and a multiple regression was run on the data recorded for this set of workers using MINITAB. Part of the MINITAB output is shown below – some of the results have been deleted for the purposes of this question.

```
The regression equation is
TIME =

Predictor       Coef      StDev        T           P
Constant       84.20      15.85                   0.000
SCORE         -0.8872     0.1581                  0.000
AGE            0.2686     0.1491                  0.105

S = 3.553   R-Sq =            R-Sq (adj) =

Analysis of Variance
Source          DF         SS        MS        F        P
Regression                3085.3    1542.6
Error                      113.6     12.6
Total                     3198.9

Source     DF     Seq SS
APT         1     3044.3
AGE         1       41.0
```

(a) Write out the estimated equation.

(b) Interpret all coefficients.

(c) Compute the appropriate actual t values.

(d) Test for the significance of b_1 and b_2 (i.e. the regression coefficients) at the 1 per cent level of significance.

(e) Is the regression as a whole significant?

(f) What are the degrees of freedom for the sum of squares explained by the regression (SSR) and the sum of squares due to error (SSE)?

(g) Compute R^2 and adjusted R^2.

(h) Interpret R^2.

(i) Estimate the time taken by a worker who is 50 years of age and has a score of 62 on the aptitude test.

(j) Should any of the variables be dropped from the model? If so which? Explain.

Chapter 20

Tests of goodness-of-fit and independence

In Chapters 18 and 19, which are concerned with simple and multiple regression and correlation, we considered situations involving relationships in which changes in one or more variables may be related to changes in another (dependent) variable. In this chapter we now turn our attention to two additional types of situation. One is measuring how well an *observed* set of data fits an *expected* set – e.g. whether an observed set of data follows a specified probability distribution (such as the normal distribution). This situation gives rise to a so-called *goodness-of-fit test*. The other situation is one in which we may wish to test whether or not a variable is influenced by (or independent of) one or more qualitative variables. For example, whether or not political allegiance may be related to factors such as social class, income status, educational attainment etc. This gives rise to so-called *tests of statistical independence*.

Both of these situations can be analysed using a test known as the *chi-square* (χ^2) *test*. Before turning to its application, it may be noted that it is especially appropriate to situations involving *categorical* variables. It will be appreciated that *measured* variables (such as income, years of education, etc.) may themselves be categorized (e.g. into groups such as high, medium or low income, below average and above average years of education, etc.). In these cases, the χ^2 test can be applied but it is generally more appropriate to use techniques such as regression or t and Z tests of differences between means in order to exploit the *numerical* nature of the data.

We now deal with the two applications of the χ^2 test – goodness-of-fit tests and tests of independence – in turn.

The following diagram provides an overview of the whole of this chapter.

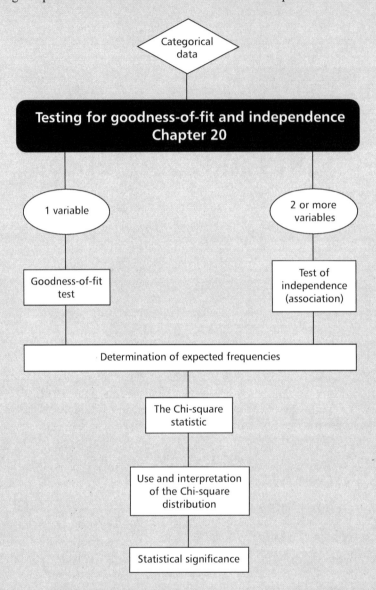

| Learning Outcomes | **After working through this chapter you will be able to:** |

- understand what is meant by goodness-of-fit and statistical independence in the context of association between variables
- appreciate the distinction between the observed and expected frequencies
- test the goodness-of-fit between sample data and an hypothesized population distribution using the chi-square (χ^2) statistic
- appreciate the conditions necessary for applying the chi-square test
- use tables of the chi-square statistic
- understand the relevance of Yates' continuity correction factor
- derive contingency tables for conducting tests of independence
- conduct tests of hypotheses concerning the independence of variables.

Principles

Goodness-of-fit tests

A car manufacturer wishes to test whether or not there is any preference amongst buyers for cars of different colours. The manufacturer intends to limit his range of colours to only three – red, blue and black – and needs to ascertain if there is likely to be a strong preference for any particular colour, otherwise the cars will be manufactured in equal numbers of each colour. A market research survey of 90 potential buyers is conducted. Prior to the survey one would expect equal numbers to favour each of three colours (i.e. 30 people) if no colour is particularly favoured. The survey reveals in fact that 45 favour red, 30 favour blue and only 15 favour black coloured cars. Do the survey results indicate significant differences in preferences or are they due to sampling error?

On the basis of the survey results we can set up a table of 'observed' frequencies (denoted O_i) and 'expected' frequencies (denoted E_i) as shown in Table 20.1. Note that the data must relate to *absolute* frequencies and not simply *relative* frequencies (e.g. percentages). Naturally, if there were no difference between the observed and expected frequencies in each category, then this would be good evidence that there is no preference. If there is a difference (as here), the question to be addressed is whether the difference has arisen merely by chance or whether it is too big to have arisen by chance alone. The null

Table 20.1 Observed and expected frequencies

Preference	Observed frequency O_i	Expected frequency E_i
Red	45	30
Blue	30	30
Black	15	30

hypothesis is that there is no particular preference – i.e. all three colours are equally liked. This hypothesis may be tested using the χ^2 *statistic* which is defined in terms of the differences between the observed and expected frequencies as follows:

χ^2 statistic

$$\chi^2 = \sum \frac{(O_i - E_i)^2}{E_i}$$

where the summation is conducted over all pairs of observed and expected frequencies $(O_i - E_i)$ relative to the expected frequency E_i for each pair.

Thus, for the example above, the χ^2 statistic is calculated as follows:

$$\chi^2 \text{ statistic} = \frac{(45 - 30)^2}{30} + \frac{(30 - 30)^2}{30} + \frac{(15 - 30)^2}{30}$$

$$= \frac{225}{30} + \frac{0}{30} + \frac{225}{20}$$

$$= 7.5 + 0 + 7.5$$

$$= 15.0$$

The significance of the χ^2 statistic is tested using a special distribution known as the χ^2 *distribution*. There is in fact a whole family of such distributions, each of which depends on the degrees of freedom available. The degrees of freedom are $k - 1$ where k is the number of categories into which each variable under consideration is sub-divided. In the example above, involving *one* variable (preferences) and *three* categories (red, blue and black), the degrees of freedom are $k - 1 = 3 - 1 = 2$. Figure 20.1 shows a 'family' of distributions for different degrees of freedom. As with the normal and t distributions,

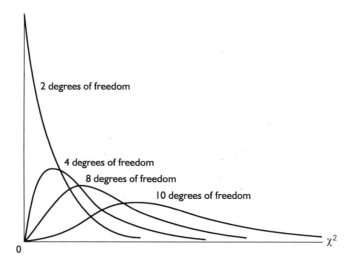

2 degrees of freedom

4 degrees of freedom

8 degrees of freedom

10 degrees of freedom

Figure 20.1 χ^2 distribution for different degrees of freedom

probabilities are measured by areas under the χ^2 curve and a χ^2 test is conducted by comparing the actual (calculated) values of the χ^2 statistic with some critical χ^2 value given in χ^2 tables (see Appendix A Table A7). It will be seen from Figure 20.1 that χ^2 distributions are always skewed to the right. Further, all have a minimum value of zero and hence all hypothesis tests are, by definition, one-tailed tests.

Having explained the χ^2 distribution and tables, we can now return to complete the example of testing car colour preferences above. If we test at the 5 per cent level of significance ($\alpha = 0.05$), the critical value of χ^2, with two degrees of freedom, equals 5.991 from Appendix A Table A7. The actual (computed) χ^2 statistic (equal to 15.0) is greater than the critical value (5.991) – this is illustrated in Figure 20.2. The conclusion is that we must reject the null hypothesis (i.e. the hypothesis of no difference between actual and expected frequencies) and conclude, therefore, that the difference in expressed preferences is statistically significant at the 5 per cent level of significance. So the manufacturer may conclude that red cars are the most preferred and black cars are the least preferred.

Conditions for applying the χ^2 test

In χ^2 tests, no value of E_i should be less than 5. If this should arise in a particular problem, then it is necessary to combine the expected frequencies in adjacent cells as appropriate – inevitably reducing the number of cells – until the combined expected frequency is at least 5. Before applying χ^2 tests, therefore, it is important to see that this condition is fulfilled. This adjustment is illustrated in Worked Example 20.2, which deals with the application of χ^2 in a goodness-of-fit test involving a hypothesized distribution. Further, when there is only one degree of freedom it is necessary to make an adjustment known as *Yates' continuity correction factor* – shown below. This is analogous to the continuity correction applied in the case of the normal approximation to the binomial distribution (see Chapter 9). It arises here because the χ^2 tables are calculated from χ^2 probability distributions which are continuous, whereas the approximations (χ^2 statistics) are discrete. This creates

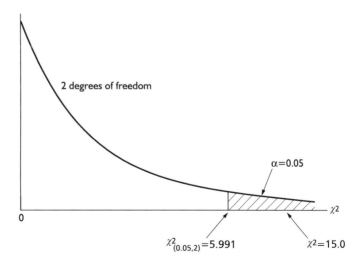

Figure 20.2 χ^2 distribution for car colour preferences

a tendency to underestimate the probability. Yates has shown that the χ^2 approximation is improved by subtracting 0.5 from the *absolute* difference between O_i and E_i. Thus, the χ^2 statistic is then given by the following expression:

Yates' continuity correction factor

$$\chi^2 = \sum \frac{(|O_i - E_i| - 0.5)^2}{E_i}$$

Note, however, that it is only necessary to introduce the correction factor when the degree of freedom is 1. Further, when the total number of observations is large, say $n > 50$, the correction factor has little effect on the outcome and can be ignored.

Tests of goodness-of-fit to hypothesized distributions

Goodness-of-fit tests using the χ^2 statistic may also be used to test whether or not a set of observed frequencies is consistent with a given probability distribution, such as the binomial, Poisson or normal distributions. It is very common to test whether a distribution is normal or not and a worked example of the application in this context is shown in Worked example 20.2 below.

Tests of independence

In goodness-of-fit tests we are concerned with a *one-dimensional* classification of a set of data. Thus, in the example of car colour preferences given above the people surveyed were classified simply on the basis of their preference for different colours, where *preference* is the only classificatory variable. The analysis may now be extended to classifications involving more than one dimension. For example, the data on car colour preferences may be further classified according to other variables such as sex of respondent, income level, age group, etc. The question of interest now is whether or not the distribution of responses (colour preferences in this case) is independent of the classificatory variables (sex, income, age, etc.). In other words, is car colour preference independent of sex, income, age, etc? To address this question, we first need to cross-classify the observations or responses to the classificatory variables of interest into a *contingency table*.

Contingency tables

A contingency table shows a cross-tabulation of the data according to classificatory variables. Thus, for example, if colour preferences are given under three catagories (red, blue and black) and further classified by sex (male, female) then we have a 3×2 contingency table. In general, an $r \times k$ contingency table corresponds to a breakdown of one variable into r rows and the other into k columns, giving rise to a total of $r \times k$ 'cells'. The number of degrees of freedom in a contingency table is equal to $(r - 1)(k - 1)$. Thus, for a 3×2 contingency table the number of degrees of freedom is $(3 - 1)(2 - 1) = 2$.

Table 20.2 Contingency table for car colour preferences

Car colour preference	Male O_i	Male E_i	Female O_i	Female E_i	Total
Red	30	$\frac{45}{90} \times 50 = 25$	15	$\frac{45}{90} \times 40 = 20$	45
Blue	15	$\frac{30}{90} \times 50 = 16.7$	15	$\frac{30}{90} \times 40 = 13.3$	30
Black	5	$\frac{15}{90} \times 50 = 8.3$	10	$\frac{15}{90} \times 40 = 6.7$	15
Total	50	50	40	40	90

Taking our earlier example of colour preferences we now extend the analysis to a *test of independence* by considering whether or not preferences are independent of sex. The test is also referred as a *test of association* (i.e. non-independence). Table 20.2 shows the cross-tabulation of observed frequencies and the expected frequencies alongside showing, for purposes of exposition, how they are derived. An explanation of their derivation follows.

The rationale for the derivation of expected frequencies is that if colour preferences are independent of sex then the distribution of preferences by males and by females would be no different from the distribution for both sexes combined (shown in the 'total' (right-hand) column of Table 20.2). Thus, to take the calculation of the expected frequency in the first cell as an example (i.e. male preferences for red cars), given that there are 50 males in the sample and 45 revealed preferences for red cars out of a total of 90 responses, the expected frequency is $(45/90) \times 50 = 25$. The other cell values are naturally calculated in a similar manner. Thus, in general, the expected frequency (E_i) for each cell in a contingency table is found as follows:

Calculation of expected cell frequencies

$$E_i = \frac{\left(\begin{array}{c}\text{total frequency of row}\\\text{in which the cell lies}\end{array}\right) \times \left(\begin{array}{c}\text{total frequency of column}\\\text{in which the cell lies}\end{array}\right)}{\text{total frequency of all cells}}$$

Hypotheses and conduct of the χ^2 test

The hypotheses to be tested in χ^2 tests of independence are

H_0: the classifications are independent

H_1: the classifications are dependent

Thus, in the example above, acceptance of H_0 would indicate that car colour preference is not related to sex. The test statistic, χ^2, is calculated as before:

$$\chi^2 = \sum \frac{(O_i - E_i)^2}{E_i}$$

where the summation is over all cells of the contingency table.

Having calculated χ^2, the test procedure is then carried out, as before, by comparing the calculated value with the critical value given in the χ^2 tables (Appendix A Table A7). H_0 is, of course, rejected if $\chi^2 > \chi^2_{0.05}$ with $(r-1)(k-1)$ degrees of freedom (df).

To illustrate the test we use the worksheet given in Table 20.2. Thus:

$$\chi^2 = \frac{(30-25)^2}{25} + \frac{(15-16.7)^2}{16.7} + \frac{(5-8.3)^2}{8.3}$$
$$+ \frac{(15-20)^2}{20} + \frac{(15-13.3)^2}{13.3} + \frac{(10-6.7)^2}{6.7}$$
$$= 1 + 0.17 + 1.31 + 1.25 + 0.22 + 1.63$$
$$= 5.58$$

$\chi^2_{0.05} = 5.991$ with $(3\text{-}1)(2\text{-}1) = 2$ degrees of freedom.

$\chi^2 < \chi^2_{0.05}$; i.e. actual χ^2 is less than critical χ^2 and therefore H_0 cannot be rejected, and we conclude that car colour preference is independent of sex at the 5 per cent level of significance. It may be noted, however, that H_0 in the example above is only *just* accepted at the 5 per cent level. At the 10 per cent level, $\chi^2_{0.10} = 4.605$ and thus H_0 would be rejected at this level.

In the Applications and Worked Examples section below we show the application of the χ^2 test to illustrate all of the principles explained in this chapter. These cover the following situations:

▪ goodness-of-fit test including Yates' continuity correction factor

▪ goodness-of-fit test for a hypothesized normal distribution

▪ test of independence for a 3×3 contingency table (the solution here is shown manually as well as using MINITAB and Excel).

It should be noted that there is a no *general* routine in MINITAB or Excel to carry out goodness-of-fit tests, although the programs could be used, of course, to carry out the arithmetic calculations required. However, it may be noted that MINITAB does include a routine for making 'normality tests' which is not based directly on the χ^2 procedure explained here. This takes us beyond the scope of this book and is not considered further.

Applications and Worked Examples

Worked example 20.1 Goodness-of-fit test using Yates' continuity correction factor

General descriptions of the MINITAB and Excel computer programs and their application are given in Appendices B and C respectively.

A company expects 20 per cent of the door-to-door calls made by a salesman to result in actual purchases. During a trial period, a newly-appointed salesman makes 40 door-to-door visits and only achieves four sales. Test whether or not the performance of the new salesman in achieving sales differs significantly from the general

Worked example 20.1 Goodness-of-fit test using Yates' continuity correction factor (continued)

performance expected by the company. (Note, since there is only one degree of freedom, i.e. $k - 1 = 1$, Yates' continuity correction factor should be applied here.)

Solution

The observed (O_i) and expected (E_i) frequencies are shown below. Note that the expected number of sales is simply given by calculating 20 per cent of 40 ($= 8$) and thus the expected number of 'no sales' equals 32.

	Sale	No sale	Total
O_i	4	36	40
E_i	8	32	40

The hypotheses to be tested are:

H_0: the performance of the new salesman does not differ from that expected by the company

H_1: the performance is significantly different from that expected.

Since there is only one degree of freedom in this case, we must introduce Yates' continuity correction factor in calculating the χ^2 statistic.
Thus the χ^2 statistic is

$$\chi^2 = \sum \frac{(|O_i - E_i| - 0.5)^2}{E_i}$$
$$= \frac{(4 - 0.5)^2}{8} + \frac{(4 - 0.5)^2}{32}$$
$$= \frac{(3.5)^2}{8} + \frac{(3.5)^2}{32}$$
$$= 1.91$$

Critical χ^2: $\chi^2_{0.05} = 3.841$.

Since $1.91 < 3.841$, we cannot reject the null hypothesis at the 5 per cent level of significance i.e. the performance of the new salesman does not differ from that expected by the company.

Worked example 20.2 Goodness-of-fit test for an hypothesized normal distribution

Examination marks for 100 students are distributed around a mean of 50 with a standard deviation of 10. The frequency distribution of the marks – giving observed frequencies O_i – is given in the table below. Test whether or not the distribution of marks differs significantly from a normal distribution.

Mark classes	Number of students (observed frequencies, O_i)
30 – <35	2
35 – <40	6
40 – <45	12
45 – <50	14
50 – <55	29
55 – <60	16
60 – <65	12
65 – <70	6
70 and over	3
Total	100

Solution

We set out the solution in full below, explaining in detail each of the required steps.

Expected frequencies

First, it is necessary to compute the expected frequencies E_i. For this purpose it is necessary to calculate the area under the normal curve, corresponding to each mark class interval to obtain probabilities for each class, using the tables of the normal distribution (see Appendix A Table A3). The expected frequencies are then obtained by multiplying these probabilities by the total sample size (i.e. 100). The calculations are set out in the table below. It should be noted that as the total expected frequency must equal the actual total frequency ($= 100$) and the normal distribution is asymptotic (see Chapter 9), it is necessary to allow for those parts of the theoretical distribution that lie in the lower and upper tails – in this case, frequencies corresponding to mark classes of *less than 35* and *70 and over*. Note also that for the expected frequencies given in the final column of the table below, the last two cells have been combined to give an expected frequency of 6.68 as each of the individual mark classes had an expected frequency of less than 5.

Mark classes	Upper class limit (x)	$Z = (x - \bar{x})/s$ $= (x - 50)/10$	Area to the left of x (from Appendix A Table A3)	Area in class interval	Expected frequency = area × 100 (E_i)
Less than 35	35	−1.5	0.0668	0.0668	6.68
35 − <40	40	−1.0	0.1587	0.0919	9.19
40 − <45	45	−0.5	0.3085	0.1498	14.98
45 − <50	50	0.0	0.5000	0.1915	19.15
50 − <55	55	0.5	0.6915	0.1915	19.15
55 − <60	60	1.0	0.8413	0.1498	14.98
60 − <65	65	1.5	0.9332	0.0919	9.19
65 − <70	70	2.0	0.9772	0.0440	4.40 ⎤
70 and over			1.0000	0.0228	2.28 ⎦ 6.68
Total				1.0000	100.00

χ^2 statistic

Having obtained expected frequencies E_i, the next step is to calculate the χ^2 statistic. The calculations are set out in the following table, showing the actual value of $\chi^2 = 13.17$.

Mark classes	O_i	E_i	$O_i - E_i$	$(O_i - E_i)^2$	$(O_i - E_i)^2/E_i$
Less than 35	2	6.68	−4.68	21.90	3.28
35 − <40	6	9.19	−3.19	10.18	1.11
40 − <45	12	14.98	−2.98	8.88	0.59
45 − <50	14	19.15	−5.15	26.52	1.38
50 − <55	29	19.15	9.85	97.02	5.07
55 − <60	16	14.98	1.02	1.04	0.07
60 − <65	12	9.19	2.81	7.90	0.86
65 and over	9	6.68	2.32	5.38	0.81
Total	100	100			$\chi^2 = 13.17$

Hypotheses

H_0: the frequency distribution of marks follows a normal distribution

H_1: the frequency distribution of marks does not follow a normal distribution

Worked example 20.2   Goodness-of-fit test for an hyphesized normal distribution (continued)

Degrees of Freedom

The degrees of freedom (df) are given by

df = number of classes − 1
      − number of estimated parameters

The number of classes is that used *after* any merging of cells as shown above – thus there are eight classes here whereas there were nine to begin with. The number of estimated parameters is two, namely the mean and standard deviation. Hence,

df = 8 − 1 − 2 = 5

In cases where the mean and standard deviation (the population parameters) are known, as opposed to being estimated from the sample data, then the degrees of freedom are equal to (number of classes −1).

Critical value of χ^2

From the χ^2 tables, we have $\chi^2_{0.05} = 11.07$ with five degrees of freedom at the 5 per cent level of significance.

Conclusion

Since the actual χ^2 statistic ($= 13.17$) is greater than the critical χ^2 value ($= 11.07$), we cannot accept the null hypothesis at the 5 per cent level of significance; i.e. the frequency distribution of marks does *not* follow a normal distribution. However, it should be noted that the null hypothesis cannot be rejected a the 1 per cent level of significance ($\chi^2_{0.01} = 15.086$). These results are illustrated in the figure below.

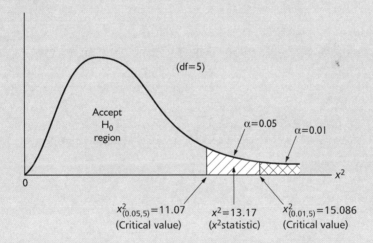

Worked example 20.3   Test of independence using a 3 × 3 contingency table

A survey of 1000 houses spread across three towns, A, B and C, has classified each property on the basis of whether or not it is *below, above* or of *average* habitable standard. The table

below summarizes the results of the survey. Does the survey provide evidence that housing standards in the three towns are significantly different?

| | | Habitable standard | | |
Town	Below	Average	Above	Total
A	15	120	330	465
B	30	140	180	350
C	20	110	55	185
Total	65	370	565	1000

Worked solution

Expected frequencies

The calculations of expected frequencies are given below. For each cell the expected frequency is obtained by multiplying the corresponding row total by the respective column total and then dividing by the total frequency.

| | | Habitable standard | | |
Town	Below	Average	Above	Total
A	30.23	172.05	262.72	465
B	22.75	129.50	197.75	350
C	12.03	68.45	104.53	185
Total	65	370	565	1000

χ^2 statistic

The value of the χ^2 statistic is calculated as follows:

$$\chi^2 = \sum \frac{(O_i - E_i)^2}{E_i}$$

$$= \frac{(15 - 30.23)^2}{30.23} + \frac{(120 - 172.05)^2}{172.05} + \frac{(330 - 262.72)^2}{262.72}$$

$$+ \frac{(30 - 22.75)^2}{22.75} + \frac{(140 - 129.50)^2}{129.50} + \frac{(180 - 197.75)^2}{197.75}$$

$$+ \frac{(20 - 12.03)^2}{12.03} + \frac{(110 - 68.45)^2}{68.45} + \frac{(55 - 104.53)^2}{104.53}$$

$$= 7.668 + 15.747 + 17.227 + 2.310 + 0.851 + 1.593$$

$$+ 5.289 + 25.221 + 23.465$$

$$= 99.373$$

Hypotheses

H_0: the habitable standard of properties is *not* statistically different in the three towns

H_1: the habitable standard of properties *is* significantly different in the three towns

Worked example 20.3 Test of independence using a 3 × 3 contingency table *(continued)*

Degrees of freedom

$$df = (r - 1) \times (k - 1)$$
$$= (3 - 1) \times (3 - 1)$$
$$= 4$$

Critical value of χ^2

From the χ^2 tables, we have $\chi^2_{0.05} = 9.488$ with 4 degrees of freedom at the 5 per cent level of significance.

Conclusion

Since $\chi^2 > \chi^2_{0.05}$, i.e. $99.374 > 9.488$, we must reject H_0 and conclude that the habitable standard of properties in the three towns is significantly different.

Computer solution using MINITAB

With the three columns of data for 'Below average' 'Average', and 'Above average' entered in C1, C2 and C3 of the MINITAB worksheet (*Note*: column and row totals should not be entered) the procedure is as follows:

Choose **Stats → Tables → Chi-Square Test**
Specify the relevant columns (C1, C2 and C3 in this case) in the **Columns containing the table** dialogue box
Click **OK**

The output is as follows:

Chi-Square Test

Contingency table (3 × 3) showing observed and expected frequencies

Expected counts are printed below observed counts

	Below	Average	Above	Total
1	15	120	330	465
	30.23	172.05	262.72	
2	30	140	180	350
	22.75	129.50	197.75	
3	20	110	55	185
	12.03	68.45	104.53	
Total	65	370	565	1000

Chi-Sq = 7.669 + 15.747 + 17.227 +
2.310 + 0.851 + 1.593 +
5.289 + 25.221 + 23.465 = 99.373 ← χ^2 statistic
DF = 4, P-Value = 0.000

Degrees of freedom   Probability

Authors' comment and interpretation: The test conclusion may be derived in two ways. The most direct way is simply to consider the probability of the result – shown by the P-value above – rather than comparing actual and critical χ^2 values: as the probability shown (less than 0.1%) is well below the specified α level of the test (5%), H_0 must be rejected and we conclude that the habitable standard of properties in the three towns is significantly different at the 5% level (and indeed at less than the 0.1% level).

The alternative, conventional way, is to compare the actual and critical values of χ^2, as in the manual example above, and this requires the further step of determining the critical value of χ^2. This can be done either by looking up the tables of χ^2 or by using MINITAB again to compute the critical value using the *inverse cdf function* (inverse cumulative probability). This procedure is shown separately below.

Choose **Calc → Probability distributions–Chi–Square**
Check the **Inverse cumulative probability** button
Specify the **degrees of freedom** (4 in this case)
Check the **Input constant** button and enter the number or constant you wish to evaluate (in this case, as $\alpha = 0.05$, we enter 0.95 – this corresponds to a CDF value of $100 - 0.05$)
Click **OK**

The output is as follows:

Inverse Cumulative Distribution Function

Chi-Square with 4 DF

P(X <= x)   x ← Critical value of χ^2 with 4 degrees of freedom and $\alpha = 0.05$.
0.9500   9.4877

Authors' comment and interpretation: As the actual value of χ^2 (99.373) > the critical value (9.4877), we reject the null hypothesis and conclude, as before, that the habitable standard of properties in the three towns is significantly different at the 5% level.

Computer solution using Excel

Use the function **CHITEST**
With actual and expected frequencies entered in the worksheet as shown below:

Click on = in the formula bar and select **CHITEST**
In **Input_Range** enter the range of cells containing the actual observations (B3:D5 in this case)

In **Expected_Range** enter the expected frequencies (B8:D10 in this case) – for details of the calculation of the expected frequencies see the manual solution to this Worked example above.

Click **OK**
This returns: 1.32902E–20 which is the probability of the χ^2 statistic.

Worked example 20.3   Test of independence using a 3 × 3 contingency table *(continued)*

Worksheet

A	B	C	D
		Actual frequencies	
	Below	Average	Above
A	15	120	330
B	30	140	180
C	20	110	55
		Expected frequencies	
A	30.23	172.05	262.725
B	22.75	129.5	197.75
C	12.02	68.45	104.525

Authors' comment: As the probability (1.32902E–20) < 0.05, the null hypothesis is rejected and thus the conclusion is that the habitable standard of properties in the three towns (A,B,C) is significantly different.

Key terms and concepts

Chi-squared (χ^2) distribution A derived family of distributions, which depend on the number of degrees of freedom available, used in tests of goodness-of-fit and independence as well as other applications.

Chi-squared (χ^2) statistic Statistic based on the differences between the observed and expected frequencies in tests of goodness-of-fit and independence and defined, in this context, as:

$$\chi^2 = \sum \frac{(O_i - E_i)^2}{E_i}$$

Contingency table A table used to summarize observed and expected frequencies for a test of independence of population characteristics.

Expected frequencies Frequencies expected, in tests of goodness-of-fit and independence, under the null hypothesis that the observed frequencies follow a certain pattern or theoretical distribution.

Goodness-of-fit test A statistical test conducted to determine whether to accept or reject a hypothesized probability distribution for a population.

Observed frequencies The actual frequencies observed in a set of data used in a test of goodness-of-fit or independence.

Test of independence A statistical test of the independence of classificatory variables.

Yates' continuity correction factor Used to adjust the value of the χ^2 statistic when there is only one degree of freedom present; thus, the expression for the χ^2 statistic is:

$$\chi^2 = \sum \frac{(|O_i - E_i| - 0.5)^2}{E_i}$$

Self-study exercises

20.1. A café recorded information about the total numbers of drinks purchased between 10.00 a.m. and 11.00 a.m. on each of the six days when the shop was open last week, as shown below. Is drink purchase independent of day of the week?

Monday	Tuesday	Wednesday	Thursday	Friday	Saturday
21	30	36	24	39	30

20.2. The births recorded in a town during one month were 63 males and 57 females. Assuming that an equal number of male and female births would be expected, test whether this result differs significantly from expectation.

20.3. In a sample of 50 firms in an industry it was found that the average number of injuries per thousand man-hours was distributed as shown in the table below with a mean of 2.35 and a standard deviation of 0.5. Test the hypothesis that the distribution is normal at the 5 per cent level of significance. (*Hint*: For ease of calculation, round the normal frequencies to whole numbers and also ensure that they sum to 50. Note that three degrees of freedom are lost.)

Average number of injuries per thousand man-hours	Number of firms
1.45 – <1.75	4
1.75 – <2.05	10
2.05 – <2.35	15
2.35 – <2.65	9
2.65 – <2.95	7
2.95 – <3.25	5

20.4. The data below give preferences for three types of drink for a random sample of 500 people who have been categorized into three age groups: teens, adults and pensioners. Is drink preference independent of age?

	Teens	Adults	Pensioners
Coffee	60	70	70
Tea	25	55	70
Soft Drink	35	55	60

20.5. A polling organization obtains the information given in the following table about voting intentions of 300 voters in three wards of a town.

Voting intention	Ward 1	Ward 2	Ward 3
Conservative	37	24	29
Labour	37	41	27
Liberal	26	55	24

(a) Is there a significant difference in voting intentions in the town as a whole?

(b) Can the differences among the wards in reported voting intentions be attributed to chance?

(c) Given that in the previous election the results were 36 per cent for Conservatives, 35 per cent for Labour, and 29 per cent for Liberal, are people showing the same voting intentions?

20.6. The table below shows the number of defective items produced by three workmen operating, in turn, three different machines, A, B and C.

Machine	Alf	Bert	Charlie
A	8	4	3
B	12	8	10
C	4	4	7

(a) Are the number of defects explained by differences:

    (i) in the performances of the workmen?

    (ii) in the performances of the machines?

(b) Is there any association between workmen and machines?

Part V

Time series and index numbers

With seasonally adjusted temperatures, you could eliminate winter in Canada.

ROBERT L. STANSFIELD

The following diagram provides an overview of the whole of this part of the book.

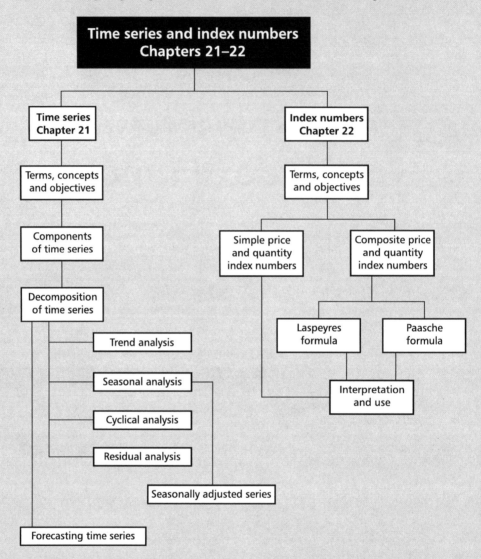

Chapter 21

Time series analysis

A sequence of observations recorded over time is known as a *time series*. Observations may be available, for example, on an hourly, daily, weekly, monthly, quarterly, etc. basis. Data of this kind are, of course, commonplace in everyday life relating to things such as hourly share prices on the Stock Exchange, weekly wages, monthly unemployment figures, quarterly electricity consumption and so on.

Time series may be affected by various factors, such as an underlying trend, seasonal factors and unusual events (e.g. an earthquake), and the objective of time series analysis is to permit the interpretation of the series taking such factors into account.

The objective of time series analysis may be contrasted with those for regression analysis, dealt with in Chapters 18 and 19. It was shown that regression models provide a means of establishing a relationship between a dependent variable and one or more independent (i.e. explanatory) variables. Regardless of whether the data for the variables are cross-sectional – measured at a point in time – or time series in nature, the aim is to derive an equation which provides the best possible explanation of the structural link between the variables under consideration. In time series analysis, however, instead of searching for relationships between two or more variables, we are only interested in studying the movements of one particular variable as it is measured across different time periods.

For practical purposes, it is useful to be able to detect any systematic movements in the time series variable under consideration and to try to extract some sense of order out of the seemingly random appearance of the information. The purpose is not only to interpret past behaviour but also to help in forecasting future events. For example, records of daily rainfall compiled over many years provide valuable information for many people such as weather forecasters and other interested parties in predicting future weather patterns. Time series analysis is particularly important in the interpretation of economic, social and business information.

A time series may be affected by four basic components: trend, seasonal variation, cyclical variation and random variation. The essence of time series analysis is to decompose the series into one or more of these four components. The nature of these four components as well as the principles of decomposition are discussed below.

After working through this chapter you will be able to:

- appreciate the purpose of time series analysis
- recognize that time series data may be subject to four influences: trend, seasonal variation, cyclical variation and random variation
- understand the meaning of the four components of a time series
- appreciate the difference between the additive and multiplicative models of time series data
- measure trends using linear and moving average methods
- measure seasonal variation using the ratio-to-moving average method
- calculate and interpret seasonally adjusted series
- measure cyclical factors and residual variation
- use the analysis of trends and seasonal variation for forecasting purposes.

Principles

A time series may be affected by four basic components:

Components of time series

- trend (denoted T)
- seasonal variation (denoted S)
- cyclical variation (denoted C)
- random variation (denoted R)

The purpose of time series analysis is to decompose the series into one or more of these four basic components. We define each of these four components below before considering the procedures for carrying out the process of decomposition in order to measure the influence of each component.

Trend (T)

A time series variable will sometimes display a steady tendency to increase or decrease over a long period of time, reflecting long-term growth or decline of the variable of interest. Such a tendency is called a *trend* (or sometimes a *secular trend*) and can be a consequence of long-term gradual changes, for example, in population, total national output, retail prices, incomes, etc. By measuring the nature of an underlying trend, we are able to make long-term projections into the future. In addition, we may wish to eliminate the trend from the data and thus highlight the influence of other components.

Seasonal variation (S)

Time series data recorded with a frequency of less than a year (e.g. monthly or quarterly) often display *seasonal variation*. Seasonal variation follows a complete cycle through the year with broadly the same general pattern repeating itself year after year. Obvious examples of such variation are sales and production of seasonal items such as ice-cream, heating fuel, ski-ing holidays and so on. Seasonal variation is not only related to climatic conditions but may also be due to social factors such as the increase in the sale of chocolate eggs at Easter time and the surge in demand for children's toys as Christmas approaches. Many time series variables contain seasonal variation, though it may not be as obvious as in these examples. Again, as with the trend, detection of seasonal effects (i.e. *seasonality*) is important in allowing us to interpret movements in the variable from one period to the next. Ice-cream sales, for example, generally fall in winter time. Therefore, in interpreting data on ice-cream sales, the question arises: is the trend of sales greater or less than would be expected, given the normal seasonal effect? To answer this question requires that the data should be adjusted to remove the effect of seasonal variation in ice-cream sales; i.e. the *seasonally unadjusted* data must be modified to a *seasonally adjusted* form. Furthermore, the analysis of seasonal variation is useful in planning production and anticipating sales and, hence, managing stock levels and staffing accordingly.

Cyclical variation (C)

In contrast to seasonal variation which, by definition, only occurs within periods of less than one year, regular variation within a time series variable in periods greater than one year is referred to as *cyclical variation*. This is commonly found in many sets of economic and business data which are affected by periodic upswings and downswings in business confidence and activity, giving rise to so-called 'business cycles' or 'trade cycles'. Such fluctuations may occur with a varying degree of regularity over a number of years, say five from the peak of one cycle to the peak of the next. Their analysis tends to be of particular interest to the government and to economists, though in some industries where the planning horizon is long (such as in the energy sectors), attention to cyclical factors, especially those that may apply to these industries in particular, may be important for long-term management decisions.

Random variation (R)

Finally, *random variation* in a time series variable arises from irregular events which cause the variable of interest to move in a completely random, unpredictable, way. Thus, it is the residual variation that remains after the *regular* effects of the trend, seasonal and cyclical factors have been removed from the data. It covers all unpredictable movements such as the effects of strikes, natural disasters such as floods and famines, the effects of the world stock market crash in October 1987, etc.

Before turning to set out the principles of time series analysis, we graphically illustrate the nature of the three regular components, T, S and C (the fourth component, R, is simply the residual variation left over after the other three have been removed and thus is not illustrated). Figure 21.1, made up of three parts, shows the trend for new car sales in part (a). As will be seen, there is a strong tendency for new car sales to increase over time, illustrated here by a linear trend. Seasonal variation of ice-cream sales is illustrated in part

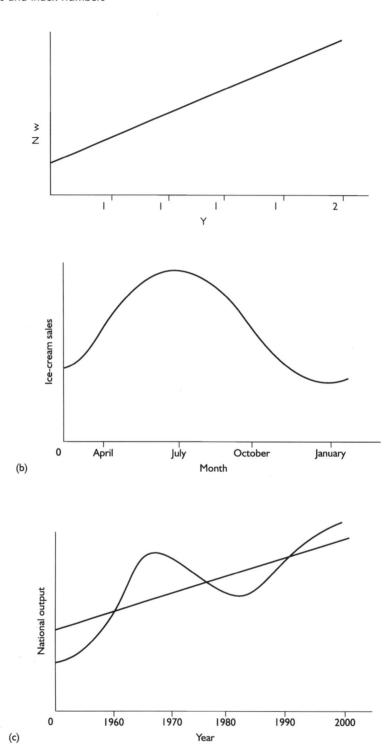

Figure 21.1 Illustrations of variation in time series data: (a) trend; (b) seasonal variation; (c) cyclical variation

(b) showing a peak for sales in the summer months and a trough in the winter. Finally, part (c) shows the cyclical variation in national output over many years, highlighting the frequency and amplitude of the business cycle, as well as the long-term trend shown by the straight line drawn through the actual data values.

We now turn to explain the principles involved in measuring the four components of time series data, separately.

Decomposition of time series

As noted earlier, the essence of time series analysis is to decompose the data for a variable into one or more of the four components. There are two fundamental approaches which may be employed for this purpose. These depend on the assumption made about the relationship between the four components: one is that the components are related in an *additive* way giving rise to a so-called *additive model*, the other is that they are related in a *multiplicative* way giving rise to a so-called *multiplicative model*. Hence, given observed data values for a variable y measured over time t, denoted y_t, the two models may be defined as follows:

Models of time series data

Additive model: $y_t = T_t + S_t + C_t + R_t$

Multiplicative model: $y_t = T_t \times S_t \times C_t \times R_t$

where the subscript t denotes that observations are recorded across a sequence of time periods.

It is common practice, however, to regard the multiplicative model as the most appropriate because it has been found to be more accurate than the additive model in describing and predicting variations in time series data. Consequently, we confine our attention to this model only. The basic principles of the analysis are the same in both cases. Note that for convenience, we drop the subscript t in the exposition given below. The principles which follow focus on the measurement of each of the four components in turn.

Measurement of trend (T)

Putting aside the possibility of 'guesstimating' a trend line from a scatter plot of the data by means of a freehand sketch, there are two ways of measuring the trend statistically. These are referred to as:

- the linear trend method
- the moving average trend method.

The linear trend method involves the use of linear regression analysis and thus is only appropriate when the time series moves up or down in a sufficiently regular way as to indicate that a straight line through the data would provide a good fit. The moving-average trend method is appropriate in situations where there is a greater degree of irregularity and especially where there are reversals of trend from time to time. The choice is determined by visual inspection of the graphed data. The first step in time series analysis, therefore, is to produce a scatterplot of the data and observe the degree of regularity or irregularity it displays.

Linear trend method

The linear trend method simply involves the application of the simple, two-variable, regression technique explained in Chapter 18, with the time series data (y) taken as the dependent variable and with time itself as the independent variable (denoted x). A linear trend for y (denoted T_y) is then defined as

Linear trend

$$T_y = a + bx$$

where T_y denotes the underlying trend values of the variable y under consideration, for each time period; x denotes the points of time (each being numbered consecutively); a is the intercept term; and b represents the change in T_y per unit of time.

The first step in fitting a linear time trend to a set of data is to number the time periods. Thus for quarterly data, for example, the four quarters in the first year would be numbered 1, 2, 3, 4, followed by 5, 6, 7, 8 for the four quarters of the next year and so on. The linear trend regression equation is derived in exactly the same way as that described in Chapter 18. Worked example 21.1 in the next section shows both the worked and computer solutions for a particular problem.

The use of the regression technique in trend analysis is particularly useful in forcasting. An example is shown in Worked example 21.2. It is worth stressing here, however, that the dangers of extrapolation, discussed in Chapter 18, are especially relevant in this context.

Moving-average trend method

As explained above, the moving-average method of fitting a trend line is used where the movement of the time series is irregular, and thus makes a linear trend inappropriate. The method provides a trend line by smoothing out fluctuations in the time series. It does so by taking rolling averages of the data over n consecutive periods – this produces, in general terms, an *n-point* moving-average series.

There are two points to note about the application of the moving-average method in practice and we consider these in turn. First, it is necessary to determine n, the time span over which the rolling averages are to be calculated. In cases where the data are recorded with a frequency of less than one year (monthly, quarterly etc.), the choice of n is clear cut. In these cases it is important to maintain the 'integrity' of the year by calculating, for example, the moving average over 12 periods for monthly data, four periods for quarterly data and so on. For annual data, the choice of n has to be decided by the researcher. The main consideration is to produce a moving-average trend line which follows the broad movements of the actual series. It is important to avoid over-smoothing the series by choosing too large a value of n. This is especially important when the data are to be used for seasonal analysis. At the same time, the larger the value of *n* then the more information on trend values we lose at the beginning and end of the data series (this point is explained and illustrated in Table 21.1 below).

The second point to note is that the moving-average trend values have to be 'centred' against the mid-point of the averaging period. The process of centring is straightforward when the time span *n* is an *odd* number. When *n* is an *even* number, however, the centring procedure is not so straightforward (since, for example, with quarterly data ($n = 4$) there is no single period which corresponds to the centre of the span!). The problem is easily solved, however, by carrying out the centring process *twice*. This point is illustrated in Table 21.2 below. It should be noted that in using this method we lose $n/2$ points at the start and end of the series. In addition, note that by virtue of smoothing out the series, the

Table 21.1 Computation of trend using moving-average method when *n* is an odd number (illustration for $n = 3$)

Year	Actual time series (y_t)	Three-point moving-average trend (T_y)
1993	10	
1994	20	$= (10 + 20 + 30)/3 = 20.0$
1995	30	$= (20 + 30 + 16)/3 = 22.0$
1996	16	$= (30 + 16 + 29)/3 = 25.0$
1997	29	etc. $= 30.0$
1998	45	$= 33.0$
1999	25	$= 40.0$
2000	50	

Table 21.2 Computation of trend using moving-average method when *n* is an even number (illustration for $n = 4$)

Year/quarter	Actual time series (y_t)	Four-quarter moving total	Four-quarter moving averages	Trend-centred moving average (T_y)
1997 Q1	6			
Q2	9			
Q3	13	40	10	10.5
Q4	12	44	11	12.0
1998 Q1	10	52	13	14.0
Q2	17	60	15	16.5
Q3	21	72	18	19.0
Q4	24	80	20	22.5
1998 Q1	18	100	25	25.0
Q2	37	100	25	26.0
Q3	21	108	27	28.5
Q4	32	120	30	30.5
2000 Q1	30	124	31	32.0
Q2	41	132	33	34.0
Q3	29	140	35	
Q4	40			

moving-average trend removes much of the random variation R present in the actual series. It is also worth noting that one disadvantage of the moving-average trend, as opposed to a linear trend, is that it is not evident how it can be projected forward reliably for prediction purposes. The MINITAB program for computing moving averages does contain a forecasting option. But the forecasts it makes simply represent the last moving average value maintained at a constant level – these are produced together with 95% prediction interval bands. A final point to note with regard to the fitting of trends and forecasting is that MINITAB contains automatic facilities for the fitting of non-linear trends but these are not considered here as this topic lies outside the scope of this book.

Having set out the principles of trend measurement, we now turn to the measurement of seasonal variation.

Measurement of seasonal variation (S)

The measurement of seasonal variation leads first to the calculation of a *seasonal index* (denoted S), consisting of n values corresponding to the n points noted above. This is useful in its own right in indicating the magnitude of seasonal effects, on average, and is thus useful for planning purposes, as explained earlier. Second, it may be used to *de-seasonalize* the observed data to produce a *seasonally adjusted series*. This permits the observation and comparison of underlying movements of the series without the distortion of seasonal effects. The seasonally adjusted series is calculated simply by dividing each observation y_t by the appropriate value of S, thus:

De-seasonalized series

$$\frac{y_t}{S}$$

Note that since we are using the multiplicative model, and thus y_t is regarded as equivalent to $T \times S \times C \times R$, we have,

$$\frac{y_t}{S} = \frac{T \times S \times C \times R}{S} = T \times C \times R$$

i.e. the resultant de-seasonalized data reflect the combined effect of trend, cyclical and random variation.

The calculation of a seasonal index simply consists of computing averages of the actual observations relative to their trend values for each quarter (or month etc.) over some given period of time. Using the multiplicative model, this is known as the *ratio-to-moving-average* method. In principle, this first involves 'de-trending' the original data, i.e. computing y_t / T_y and secondly, averaging these de-trended values over all corresponding time periods (this averages out the random variation component of the data) to produce the required seasonal index S.

It should be appreciated that the use of a moving-average trend, as opposed to a linear trend, picks up the influence of cyclical factors as well as the trend itself. Thus, the procedure for computing the ratio-to-moving average for any time period is:

$$\frac{\text{actual value}}{\text{centred moving average}}$$

The rationale of the procedure may be expressed as follows, using the usual notation for the multiplicative model:

$$\frac{y_t}{T \times C} \text{ is equivalent to } \frac{T \times S \times C \times R}{T \times C} = S \times R$$

Hence, by de-trending the data using the moving-average trend method (and assuming a multiplicative time series model) means that the de-trended series effectively consists of seasonal and random movements only. To isolate S requires that the random component R be removed and this is achieved as explained above (by taking moving averages).

The procedural steps for computing a seasonal index may be summarized as follows:

Ratio-to-moving average method for computing a seasonal index

1. de-trend the series, y by computing y_t/T_y;

2. average the de-trended values for the corresponding time periods (e.g. corresponding quarters of each year, etc.);

3. if necessary, scale the averages corresponding to each time period to ensure that they sum to 400 (for quarterly data) or 1200 (for monthly data), etc.

An illustration of the calculation of a seasonal index and a seasonally adjusted series is shown in Worked examples 21.3 and 21.4 below respectively together with examples of the application of the MINITAB program which uses an adaptation of the traditional procedure.

Measurement of cyclical variation (C)

The measurement of cyclical variation requires the isolation of the C component in the time series variable y_t – in other words, the removal of T, S and R from the series. Thus, given that a de-seasonalized series reflects the effects of trend, cycles and random events, dividing such a series by the trend leaves a series consisting of $C \times R$. Thus, symbolically, de-trending after de-seasonalizing produces:

$$\frac{T \times C \times R}{T} = C \times R$$

An important point to note here, however, is that T must be found using the linear trend method since, as shown earlier, the alternative moving-average method produces a measure of $T \times C$ combined – hence dividing by a moving-average trend at this stage would leave only random effects! Hence, the isolation of C depends either on the initial calculation of T using only the linear trend method or, alternatively, on fitting a trend line to the de-seasonalized series.

Finally, to isolate C requires the removal of as much of the random variation as possible. This is done by using the moving-average technique again. The moving averages in this case constitute the required cyclical index. Again, care is required to average the series over an appropriate number of periods, n. Too large a value of n will oversmooth the series and obscure the cyclical movements it is desired to isolate. Likewise, too small a value of n will lead to inadequate smoothing and hence not reveal the cyclical movements unobscured by random movements.

The procedural steps involved in cyclical analysis may be summarized as follows:

Computation of cyclical variation

1. using the linear trend method, de-trend the de-seasonalized data, i.e. calculate

$$\frac{T \times C \times R}{T} \text{ to give } C \times R$$

2. compute moving averages of the $C \times R$ values for the corresponding time periods to remove as much of the random variation R as possible. This allows the identification of cycle peaks and troughs.

Measurement of random variation (R)

The term random variation is used to cover all types of variation other than trend, seasonal and cyclical movements. It can be measured as the residual after the regular (trend, seasonal and cyclical) factors have been removed from a series, but there is little or no interest in such residuals in their own right and we devote no further attention to them.

Applications and Worked Examples

General descriptions of the MINITAB and Excel computer programs and their application are given in Appendices B and C respectively.

Before showing worked examples, it should be noted that MINITAB now contains extensive facilities for time series analysis which go beyond the scope of this book. Wherever appropriate we show the applications of MINITAB and comment on those features of the program which represent more advanced treatments. Excel has only limited features for time series analysis; these are also illustrated below.

Worked example 21.1   Estimating a linear trend

The data below show annual new car sales (y_t) for a company over time (x_t) from 1980 to 1999. Note that the time variable x_t is numbered consecutively with $1980 = 1$, $1981 = 2$, etc. (Numbering does not, of course, have to start at 1 – it could start at 0 or any other number.)

Estimate a linear trend for the data using regression analysis. Plot a scatter diagram of actual sales and plot the linear trend.

Year	Time (x_t)	New car sales (y_t) (000s)
1980	1	2.6
1981	2	2.5
1982	3	2.3
1983	4	3.2
1984	5	2.9
1985	6	3.2
1986	7	3.3
1987	8	3.6
1988	9	3.8
1989	10	3.7
1990	11	3.4
1991	12	3.6
1992	13	4.0
1993	14	4.2
1994	15	4.5
1995	16	4.6
1996	17	4.4
1997	18	4.2
1998	19	3.8
1999	20	4.8

Worked solution

$\Sigma x = 210.0$ $\Sigma x^2 = 2870.0$ $\Sigma xy = 835.20$
$\Sigma y = 72.6$ $\Sigma y^2 = 273.22$

Regression coefficients

$$b = \frac{n\Sigma xy - \Sigma x \Sigma y}{n\Sigma x^2 - (\Sigma x)^2}$$

$$= \frac{20(835.20) - (210.0)(72.6)}{20(2870) - (210.0)^2}$$

$$= 0.11 \text{ (approximately)}$$

$$a = \bar{y} - b\bar{x} = 3.63 - 0.11(10.5) = 2.48.$$

Regression equation

$$y_c = a + bx = 2.48 + 0.11x$$

Substituting for time (i.e. $x = 1-20$) in this equation gives the linear trend values for sales (i.e. values of y_c) shown in the following table.

1980	2.59	1990	3.69
1981	2.70	1991	3.80
1982	2.81	1992	3.91
1983	2.92	1993	4.02
1984	3.03	1994	4.13
1985	3.14	1995	4.24
1986	3.25	1996	4.35
1987	3.36	1997	4.46
1988	3.47	1998	4.57
1989	3.58	1999	4.68

The following figure shows the scatter plot and linear trend of car sales.

Worked example 21.1   Estimating a linear trend (continued)

Computer solution using MINITAB

Choose   **Stat → Time   Series → Trend analysis**

In **Variable**, enter the column containing the time series observations (C1 in this case) [*Note: there is no need to enter time as such*]

Under **Model type** choose **Linear**

Click **OK**

The output is as follows:

Trend Analysis

Data	Sales (y_t)
Length	20.0000
NMissing	0

Fitted Trend Equation

$$Y_t = 2.47895 + 0.109624*t$$

Accuracy Measures

MAPE:	6.66079
MAD:	0.232632
MSD:	0.0845203

Authors' comment: Explanation of these 'Accuracy Measures' is given in the text below the graphical output.

Worksheet

C1 Sales (y_t)	C2 FITS1
2.6	2.58857
2.5	2.69820
2.3	2.80782
3.2	2.91744
2.9	3.02707
3.2	3.13669
3.3	3.24632
3.6	3.35594
3.8	3.46556
3.7	3.57519
3.4	3.68481
3.6	3.79444
4.0	3.90406
4.2	4.01368
4.5	4.12331
4.6	4.23293
4.4	4.34256
4.2	4.45218
3.8	4.56180
4.8	4.67143

Authors' comment: Predicted y_t values (y_c) – these are inserted into the worksheet automatically.

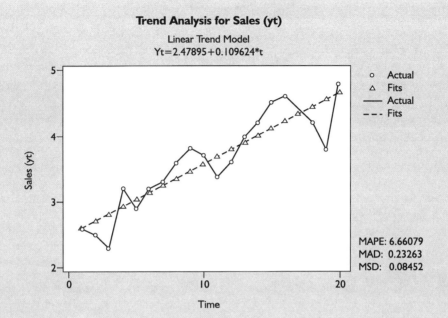

Trend Analysis for Sales (yt)

Linear Trend Model
$Yt = 2.47895 + 0.109624 \cdot t$

○ Actual
△ Fits
— Actual
--- Fits

MAPE: 6.66079
MAD: 0.23263
MSD: 0.08452

Authors' comment and interpretation: Accuracy Measures The Accuracy Measures (MAPE, MAD and MSD) are computed by MINITAB automatically. All are measures based on the 'errors' (deviations) between the actual and the fitted values of y_t. Their use is to compare different methods of modelling time series. Their interpretation, in each case, is that the smaller the value the better the fit of the model. The definitions of the measures are:

$$\text{MAPE (Mean Absolute Percentage Error)} = \frac{\sum_{t=1}^{n} |(y_t - y_c)/y_t|}{n} \times 100 \; (y_t \neq 0)$$

$$\text{MAD (Mean Absolute Deviation)} = \frac{\sum_{t=1}^{n} |y_t - y_c|}{n}$$

$$\text{MSD (Mean Squared Deviation)} = \frac{\sum_{t=1}^{n} (y_t - y_c)^2}{n}$$

Computer solution using Excel

Estimation of the linear trend equation requires the use of **Regression** in the **Data Analysis Tools**. But if a plot of the data and the trend line only are required, a simpler procedure is to use the **Chart Wizard** together with an option for plotting a trend line automatically. We illustrate both procedures here.

Worked example 21.1   Estimating a linear trend (continued)

(a) Regression analysis

With the x variable (time numbered from 1 to 20) entered in cells A1 : A20 of the worksheet and the y variable (Sales) entered in cells B1 : B20:

Choose **Tools** → **Data Analysis. . .** → **Regression**
In **Input Y Range** enter y values by highlighting cells B1 : B20 in the worksheet
In **Input X Range** enter x by highlighting cells A1 : A20 in the worksheet
Under **Residuals** check the **Line Fit Plots** box
Click **OK**

Authors' comment: Checking *Line Fit Plots* includes in the output a list of all the estimated values of y (y_c) from the regression plus a plot of the data and trend line – as shown below.

The output is as follows:

SUMMARY OUTPUT

Regression Statistics	
Multiple R	0.90851932
R Square	0.82540735
Adjusted R Square	0.81570776
Standard Error	0.30644974
Observations	20

ANOVA

	df	SS	MS	F	Significance F
Regression	1	7.991593985	7.991594	85.09712	3.0457E-08
Residual	18	1.690406015	0.093911		
Total	19	9.682			

	Coefficients	Standard Error	t Stat	P-value	Lower 95%	Upper 95%
Intercept	2.47894737	0.142355546	17.41377	1.04E-12	2.17986923	2.778026
X Variable 1	0.10962406	0.011883611	9.22481	3.05E-08	0.0846575	0.134591

Regression coefficients

RESIDUAL OUTPUT

Predicted values of y

Observation	Predicted Y	Residuals
1	2.58857143	0.011428571
2	2.69819549	−0.198195489
3	2.80781955	−0.507819549
4	2.91744361	0.282556391
5	3.02706767	−0.127067669

6	3.13669173	0.063308271
7	3.24631579	0.053684211
8	3.35593985	0.24406015
9	3.46556391	0.33443609
10	3.57518797	0.12481203
11	3.68481203	−0.28481203
12	3.79443609	−0.19443609
13	3.90406015	0.09593985
14	4.01368421	0.186315789
15	4.12330827	0.376691729
16	4.23293233	0.367067669
17	4.34255639	0.057443609
18	4.45218045	−0.252180451
19	4.56180451	−0.761804511
20	4.67142857	0.128571429

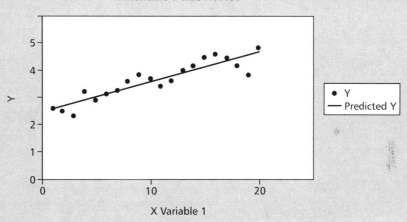

(b) Scatter plot and trend line

Use the **Chart Wizard** to produce a **scatter plot** (as described in the Worked examples for Chapter 2 – see, in particular, Worked example 2.1). Then, to plot a trend line automatically:

Click the **data series** to which you want to add a **trendline**
On the **Chart** menu, click **Add Trendline**
On the **Type** tab, click the type of regression trendline you want (**linear** in this case)
Click **OK**

The plot output (plus the actual sales data) is shown on the next page:

Worked example 21.1   Estimating a linear trend *(continued)*

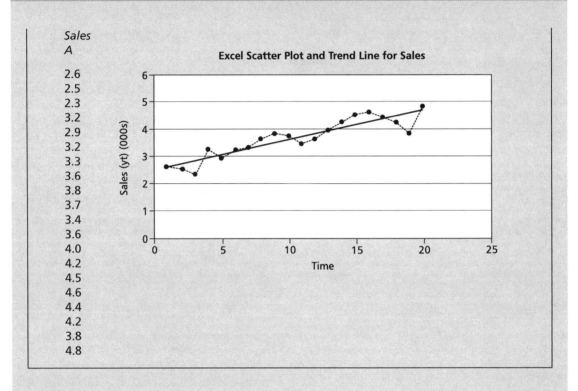

Sales
A

2.6
2.5
2.3
3.2
2.9
3.2
3.3
3.6
3.8
3.7
3.4
3.6
4.0
4.2
4.5
4.6
4.4
4.2
3.8
4.8

Worked example 21.2   Forecasting using a linear trend

Using the linear trend equation computed in Worked example 21.1, forecast a new car sales (y^f) for the years 2000–3.

Worked solution

Given $y^f = 2.48 + 0.11x$, then

2000 $x = 21$: $y^f_{2000} = 2.48 + 0.11(21) = 4.79$ (000s)

2001 $x = 22$: $y^f_{2001} = 2.48 + 0.11(22) = 4.90$ (000s)

2002 $x = 23$: $y^f_{2002} = 2.48 + 0.11(23) = 5.01$ (000s)

2003 $x = 24$: $y^f_{2003} = 2.48 + 0.11(24) = 5.12$ (000s)

Computer solution using MINITAB

In addition to the procedure under Worked example 21.1 above:

Check **Generate forecasts**

In **Number of forecasts** enter the number of time periods for which forecasts are required (4 in this case)

In **Starting from origin** enter the time period preceding the first period for which the new forecasts are required (i.e. 20 in this case)

Click **OK**

The output is as follows:

Trend Analysis

Data	Sales (y_t)
Length	20.0000
NMissing	0

Fitted Trend Equation

$Y_t = 2.47895 + 0.109624 * t$

Accuracy Measures

MAPE:	6.66079
MAD:	0.232632
MSD:	0.0845203

Row	Period	Forecast
1	21	4.78105
2	22	4.89068
3	23	5.00030
4	24	5.10992

Forecast values for the years 2000–2003. Differences between these and the manual results are due to rounding errors (fewer decimal places being used in the manual calculations). The graph on the following page shows the forecasts as well as the actual data and fitted values.

Worked example 21.2 Forecasting using a linear trend *(continued)*

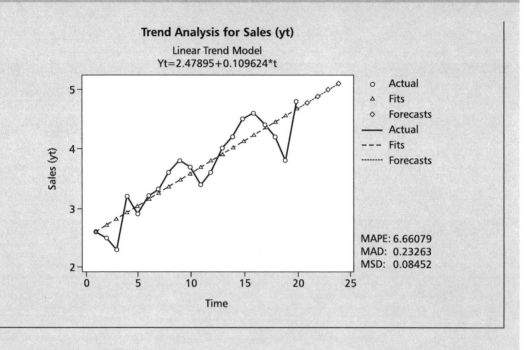

Trend Analysis for Sales (yt)

Linear Trend Model
$Yt = 2.47895 + 0.109624*t$

MAPE: 6.66079
MAD: 0.23263
MSD: 0.08452

Computer solution using Excel

Forecast values of the trend can be obtained by using the fill handle facility to fill in the required values by projecting the series forwards or backwards as follows:

Select at least two cells that contain the starting values for the trend (to increase the accuracy of the trend series, select additional starting values)
Drag the fill handle in the direction you want to fill with increasing or decreasing values.

	A	B
	Time	Predicted by
1	19	4.5 1805
2	20	4..71429
3	21	4.78105
4	22	4.890 77
5	2	5.000 01
6	24	5.109925

Last two values of the predicted series from the Excel regression output in Worked example 21.1 above.

Values projected forwards by dragging the fill handle on cell B2 down after highlighting cells B1 and B2. The slight differences between these values and those in the manual worked example are due to rounding in the manual example.

Worked example 21.3   Seasonal index using ratio-to-moving-average method

Using the data shown in Table 21.2, compute a seasonal index based on the ratio-to-moving-average method.

Worked solution

We reproduce below the relevant information from Table 21.2 i.e. the actual time series (y_t) and the trend values (T_y), from which we can calculate the seasonal index.

Year/quarter		Actual time series (y_t)	Trend-centred moving average (T_y)	Ratio $(y_t/T_y) \times 100$
1997	Q1	6		
	Q2	9		
	Q3	13	10.5	123.8
	Q4	12	12.0	100.0
1998	Q1	10	14.0	71.4
	Q2	17	16.5	103.0
	Q3	21	19.0	110.5
	Q4	24	22.5	106.7
1999	Q1	18	25.0	72.0
	Q2	37	26.0	142.3
	Q3	21	28.5	73.7
	Q4	32	30.5	104.9
2000	Q1	30	32.0	93.8
	Q2	41	34.0	120.6
	Q3	29		
	Q4	40		

Calculation of seasonal index

Year	Q1	Q2	Q3	Q4	
	Ratios: $(y_t/T_y) \times 100$				
1997	—	—	123.8	100.0	
1998	71.4	103.0	110.5	106.7	
1999	72.0	142.3	73.7	104.9	
2000	93.8	120.6	—	—	
Average	79.1	122.0	102.7	103.9	(Total = 407.7)
Adjusted average (seasonal index, *S*)	77.6	119.7	100.8	101.9	(Total = 400)

Worked example 21.3   Seasonal index using ratio-to-moving-average method
                         (continued)

Note that in calculating the seasonal index, adjustment of the average is required because the four quarterly seasonal indexes should sum to 400. Thus:

$$\text{Seasonal indexes} = \frac{\text{Average for corresponding quarters}}{407.7} \times 400$$

Computer solution using MINITAB

Note on the use of MINITAB

MINITAB will compute moving averages and we show the procedure below. It will also compute seasonal indices but it uses a different method to do so than the one we have shown here. The method set out above is the traditional method of computing a seasonal index, but with the advent of computers alternative and more sophisticated methods have been introduced and these are followed in MINITAB. We show first how MINITAB may be used to produce the results shown above. We then give another example to explain and illustrate the fully automatic procedure in MINITAB – Worked example 21.4.

First, MINITAB is used to compute a moving average and then the required ratios between the actual time series data and the moving average trend are calculated (the de-trended series) as follows:

Choose   **Stat → Time Series → Moving Average**

In **Variable** enter the location of the time series (C1 in this case, or, if named, the name of the series)

In **Moving average length** enter '4' (because the data are quarterly)

Check **Centre the moving averages**
Choose **Results**
Under **Graphics** check **Do not display graph** and under **Output** check **Summary table**
Click **OK**
Select **Storage** and check **Moving averages** [this command puts the moving average trend into the next available blank column in the worksheet – C2 in this case – see below]
Click **OK** twice

The output is as follows (recall that the moving average values are placed in C2 of the worksheet):

Moving average

Data	Time series
Length	16.0000
NMissing	0

Moving Average
Length: 4

Accuracy Measures
MAPE: 25.2502
MAD: 7.2273
MSD: 76.2955

With the moving average values in C2 of the worksheet, the rest of the procedure involves the use of MINITAB to mirror the manual calculations. First it is necessary to calculate the detrended series (ratios to trend) using the MINITAB calculator as follows:

Choose **Calc → Calculator**
In **Store result in variable** enter C3 (or the name of the de-trended series: 'Ratios')
In **Expression** enter (c1/c2)*100
Click **OK**

The next step is to reproduce the data in the table above, which shows the calculation of the seasonal indices; this may be reproduced in MINITAB by unstacking C3 (in order to sort the data into Q1 values, Q2 values etc.), but it is first necessary to code the ratios in C3 according to the quarter to which they relate as follows:

In C4 of worksheet enter codes 1,2,3 or 4 as appropriate to represent the quarter to which each of the ratios relate.

Next, Choose **Manip → Stack/Unstack → Unstack one column**
In **Unstack the data in** enter C3
In **Store the stacked data in** enter C5, C6, C7, C8
In **Using subscripts in** enter C4 [together with the last command, this puts the Q1 ratios in C5, the Q2 ratios in C6 etc.]
Click **OK** [the results are shown in cols C4–C8 of the worksheet reproduced below].

Next it is necessary to calculate the means of cols C5–C8 and adjust them to ensure that they sum to 400 as follows:

Choose **Calc → Calculator**
In **Expression** enter Mean(Q1) [or, alternatively Mean(C5)] and in **Store result in variable** enter C9; repeat for the remaining 3 quarters storing the results in C10–C12 [these results are shown in Columns C9–C12 of the worksheet below].

Finally the means need to be adjusted to ensure that they sum to 400; this is done as follows and the result stored in K1–K4:

Choose **Calc → Calculator**
In **Expression** enter RSUM(C9,C10,C11, C12) and in **Store results in variable** enter K5 [this sums the means across the four columns and enters the sum in K5.

[Note: to see the contents of K5 Choose **Manip → Display data** and enter K5 – the content of K5 is then displayed in the Session Window.]

Choose **Calc → Calculator** again
In **Expression** enter C9/K5*400. Click **OK** and in **Store results in variable** enter K1. Repeat for columns C10–C12 storing the results in K2–K4 respectively
Choose **Manip → Display data** and enter K1–K4
Click **OK**

The output is as follows:

Data Display

K1	77.5911
K2	119.710
K3	100.766
K4	101.933

Seasonal index for Q1–Q4.
Compare with the results in the Worked solution shown above.

Worked example 21.3 Seasonal index using ratio-to-moving-average method (continued)

Worksheet

C1	C2	C3	C4	C5	C6	C7	C8	C9	C10	C11	C12
T								Mean	Mean	Mean	Mean
series	AVER1	Ratios	Qtrs	Q1	Q2	Q3	Q4	Q1	Q2	Q3	Q4
6	*	*	*	71.429	103.030	123.810	100.000	79.060	121.975	102.673	103.862
9	*	*	*	72.000	142.308	110.526	106.667				
13	10.5	123.81	3	93.750	120.588	73.684	104.918				
12	12	100	4								
10	14	71.429	1								
17	16.5	103.03	2								
21	19	110.526	3								
24	22.5	106.667	4								
18	25	72	1								
37	26	142.308	2								
21	28.5	73.684	3								
32	30.5	104.918	4								
30	32	93.75	1								
41	34	120.588	2								
29	*	*	3								
40	*	*	4								

Computer solution using Excel

There is no pre-defined tool in Excel for solving this problem. Excel does contain a data analysis tool for computing **moving averages** automatically, but it is not suitable for our purposes here because it does not centre the moving average values. The moving average values could be manipulated arithmetically in Excel, using appropriate formulae, to generate properly centred values and then used to construct a seasonal index but as the procedure simply mirrors that in the manual example it is not repeated here.

Worked example 21.4 Seasonal index using the MINITAB decomposition method

The differences between the seasonal analysis procedure used in Worked example 21.3 and the MINITAB decomposition method set out here are:

1. Use of a linear, rather than a moving average, trend.

2. Introduction of an additional step after de-trending, namely smoothing of the *de-trended* data using a moving average and calculating the initial seasonal indices ('raw seasonals') as ratios between the de-trended data and the *smoothed de-trended data* (the moving average).

3. Use of the median, rather than the arithmetic mean, in averaging the ratios ('raw seasonals').

We first illustrate the method manually, using the same data as in Worked example 21.3, and then using MINITAB. As before, the multiplicative model only is used.

Worked solution

The decomposition method involves the following steps:

1. Fit a trend line to the data using least squares regression.
2. Detrend the data by dividing them by the trend component (as the multiplicative model is being used).
3. Smooth the de-trended data using a centred moving average with a length equal to the length of the seasonal cycle.

4. Divide the moving average values into the detrended data to obtain the 'raw seasonals'.
5. For each seasonal period, find the median value of the raw seasonals. Adjust the medians, by scaling arithmetically, so that their mean is one (*note that, in contrast to the previous example where we scaled to a sum of 400, MINITAB computes ratios rather than indexes and scales to a mean value of one rather than a sum of 4*). These adjusted medians constitute the seasonal indices.

The table below reproduces the actual time series in column (A). The linear trend equation fitted to these data is: $y_c = 4.3 + 2.14118y_t$. The fitted trend values are given in column (B). The rest of the table follows the steps set out above and is self-explanatory.

Time		Time series y_t (A)	Linear Trend y_c (B)	De-trended (C) = (A)/(B)	Centred moving average of de-trended data (D) = moving ave of (C)	'Raw seasonals' (E) = (C)/(D)
1997	q1	6	6.441	0.932		
	q2	9	8.582	1.049		
	q3	13	10.724	1.212	0.998	1.215
	q4	12	12.865	0.933	0.958	0.974
1998	q1	10	15.006	0.666	0.935	0.713
	q2	17	17.147	0.991	0.943	1.051
	q3	21	19.288	1.089	0.979	1.112
	q4	24	21.429	1.120	1.047	1.070
1999	q1	18	23.571	0.764	1.061	0.720
	q2	37	25.712	1.439	1.013	1.421
	q3	21	27.853	0.754	1.027	0.734
	q4	32	29.994	1.067	1.018	1.048
2000	q1	30	32.135	0.934	0.993	0.940
	q2	41	34.277	1.196	0.995	1.203
	q3	29	36.418	0.796		
	q4	40	38.559	1.037		

Worked example 21.4   Seasonal index using the MINITAB decomposition method
(continued)

	Raw Seasonals				
	q1	q2	q3	q4	
1997			1.215	0.974	
1998	0.713	1.051	1.112	1.070	
1999	0.720	1.421	0.734	1.048	
2000	0.940	1.203			
					Means
Medians	0.720	1.203	1.112	1.048	1.021
Adjusted Medians (Seasonal indices)	0.705	1.178	1.090	1.027	1.000

The adjusted medians are calculated as: Median$\times(1/1.021)$. Expressed in a comparable way to the seasonal indices calculated in Worked example 21.3, where they are expressed as percentages, the seasonal indices would be: 70.5 (q1), 117.8 (q2), 109.0 (q3) and 102.7 (q4).

It will be appreciated that while, in principle, there is a good argument in favour of the median, rather than the mean, in calculating the seasonal indices, it is not satisfactory in conjunction with such a small data set. Indeed reasonably reliable seasonal indices can only be calculated using much larger data sets, regardless of the method of measurement. The examples used here were chosen merely for illustrative purposes and ease of calculation. It should also be noted that the use of a linear trend may not be appropriate (although in this case it will be shown later that it has a better fit than the moving average trend used in Worked example 21.1).

Computer solution using MINITAB

We illustrate here the use of the MINITAB Decomposition procedure and the output. With the time series in column C1 of the MINITAB worksheet, the procedure is as follows:

Choose **Stat** → **Time Series** → **Decomposition**
In **Variable** enter C1 and in **Seasonal length** enter '4'
Under **Model type** check **multiplicative**
Under **Model Components** check **Trend plus seasonal**
In **First obs is in seasonal period** enter '1'
Choose **Results**
Under **Graphics** check **Display plots**
Under **Output** check **Summary table and results table**
Click **OK**

The output is as follows:

Time Series Decomposition

Data	C1
Length	16.0000
NMissing	0

Trend Line Equation

$Yt = 4.3 + 2.14118*t$

Seasonal Indices

Period	Index
1	0.705154
2	1.17832
3	1.08977
4	1.02675

Required seasonal index ←

Accuracy of Model

MAPE:	14.2398
MAD:	3.0365
MSD:	20.3462

Authors' comment and interpretation: Comparison of these measures of accuracy with those for the moving average trend (Worked example 21.3) shows that the linear trend provides a better fit. The corresponding measures for the moving average trend are: MAPE = 25.2502; MAD = 7.2273 and MSD = 76.2955.

Authors comment: Textual output is confined to that shown here when the option of 'Summary Table' is chosen. Choice of 'Summary table and table of results' produces, in addition, the table shown below. A set of three charts is also produced automatically by MINITAB unless graphical output is suppressed. These charts are also shown below.

Row	Actual time series	Linear trend	Seasonal index	(C1)/(C2)	Seasonally adjusted series = (C1)/(C3)	Predicted values = seasonalized trend = (C2)×(C3)	Residuals = (C1)–(C6)
	TS (C1)	Trend (C2)	Seasonal (C3)	De-trend (C4)	Deseason (C5)	Model (C6)	Error (C7)
1	6	6.4412	0.70515	0.93151	8.5088	4.5420	1.4580
2	9	8.5824	1.17832	1.04866	7.6380	10.1128	−1.1128
3	13	10.7235	1.08977	1.21229	11.9291	11.6862	1.3138
4	12	12.8647	1.02675	0.93278	11.6873	13.2089	−1.2089
5	10	15.0059	0.70515	0.66641	14.1813	10.5815	−0.5815
6	17	17.1471	1.17832	0.99142	14.4273	20.2048	−3.2048
7	21	19.2882	1.08977	1.08875	19.2701	21.0197	−0.0197
8	24	21.4294	1.02675	1.11996	23.3747	22.0027	1.9973
9	18	23.5706	0.70515	0.76366	25.5263	16.6209	1.3791
10	37	25.7118	1.17832	1.43903	31.4006	30.2968	6.7032
11	21	27.8529	1.08977	0.75396	19.2701	30.3533	−9.3533
12	32	29.9941	1.02675	1.06688	31.1662	30.7965	1.2035
13	30	32.1353	0.70515	0.93355	42.5439	22.6603	7.3397
14	41	34.2765	1.17832	1.19616	34.7952	40.3887	0.6113
15	29	36.4176	1.08977	1.79632	26.6111	39.6869	−10.6869
16	40	38.5588	1.02675	1.03738	38.9578	39.5904	0.4096

Worked example 21.4 Seasonal index using the MINITAB decomposition method
(continued)

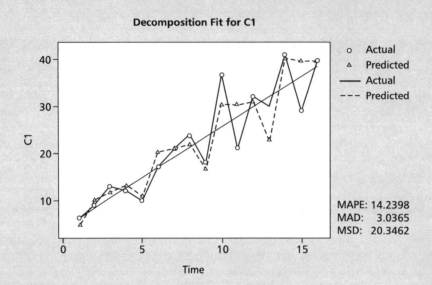

Decomposition Fit for C1

o Actual
△ Predicted
— Actual
--- Predicted

MAPE: 14.2398
MAD: 3.0365
MSD: 20.3462

Authors' comment: The 'Decomposition Fit for C1', shown above, plots the actual time series (C1) against the predicted values (C6) and the linear trend (C2).

Co ponent nal sis for C1

riginal Data

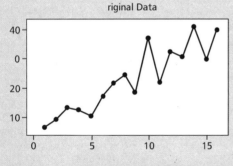

Detrended Data

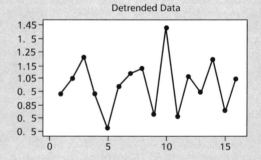

Seasonall d usted Data

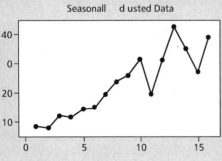

Seasonall d usted and Detrended Data

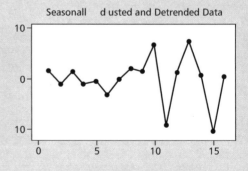

Authors' comment: The 'Component Analysis for C1', shown above, is self-explanatory – plotting the actual (C1) and seasonally-adjusted (C5) series, on the left, and the de-trended series (C4) and the residual errors (C7) on the right. This set of plots is intended to show how adjusting the series for different components affects the data.

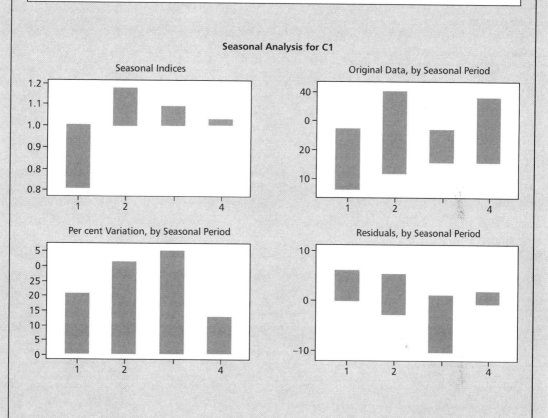

Seasonal Analysis for C1

Authors' comment: The 'Seasonal Analysis for C1' shows charts of seasonal indices and per cent variation within each season relative to the sum of variation by season and boxplots of the data and of the residuals by seasonal period.

Computer solution using Excel

There is no pre-defined tool in Excel for solving this problem.

Worked example 21.5 Computing a seasonally adjusted series

Seasonally adjust the original time series data y_t shown in Table 21.2 using the seasonal index derived in Worked example 21.3 above.

Worked solution

Seasonal adjustment is carried out by dividing the actual time series y_t by the corresponding seasonal index S for each time period. The results are shown in the final column of the table below.

Year/quarter		Actual time series (y_t)	Seasonal index (S_t)	Seasonally adjusted series $(y_t/S_t) \times 100$
1997	Q1	6	77.6	7.73
	Q2	9	119.7	7.52
	Q3	13	100.8	12.90
	Q4	12	101.9	11.78
1998	Q1	10	77.6	12.89
	Q2	17	119.7	14.20
	Q3	21	100.8	20.83
	Q4	24	101.9	23.55
1999	Q1	18	77.6	23.20
	Q2	37	119.7	30.91
	Q3	21	100.8	20.83
	Q4	32	101.9	31.40
2000	Q1	30	77.6	38.66
	Q2	41	119.7	34.25
	Q3	29	100.8	28.77
	Q4	40	101.9	39.25

Computer solution using MINITAB

Seasonal adjustment using MINITAB is carried out automatically as part of the MINITAB Decomposition Method illustrated above in Worked example 21.4 where the seasonally adjusted series is shown in column C4 in the MINITAB tabular output. It differs from the series shown above because of the difference in methodology – as explained in Worked example 21.4.

Worked example 21.6 Deriving seasonalized forecasts

In Worked example 21.2 we used a linear trend fitted to annual data to derive forecasts for subsequent years. When using non-annual data (e.g. quarterly or monthly, etc.) for forecasting purposes we need to allow for seasonal effects in order to produce seasonalized forecasts.

Given the information in the table below, which shows passenger-miles (millions) flown by a major airline each quarter from 1997 to 1999, derive forecasts for each quarter of 2000.

	Years		
Quarter	1997	1998	1999
Q1	77	99	113
Q2	166	191	229
Q3	252	287	316
Q4	104	123	156

Worked solution

The table below shows the calculation of a linear trend and the ratios of actual (y_t) to trend (T_y) values – i.e. y_t/T_y. The calculation of a seasonal index using these ratios is shown below the table. Finally, we also show the forecast trend values (T_y^f) and the corresponding seasonalized forecasts.

Linear trend and ratios of actual to trend

Date		Time x_t	Miles y_t	Trend T_y $(132 + 8x_t)$	Ratio $(y_t/T_y) \times 100$
1997	Q1	0	77	132	58.3
	Q2	1	166	140	118.6
	Q3	2	252	148	170.3
	Q4	3	104	156	66.7
1998	Q1	4	99	164	60.4
	Q2	5	191	172	111.0
	Q3	6	287	180	159.4
	Q4	7	123	188	65.4
1999	Q1	8	113	196	57.7
	Q2	9	229	204	112.3
	Q3	10	316	212	149.1
	Q4	11	156	220	70.9

Calculation of a seasonal index

Year	Ratio $(y_t/T_y) \times 100$				
	Q1	Q2	Q3	Q4	
1997	58.3	118.6	170.3	66.7	
1998	60.4	111.0	159.4	65.4	
1999	57.7	112.3	149.1	70.9	
Average (seasonal index, S)	58.80	113.97	159.60	67.67	(Total = 400)

383

Worked example 21.6   Deriving seasonalized forecasts *(continued)*

Forecast trend values and seasonal forecasts

Date	Time x_t	Forecast trend T_y^f	Seasonal index S	Seasonalized forecasts $(T_y^f \times S)/100$
2000 Q1	12	228	58.80	134
Q2	13	236	113.97	269
Q3	14	244	159.60	389
Q4	15	252	67.67	171

Computer solution using MINITAB

Seasonalized forecasts are computed automatically in MINITAB as part of the MINITAB Decomposition Method explained in Worked example 21.4. In addition to the procedure set out there it is merely necessary to add the following commands:

Check **Generate forecasts** and in **Number of forecasts** enter '4' and in **Starting at origin** enter 12 (this is the period immediately preceding the first forecast period). The number assigned to this period differs from that shown in the manual example because MINITAB automatically starts the numbering of the time periods at '1' not '0', as in that table.

Time Series Decomposition

Data	C1
Length	12.000
NMissing	0

Trend Line Equation

Yt = 124.470 + 7.94056*t

Seasonal Indices

Period	Index
1	0.592778
2	1.13402
3	1.62072
4	0.652479

Accuracy of Model

MAPE:	3.2477
MAD:	6.5268
MSD:	95.9720

Forecasts

Row	Period	Forecast
1	13	134.974
2	14	267.218
3	15	394.772
4	16	164.111

Required forecasts. These differ from the manual forecasts above because of the differences in methodology explained in Worked example 21.4

Computer solution using Excel

There is no pre-defined tool in Excel for solving this problem. Part of the solution – forecasts of linear trend values – may be done automatically in Excel (as in Worked example 21.2 above) but the required seasonal index would need to be calculated manually, as before. Given a seasonal index, the procedure to compute seasonalized forecasts could be carried out as a simple spreadsheet procedure.

Worked example 21.7　Cyclical analysis

The data for annual new car sales, given in Worked example 21.1 above, show fairly marked cyclical variations – see the graph on p. 365. By making appropriate calculations, determine how many cycles occurred over the period and indicate the interval between successive peaks and troughs.

Worked solution

The first step is to de-trend the data by computing the ratio y_t/T_y. The result of this step is shown below:

Year	Time (x_t)	Sales (y_t)	Trend (T_y)	Ratio (y_t/T_y)
1980	1	2.6	2.59	1.00
1981	2	2.5	2.70	0.93
1982	3	2.3	2.81	0.82
1983	4	3.2	2.92	1.10
1984	5	2.9	3.03	0.96
1985	6	3.2	3.14	1.02
1986	7	3.3	3.25	1.02
1987	8	3.6	3.36	1.07
1988	9	3.8	3.47	1.10
1989	10	3.7	3.58	1.03
1990	11	3.4	3.69	0.92
1991	12	3.6	3.80	0.95
1992	13	4.0	3.91	1.02
1993	14	4.2	4.02	1.05
1994	15	4.5	4.13	1.09
1995	16	4.6	4.24	1.09
1996	17	4.4	4.35	1.01
1997	18	4.2	4.46	0.94
1998	19	3.8	4.57	0.83
1999	20	4.8	4.68	1.03

The de-trended series y_t/T_y still contains the effect of random factors. To remove these, we compute a moving average of the de-trended series to smooth out the influence of these factors. For this purpose, it is necessary to take a moving average over a fairly short period in order to guard against the danger

Worked example 21.7   Cyclical analysis *(continued)*

of oversmoothing. In the example here, we use a three-period moving average. The calculations are shown in the table below. It will be seen from the moving average of the de-trended car sales data that there are three cycles defined by sales peaks in 1984, 1988, and 1994, with corresponding sales troughs in 1985, 1991 and 1997. The figure below illustrates these results.

Year	Ratio (y_t/T_y)	Three-period moving average	
1980	1.00	—	
1981	0.93	0.92	
1982	0.82	0.95	
1984	1.10	0.96	
1984	0.96	1.03	←Peak
1985	1.02	1.00	←Trough
1986	1.02	1.04	
1987	1.07	1.06	
1988	1.10	1.07	←Peak
1989	1.03	1.02	
1990	0.92	0.97	
1991	0.95	0.96	←Trough
1992	1.02	1.01	
1993	1.05	1.05	
1994	1.09	1.08	←Peak
1995	1.09	1.06	
1996	1.01	1.01	
1997	0.94	0.93	←Trough
1998	0.83	0.93	
1999	1.03	—	

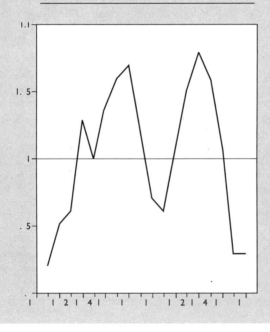

Computer solutions

There are no automatic procedures in MINITAB or Excel for cyclical analysis.

Key terms and concepts

Additive model

$$y_t = T_t + S_t + C_t + R_t$$

Cyclical component (C) The component of a time series that results in periodic above-trend and below-trend movements lasting more than 1 year.

Irregular (random) component (R) The component of a time series that reflects the variation of the series other than that which can be explained by trend, cyclical and seasonal components.

Linear trend method A method of measuring the trend (T_y) which applies regression analysis to the time series data with time (x) taken as the independent (i.e. explanatory) variable.

$$T_y = a + bx$$

Moving average method A method of smoothing a time series by averaging successive groups of the observations.

Multiplicative model

$$y_t = T_t \times S_t \times C_t \times R_t$$

Seasonal component (S) The component of a time series that shows a regular pattern within one calendar year.

Seasonal index An index which measures the average magnitude of seasonal effects in a particular time series.

Seasonally adjusted series Time series in which the average effect of the seasons has been removed.

Time series A set of observations measured at successive points in time or over successive periods of time.

Trend (T) The long-run movement in a time series.

Self-study exercises

21.1. What is meant by the decomposition of time series?

21.2. With which characteristic movement of a time series would you mainly associate each end of the following:

(a) the change in the population since 1900?

(b) an era of prosperity?

(c) a strike?

(d) monthly electricity consumption?

    (e) Christmas sales revenue?

    (f) a prolonged heat-wave?

21.3. The table below shows quarterly beer consumption statistics (million bulk barrels). Calculate a linear trend equation.

	Quarter			
Year	Q1	Q2	Q3	Q4
1998	8	9	10	9
1999	8	10	11	12
2000	9	10	12	13

21.4. The figures below show quarterly butter production (thousands of tonnes) over a four-year period.

	1997	1998	1999	2000
January–March	2	7	10	14
April–June	8	14	17	20
July–September	2	9	12	15
October–December	2	8	9	12

    (a) Derive a moving-average trend.

    (b) Was the rise in production in the second quarter of 2000 more or less than could have been anticipated on seasonal grounds?

    (c) Comment on the most notable features of the data and offer some explanation for them.

21.5. The table below shows a quarterly time series of unemployment together with its estimated trend in brackets. Derive a seasonal index and determine whether a value of 50.4 (000s) unemployed for Spring 2001 represents, after seasonal adjustment, a rise or a fall on Winter 2000.

	Unemployment (000s)			
Year	Spring	Summer	Autumn	Winter
1997	45.8	43.6	45.5	47.5
	(45.6)	(45.2)	(45.3)	(45.7)
1998	47.2	44.6	46.3	46.0
	(46.0)	(45.7)	(45.8)	(45.9)
1999	45.4	40.4	46.1	47.5
	(45.8)	(44.7)	(45.0)	(45.5)
2000	43.9	43.4	50.1	53.0
	(45.2)	(44.8)	(45.9)	(47.3)

21.6. Produce a seasonally adjusted series for the unemployment data given in question 21.5 above.

21.7. The data shown below give quarterly sales of electricity (ELEC) in billions of kilowatt hours.

Year	Q1	Q2	Q3	Q4
		Quarter		
1997	6.00	4.80	4.96	5.70
1998	4.46	4.10	4.60	5.64
1999	4.90	5.00	4.82	7.32
2000	6.36	6.20	7.40	9.40

With TIME measured from 1997 Q1 = 1, 1997 Q2 = 2, etc., the estimated regression equation for these data is:

ELEC = 4.12 + 0.189 TIME.

(a) Use the linear trend to calculate the seasonal index.

(b) By how much is electricity sales normally expected to rise or fall in the first quarter on account of seasonal factors?

(c) De-seasonalize the data and explain why the de-seasonalized data are not equal to the linear trend.

(d) Forecast quarterly electricity sales for 2001 and explain the problems that might result if the forecasts were extended to 2009.

21.8. The sales of a product over the period 1985–2000 have shown cyclical variations around a strong underlying linear trend. Actual sales and trend values are reported in the table below. By making appropriate calculations, say how many cycle peaks and troughs occurred in this period and indicate the interval between successive peaks and troughs.

Year	Actual	Trend
	Sales (000s)	
1985	1.0	0.8
1986	1.8	1.0
1987	2.1	1.2
1988	2.0	1.4
1989	1.8	1.6
1990	1.9	1.8
1991	2.3	2.0
1992	3.0	2.2
1993	3.3	2.4
1994	3.6	2.6
1995	3.6	2.8
1996	3.2	3.0
1997	2.9	3.2
1998	3.6	3.4
1999	3.9	3.6
2000	3.8	3.8

| Chapter 22 | Index numbers |

Index numbers

Aims In the last chapter we explored how time series data can be broken down into four components and how they may be used for forecasting purposes. We now use time series data in this chapter to construct index numbers as a way of measuring changes in prices or quantities over time (or space).

Index numbers are a set of percentage values with reference to a base period, designated as 100. Thus, an index number with a value exceeding 100 indicates an increase in the variable under consideration relative to the value in the base period. Similarly, a value less than 100 indicates a decrease relative to the base period value. Index numbers may be used to measure differences in the magnitude of a single variable (producing a *simple* index) or a group of related variables (producing a *composite* index) over time. In addition, index numbers may also be used to measure differences between spatial locations (i.e. at a single point in time); e.g. an index could be constructed to show differences between average house prices in Scotland and Wales in a particular time period.

Index numbers are used widely, especially by the government for monitoring economic and social trends and by management for monitoring business performance. The most well-known example of a *price index* in the UK is the retail prices index (RPI) which measures inflation in retail prices. It is generally a prominent news item every month, especially when prices are rising fairly quickly. What the index does is to express in a single summary measure exactly how much prices in general have changed for a broad range of goods and services over some period of time. This is frequently a crucial element, for example, in wage bargaining (as workers attempt to maintain the purchasing power of their wages) and in the index-linking of pensions. In many countries the comparable measure is called the consumer price index (CPI). Another well-known index in the UK is the index of the output of production industries which measures the change in the output of the broad range of manufacturing and other industries. This index – in contrast to the RPI – is a *quantity index* and is an important measure of economic growth.

In this chapter we illustrate, first, the basic principles of index number construction with reference to price changes over time. We then show how the same principles may be applied in measuring quantity changes. The chapter deals with the following aspects of index numbers in turn:

- simple price indexes
- composite price indexes:

  weighted averages of simple price indexes

  weighted aggregate price indexes
- a summary and comparison of price index formulae
- quantity indexes:

  weighted averages of simple quantity indexes

  weighted aggregate quantity indexes
- considerations in the construction of index numbers

Learning Outcomes   After working through this chapter you will be able to:

- appreciate the meaning and uses of index numbers of prices and quantities
- comprehend the difference between simple index numbers and composite index numbers
- understand the notation and formulae concerning the use and construction of index numbers
- recognize the difference between base weights and current weights
- understand the meaning of the base of an index
- compute simple price and quantity indexes
- construct composite (aggregate) price index numbers using the Laspeyres and Paasche formulae
- construct composite (aggregate) quantity index numbers using the Laspeyres and Paasche formulae
- appreciate the distinction between, and use of, current and constant prices
- use a price index to deflate current prices to constant prices
- appreciate the technical considerations which underlie the construction of index numbers generally.

Principles

Simple price indexes

A *simple price index* – often referred to technically as a *price relative* – expresses the price of an item in a particular time period as a percentage of its price in a chosen base time period. Thus, if the price of a packet of sugar was £1 on 1 January 1990 and the price on 1 January 1999 was £1.45, then a simple price index for 1 January 1999, taking 1 January 1990 as the base, is given by

$$\frac{P_{\text{January 1999}}}{P_{\text{January 1990}}} \times 100 = \frac{£1.45}{£1.00} \times 100 = 145$$

where P represents the price (on either 1 January 1999 or 1 January 1990).

Thus, the index number 145 indicates that by 1 January 1999 the price of sugar had risen by 45 per cent since the base period, 1 January 1990.

Notation

At this point it is convenient to introduce notation which will be used throughout the rest of this chapter. We use the letters p for prices and q for quantities and subscripts n for the *current* time period, o for the *base* time period and t for *any* time period. Thus, p_n refers to current period prices and p_o to base period prices (and so on for quantities). In addition, PI denotes a price index and QI a quantity index. Hence, using this notation, a simple price index for the time period n, relative to a base period o, is defined as:

Simple price index

$$\text{PI}_n = \frac{p_n}{p_o} \times 100$$

Composite price indexes

In contrast to a simple price index, which measures the change in price of a *single* item, a *composite price index* provides a summary measure for the average price change of a *group* of items. This is a more complex problem and it is this problem that lies at the centre of index number theory. The RPI referred to above, which covers a whole group of commodities – the typical 'basket' of goods and services – consumed by households in general, is a good example of such an index. Obviously, it would be possible to take an average of the simple price indexes for each constituent item in the typical basket of goods and services. However, this has the major disadvantage of giving equal weight to each item even though consumers spend more on some items than on others. What is needed, therefore, is a *weighted average* which takes account of the relative importance of each item in the basket.

Weighted averages can be applied in two ways to produce composite indexes. These are:

▪ weighted averages of simple price indexes
▪ weighted aggregate price indexes

We deal with each in turn.

Weighted averages of simple price indexes

Consider the data in Table 22.1 below which show details of just three commodities bought by a consumer: sugar, milk and eggs in 1990 and 1999.

The table gives the price p of each commodity together with the quantity q purchased and the value of expenditure (pq) on each item for each year.

Simple price indexes (i.e. price relatives) for these three commodities showing the price increase over the period 1990–9 are given in Table 22.2 below.

It will be seen that over the period 1990–9 each of the individual commodities has increased in price by a different percentage – sugar by 30 per cent, milk by 20 per cent and eggs have not risen at all. Thus, the simple price indexes (with base 1990 = 100) are 130, 120 and 100 respectively.

To produce a summary – composite – measure of the price increase for these three commodities requires the computation of a weighted average of the three simple price indexes.

Denoting the weights for each of the i commodities as W_i we have:

$$PI_n = \frac{\Sigma(\text{simple price indexes} \times W_i)}{\Sigma W_i}$$

This is equivalent to

$$PI_n = \frac{\Sigma(p_{in}/p_{io})W_i}{\Sigma W_i} \times 100$$

The W_i weights should reflect the relative importance of the purchases of the three commodities. This is reflected both in the quantities bought and also in the value of expenditure on each commodity – as shown in Table 22.1. The use of quantity weights to

Table 22.1 Prices, quantities and value of expenditure on selected food items

Commodity	Unit	Unit prices (£)		Quantities bought		Value of expenditure (£)	
		1990 p_o	1999 p_n	1990 q_o	1999 q_n	1990 $p_o q_o$	1999 $p_n q_n$
Sugar	kilo	1.00	1.30	2	3	2.00	3.90
Milk	litre	0.50	0.60	10	10	5.00	6.00
Eggs	dozen	2.00	2.00	4	5	8.00	10.00

Table 22.2 Simple price indexes for three commodities

Commodity	Price relatives, p_n/p_o 1999/1990	Simple price index (1990 = 100)
Sugar	1.30/1.00 = 1.30	1.30×100 = 130
Milk	0.60/0.50 = 1.20	1.20×100 = 120
Eggs	2.00/2.00 = 1.00	1.00×100 = 100

produce a weighted average of simple price indexes is not appropriate, however, because the units of measurement (kilos, litres and dozens) are different and therefore cannot be added together to give the total weight ΣW_i required as the denominator for the above expression. The only common unit of measurement is money – hence, the appropriate weights in this case are the *values* of expenditure on each commodity. These values (i.e. prices × quantities) are also shown in Table 22.1.

It should be appreciated that there is a choice of weights to be made: W_i can either be taken as the value of expenditure in the base year (i.e. $p_{io}q_{io}$) or in the current year (i.e. $p_{in}q_{in}$). These two choices produce *base-weighted* and *current-weighted* indexes respectively, defined formally as follows:

Base-weighted average of simple price indexes

$$\text{PI}_n = \frac{\Sigma(p_{in}/p_{io})W_i}{\Sigma W_i} \times 100$$

where $W_i = p_{io}q_{io}$ for each of the i commodities.

Current-weighted average of simple price indexes

$$\text{PI}_n = \frac{\Sigma(p_{in}/p_{io})W_i}{\Sigma W_i} \times 100$$

where $W_i = p_{in}q_{in}$ for each of the i commodities.

Table 22.3 shows the data for the calculations of these two indexes using the information given in the example above (see Table 22.1). The actual calculations (shown overleaf) give a base-weighted composite index of 110.7; i.e. an average increase for prices in general over the period 1990–9 of 10.7 per cent. The current-weighted composite index shows an

Table 22.3 Calculation of weighted average of simple price indexes

Commodity	Simple price index $(p_{in}/p_{io}) \times 100$ (a)	Value weights 1990 $(p_{io}q_{io})$ (b)	Value weights 1999 $(p_{in}q_{in})$ (c)	Weighted simple price indexes 1990 (a)×(b)	Weighted simple price indexes 1999 (a)×(c)
Sugar	130	2.0	3.9	260	507
Milk	120	5.0	6.0	600	720
Eggs	100	8.0	10.0	800	1000
Summations (Σ)		15.0	19.9	1660	2227

equivalent rise of 11.9 per cent. The reason for this difference is due, obviously, to the fact that different weights are used in each case. The pros and cons for each are discussed in the later section 'Considerations in the construction of index numbers'.

Calculation of price indexes

(a) Base-weighted average of simple price indexes

$$\text{PI}_n = \frac{\Sigma(p_{in}/p_{io})p_{io}q_{io}}{\Sigma p_{io}q_{io}} \times 100 = \frac{1660}{15.0} = 110.7 \text{ (approx).}$$

(b) Current-weighted average of simple price indexes

$$\text{PI}_n = \frac{\Sigma(p_{in}/p_{io})p_{in}q_{in}}{\Sigma p_{in}q_{in}} \times 100 = \frac{2227}{19.9} = 111.9 \text{ (approx).}$$

Weighted aggregate price indexes

In contrast to the calculation of weighted averages of simple price indexes which we have just dealt with, an alternative is to compute weighted aggregates of prices × quantities $(p_i q_i)$, in which the quantities q_i are used as weights. Thus, $p_o q_o$ gives the value of base period quantities q_o at base period prices p_o and $p_i q_o$ gives the value of base period quantities q_o at current period prices p_i. Comparison of these two values gives a measure of the general change in price for a particular basket of goods relevant to the base period, producing a *base-weighted aggregate index*. This type of index is referred to as a *Laspeyres index*.

Likewise, a *current-weighted aggregate index* may be defined in which q_{in} weights are used instead of q_{io} weights. This type of index is referred to as a *Paasche index*.

Thus we have the following formulae for each type.

Laspeyres (base-weighted) aggregate price index

$$\text{PI}_n = \frac{\Sigma p_{in}q_{io}}{\Sigma p_{io}q_{io}} \times 100$$

Paasche (currrent-weighted) aggregate price index

$$\text{PI}_n = \frac{\Sigma p_{in}q_{in}}{\Sigma p_{io}q_{in}} \times 100$$

where p_{io} and p_{in} represent the prices and q_{io} and q_{in} represent the quantities of each good i bought in the base period o and current period n respectively.

Table 22.4 shows the calculation of both types of weighted aggregate price indexes using the same data as before (see Table 22.1).

Table 22.4 Calculation of Laspeyres and Paasche aggregate price indexes

| | Unit prices | | Quantities bought (weights) | | Values (£) – Aggregates | | | |
| | | | | | Base weights | | Current weights | |
Commodity	1990 p_{io}	1999 p_{in}	1990 q_{io}	1999 q_{in}	$p_{io}q_{io}$	$p_{in}q_{io}$	$p_{io}q_{in}$	$p_{in}q_{in}$
Sugar	1.00	1.30	2	3	2.00	2.60	3.00	3.90
Milk	0.50	0.60	10	10	5.00	6.00	5.00	6.00
Eggs	2.00	2.00	4	5	8.00	8.00	10.00	10.00
Summations (Σ)					15.00	16.60	18.00	19.90

Calculation of price indexes

(a) Laspeyres (base-weighted) aggregate price index

$$\text{PI}_n = \frac{\Sigma p_{in}q_{io}}{\Sigma p_{io}q_{io}} \times 100 = \frac{16.60}{15.00} \times 100 = 110.7 \text{ (approx)}.$$

(b) Paasche (current-weighted) aggregate price index

$$\text{PI}_n = \frac{\Sigma p_{in}q_{in}}{\Sigma p_{io}q_{in}} \times 100 = \frac{19.90}{18.00} \times 100 = 110.6 \text{ (approx)}.$$

Summary and comparison of price index formulae

It should be noted that in the case of base-weighted indexes, the index calculated as a weighted average of simple price indexes (see Table 22.3) is the same as that calculated as the weighted aggregate index shown above (Table 22.4) – both are 110.7 (rounded to one decimal place). This will always be the case because the formulae for the two calculations are equivalent; i.e.:

Base-weighted average of simple price indexes		Base-weighted aggregate price index
$\dfrac{\Sigma(p_{in}/p_{io})p_{io}q_{io}}{\Sigma p_{io}q_{io}}$	$\equiv$	$\dfrac{\Sigma p_{in}q_{io}}{\Sigma p_{io}q_{io}}$

However, in the case of the current-weighted indexes, the index calculated as a weighted average of simple price indexes (see Table 22.3) is not the same as a weighted aggregate price index (Table 22.4) – the former is 111.9 while the latter is 110.6. This is because the formulae are *not* equivalent; i.e.

Current-weighted average of simple price indexes		Current-weighted aggregate price index
$\dfrac{\Sigma(p_{in}/p_{io})p_{in}q_{in}}{\Sigma p_{in}q_{in}}$	$\neq$	$\dfrac{\Sigma p_{in}q_{in}}{\Sigma p_{io}q_{in}}$

A summary of the four price indexes we have considered in this chapter – weighted averages of simple price indexes and weighted aggregate indexes each weighted using base weights and current weights respectively – is shown in Table 22.5.

Table 22.5 Summary of price index formulae

Type of price index	Formula	Fixed weights used
(a) Weighted average of simple price indexes		
Base-weighted	$\dfrac{\Sigma(p_{in}/p_{io})p_{io}q_{io}}{\Sigma p_{io}q_{io}}$	Base weights $(p_{io}q_{io})$
Current-weighted	$\dfrac{\Sigma(p_{in}/p_{io})p_{in}q_{in}}{\Sigma p_{io}q_{in}}$	Current weights $(p_{in}q_{in})$
(b) Weighted aggregate price indexes		
Base-weighted (Laspeyres)	$\dfrac{\Sigma p_{in}q_{io}}{\Sigma p_{io}q_{io}}$	Base weights (q_{io})
Current-weighted (Paasche)	$\dfrac{\Sigma p_{in}q_{in}}{\Sigma p_{io}q_{in}}$	Current weights (q_{in})

As we noted earlier, the use of base-weighted and current-weighted formulae produces different index numbers. Care is required, therefore, in their interpretation. It should be appreciated that the different index numbers provide answers to different questions. Base-weighted indexes indicate the general rise in prices if the base period pattern of expenditure is maintained, while current-weighted indexes indicate the general rise in prices if the pattern of expenditure in the base period had been the same as in the current period. Thus, the choice of formula to use depends on the nature of the question being addressed. For most practical purposes, base-weighted indexes are most common because they do not require the determination of new weights in each time period, as is the case with current-weighted indexes. They are therefore more convenient and less costly to produce.

Quantity indexes

So far we have dealt with the derivation of *price* index numbers. We now turn to consider the construction of *quantity* index numbers. A *quantity index* (denoted QI) is a composite measure of the change in output or sales, etc. of a group of different commodities. An example of an official quantity index in the UK is the index of the output of production industries which, as the name implies, measures changes in the volume of output of manufacturing and other production industries. The principles of the methodology for computing quantity indexes are the same as those described above for computing price indexes. Consequently, they can be dealt with very briefly.

As before, a quantity index may be constructed using one of two methods, namely:

- the weighted average of simple quantity indexes method
- the weighted aggregate of quantities method

We describe each in turn below.

Weighted average of simple quantity indexes

Just as price indexes may be calculated as weighted averages of simple price indexes, so quantity indexes may be calculated as weighted averages of simple quantity indexes. As before, the weights used in these cases must be *value* weights (i.e. prices × quantities).

Using the same data as before involving the prices and quantities bought of three commodities – sugar, milk, eggs – shown in Table 22.1, we first need to compute the simple quantity indexes (i.e. *quantity relatives*) in the same way as we previously computed simple price indexes. We also need to compute base and current period value weights (i.e. expenditure $p_{io}q_{io}$ and $p_{in}q_{in}$ respectively).

Base-weighted and current-weighted composite quantity indexes may be expressed in general terms as

$$QI = \frac{\Sigma(\text{simple quantity indexes}) \times \text{weight}}{\Sigma \text{weights}}$$

The formulae for both types of composite index are as follows:

Weighted average of simple quantity indexes

Base-weighted: $QI_n = \dfrac{\Sigma(q_{in}/q_{io})p_{io}q_{io}}{\Sigma p_{io}q_{io}} \times 100$

Current-weighted: $QI_n = \dfrac{\Sigma(q_{in}/q_{io})p_{in}q_{in}}{\Sigma p_{in}q_{in}} \times 100$

The computations using these formulae, relating to the example above involving three commodities, sugar, milk and eggs, are set out in Table 22.6.

Table 22.6 Calculation of weighted average of simple quantity indexes

Commodity	Simple quantity index $(q_{in}/q_{io}) \times 100$ (a)	Value weights		Weighted simple quantity indexes	
		1990 $(p_{io}q_{io})$ (b)	1999 $(p_{in}q_{in})$ (c)	1990 (a)×(b)	1999 (a)×(c)
Sugar	150	2.00	3.90	300	585
Milk	100	5.00	6.00	500	600
Eggs	125	8.00	10.00	1000	1250
Summations (Σ)		15.00	19.90	1800	2435

Calculation of quantity indexes

(a) Base-weighted average of simple quantity indexes

$$QI_n = \frac{\Sigma(q_{in}/q_{io})p_{io}q_{io}}{\Sigma p_{io}q_{io}} \times 100 = \frac{1800}{15.00} = 120.0$$

(b) Current-weighted average of simple quantity indexes

$$QI_n = \frac{\Sigma(q_{in}/q_{io})p_{in}q_{in}}{\Sigma p_{in}q_{in}} \times 100 = \frac{2435}{19.90} = 110.7 \text{ (approx)}$$

Weighted aggregate quantity indexes

Weighted aggregate quantity indexes involve measuring the change in the value of quantities produced (or sold etc.) over a period of time in terms of unchanged (i.e. constant) prices. In effect, therefore, weighted aggregates are computed in which the weights are *prices* (contrast this with the corresponding price indexes in which the weights are *quantities* – in other words, prices and quantities are simply switched around).

Again, Laspeyres (base-weighted) indexes and Paasche (current-weighted) indexes may be calculated using the following formulae:

Weighted aggregate quantity indexes

Laspeyres (base-weighted): $\dfrac{\Sigma q_{in}p_{io}}{\Sigma q_{io}p_{io}} \times 100$

Paasche (current-weighted): $\dfrac{\Sigma q_{in}p_{in}}{\Sigma q_{io}p_{in}} \times 100$

Using the same data as before, the calculations needed to construct the resultant Laspeyres and Paasche indexes are shown in Table 22.7.

Table 22.7 Calculation of Laspeyres and Paasche aggregate quantity indexes

Commodity	Unit prices (£) 1990 p_{io} (a)	Unit prices (£) 1999 p_{in} (b)	Quantities bought 1990 q_{io} (c)	Quantities bought 1999 q_{in} (d)	Values (£) Base weights $q_{io}p_{io}$ (c)×(a)	Values (£) Base weights $q_{in}p_{io}$ (d)×(a)	Values (£) Current weights $q_{io}p_{in}$ (c)×(b)	Values (£) Current weights $q_{in}p_{in}$ (d)×(b)
Sugar	1.00	1.30	2	3	2.00	3.00	2.60	3.90
Milk	0.50	0.60	10	10	5.00	5.00	6.00	6.00
Eggs	2.00	2.00	4	5	8.00	10.00	8.00	10.00
Summations (Σ)					15.00	18.00	16.60	19.90

Calculation of quantity indexes

(a) Laspeyres (base-weighted) aggregate quantity index

$$QI_n = \frac{\Sigma q_{in}p_{io}}{\Sigma q_{io}p_{io}} \times 100 = \frac{18.00}{15.00} \times 100 = 120.0$$

(b) Paasche (current-weighted) aggregate quantity index

$$QI_n = \frac{\Sigma q_{in}p_{in}}{\Sigma q_{io}p_{in}} \times 100 = \frac{19.90}{16.00} \times 100 = 119.9 \text{ (approx)}$$

Considerations in the construction of index numbers

To conclude this chapter on index numbers we consider briefly a number of issues that arise in practice. It is also worth noting in passing that the theory relating to index number construction is much more complex than that presented here. Attention has been confined here to the basic statistical principles of index number construction. There are several factors to consider in the construction of index numbers in practice:

▪ choice of base period

▪ purpose of measurement

▪ representativeness

▪ changing composition

▪ choice of computational method

We discuss each of these in turn.

Choice of base period

It is always desirable to choose as the base or 'reference' period, for the index, a period of time which is not in some sense abnormal, otherwise the index can give a misleading impression. For example, basing an output index on a period badly affected by strikes and thus when volumes of output are low, exaggerates the increase in output afterwards. Likewise, basing a price index at the peak of an inflationary period tends to exaggerate the subsequent moderation of price movements. The choice of base period, therefore, should reflect some sort of norm – a 'typical' period (as far as possible) so that comparisons of prices or quantities made against it are not distorted.

Purpose of measurement

Needless to say, the construction of the index needs to match the purposes for which it is required. For example, the adjustment of old-age pensions in order to maintain their purchasing power over time requires that they be adjusted by an index of prices which measures the change in prices for goods and services typically bought by old-age pensioners rather than prices in general (i.e. relevant to *all* households). In other words, a special pensioner price index is required for this purpose rather than a general one. Likewise, if the purpose is to construct an index of wholesale prices, it would be inappropriate to consider the retail prices of goods for inclusion. It is important, therefore, to be clear about the purpose of measurement before deciding on the relevant items to be included.

Representativeness

The items selected for inclusion in an index naturally need to be representative of the underlying population from which the included sample items are drawn. For example, in constructing a consumer price index, the 'basket' of goods and services re-priced in each time period must be representative of the range of goods and services bought by the consumers in question. At the same time, detailed specifications for the items must be defined and adhered to in collecting the relevant prices in each time period. In addition, a sufficiently large sample has to be collected from different outlets to ensure representativeness.

Changing composition

A major problem in practice for indexes maintained over a long period of time is that new products come into use while others become obsolete. This problem has to be taken account of by some method of revising the composition of the index from time to time and chaining the indexes before and after to produce a continuous series. At the same time, changes in the quality of goods are taking place all the time and adjustments need to be made to allow for this if the series of index numbers is to remain unbiased.

Choice of computational method

As we showed earlier, alternative formulae may be used for computing index numbers but these produce different values. A choice has to be made, therefore, concerning the appropriate computational method. This must depend essentially upon the purpose for which the index is required. For example, a base-weighted price index measures, by definition, the change in prices of the basket of goods that was typical for the base period. If the focus of interest, however, is on current, rather than historic, consumption patterns, then a current-weighted index is appropriate.

Applications and Worked Examples

The data given below show the prices, quantities and values of purchases of three chemicals used by a firm in the production of fertilizers for the years 1990 and 2000.

Chemical	Prices (£ per kilogram)		Quantities (kilograms)		Values (£)	
	1990	2000	1990	2000	1990	2000
A	2	3	40	50	80	150
B	4	5	20	25	80	125
C	10	12	10	15	120	180

Worked examples 22.1–22.4 below refer to these data. Note that computer solutions are not given here because there are no automatic routines for index numbers in MINITAB or Excel (although the programs could, of course, be used to do the arithmetic calculations).

Worked example 22.1　Simple price indexes

Using the data above, calculate simple price indexes for the three chemicals for 2000 with a base year of 1990 = 100.

Solution

$$A: \text{PI}_{2000} = \frac{P_{2000}}{P_{1990}} \times 100 = \frac{3}{2} \times 100 = 150.0$$

$$B: \text{PI}_{2000} = \frac{P_{2000}}{P_{1990}} \times 100 = \frac{5}{4} \times 100 = 125.0$$

$$C: \text{PI}_{2000} = \frac{P_{2000}}{P_{1990}} \times 100 = \frac{12}{10} \times 100 = 120.0$$

Worked example 22.2　Composite price indexes

Using the information in Worked example 22.1 above, compute the following:

(i) a base-weighted average of simple price indexes for 2000 with a base year of 1990 = 100;

(ii) a current-weighted average of simple price indexes for 2000 with a base year of 1990 = 100;

(iii) a base-weighted (Laspeyres) aggregate price index for 2000 with a base year of 1990 = 100;

(iv) a current-weighted (Paasche) aggregate price index for 2000 with a base year of 1990 = 100.

Worked example 22.2   Composite price indexes *(continued)*

Solutions

The solutions to these problems are obtained using the worksheet set out below.

Chemical	Prices (£ per kilogram)		Quantities (kilograms)		Values (£)			
	p_{1990}	p_{2000}	q_{1990}	q_{2000}	$p_{1990}q_{1990}$	$p_{2000}q_{2000}$	$p_{2000}q_{1990}$	$p_{1990}q_{2000}$
A	2	3	40	50	80	150	120	100
B	4	5	20	25	80	125	100	100
C	10	12	10	15	100	180	120	150

(i) Base-weighted average of simple price indexes for 2000 with a base year of 1990 = 100. Using value weights:

$$PI_{2000} = \frac{\Sigma(p_{i\,2000}/p_{i\,1990})W_i}{\Sigma W_i} \times 100$$

where $W_i = p_{i1990}q_{i1990}$

$$PI_{2000} = \frac{(3/2) \times 80 + (5/4) \times 80 + (12/10) \times 100}{80 + 80 + 100} \times 100$$

$$= \frac{340}{260} \times 100$$

$$= 130.77$$

(ii) Current-weighted average of simple price indexes for 2000 with a base year of 1990 = 100.

$$PI_{2000} = \frac{\Sigma(p_{i\,2000}/p_{i\,1990})W_i}{\Sigma W_i} \times 100$$

where $W_i = p_{i\,2000}q_{i\,2000}$

$$PI_{2000} = \frac{(3/2) \times 150 + (5/4)125 + (12/10) \times 180}{150 + 125 + 180} \times 100$$

$$= \frac{597.25}{455} \times 100$$

$$= 131.26$$

(iii) Base-weighted (Laspeyres) aggregate price index for 2000 with a base year of 1990 = 100.

$$PI_{2000} = \frac{\Sigma p_{i\,2000}q_{i\,1990}}{\Sigma p_{i\,1990}q_{i\,1990}} \times 100$$

$$= \frac{120 + 100 + 120}{80 + 80 + 100} \times 100$$

$$= 130.77$$

(iv) Current-weighted (Paasche) aggregate price index for 2000 with a base year of 1990 = 100.

$$PI_{2000} = \frac{\Sigma p_{i\,2000} q_{i\,2000}}{\Sigma p_{i\,1990} q_{i\,2000}} \times 100$$

$$= \frac{150 + 125 + 180}{100 + 100 + 150} \times 100$$

$$= 130.0$$

Worked example 22.3   Simple quantity indexes

Using the data above, calculate simple quantity indexes for the three chemicals for 2000 with a base of 1990 = 100.

Solution

$$A:\ QI_{2000} = \frac{Q_{2000}}{Q_{1990}} \times 100 = \frac{50}{40} \times 100 = 125$$

$$B:\ QI_{2000} = \frac{Q_{2000}}{Q_{1990}} \times 100 = \frac{25}{20} \times 100 = 125$$

$$C:\ QI_{2000} = \frac{Q_{2000}}{Q_{1990}} \times 100 = \frac{15}{10} \times 100 = 150$$

Worked example 22.4   Composite quantity indexes

Using the data above, compute the following:

(i) a base-weighted average of simple quantity indexes for 2000 with a base year of 1990 = 100;

(ii) a current-weighted average of simple quantity indexes for 2000 with a base year of 1990 = 100;

(iii) a base-weighted (Laspeyres) aggregate quantity for 2000 with a base year of 1990 = 100;

(iv) a current-weighted (Paasche) aggregate quantity indexes for 2000 with a base year of 1990 = 100.

Solutions

The solutions to these problems are obtained using the worksheet set out in Worked example 22.2 above.

(i) Base-weighted average of simple quantity indexes for 2000 with a base year of 1990 = 100.

$$QI_{2000} = \frac{\Sigma(q_{i\,2000}/q_{i\,1990})W_i}{\Sigma W_i} \times 100$$

Worked Example 22.4 Composite quantity indexes *(continued)*

where $W_i = p_{i\,1990}q_{i\,1990}$

$$QI_{2000} = \frac{(50/40) \times 80 + (25/20) \times 80 + (15/10) \times 100}{80 + 80 + 100} \times 100$$

$$= \frac{350}{260} \times 100$$

$$= 134.62$$

(ii) Current-weighted average of simple quantity indexes for 2000 with a base year of 1990 = 100.

$$QI_{2000} = \frac{\Sigma(q_{i\,2000}/q_{i\,1990})W_i}{\Sigma W_i} \times 100$$

where $W_i = p_{i\,2000}q_{i\,2000}$

$$QI_{2000} = \frac{(50/40) \times 150 + (25/20) \times 125 + (15/10) \times 180}{150 + 125 + 180} \times 100$$

$$= \frac{613.75}{455} \times 100$$

$$= 134.89$$

(iii) Base-weighted (Laspeyres) aggregate quantity index for 1990 with a base year of 1990 = 100.

$$QI_{2000} = \frac{\Sigma q_{i\,2000}p_{i\,1990}}{\Sigma q_{i\,1990}p_{i\,1990}} \times 100$$

$$= \frac{100 + 100 + 150}{80 + 80 + 100} \times 100$$

$$= \frac{350}{260} \times 100$$

$$= 134.62$$

(iv) Current-weighted (Paasche) aggregate quantity index for 1990 with a base year of 1990 = 100.

$$QI_{2000} = \frac{\Sigma q_{i\,2000}p_{i\,2000}}{\Sigma q_{i\,1990}p_{i\,2000}} \times 100$$

$$= \frac{150 + 125 + 180}{120 + 100 + 120} \times 100$$

$$= \frac{455}{340} \times 100$$

$$= 133.82$$

Worked example 22.5 Use of price indexes – calculating 'real' percentage changes

Over the period from 1990 to 2000 the price index for materials purchased by a firm increased from 100 to 189 while the value of materials purchased rose from £280,000 to £950,000. By what percentage did the 'real' value of materials purchased increase?

Solution

The value in real terms represents the values at *constant* prices. At 1990 price levels these are:

$$\text{1990 value:} \quad \frac{£280,000}{100} \times 100 = £280,000$$

$$\text{2000 value:} \quad \frac{£950,000}{189} \times 100 = £502,645.5$$

Expressing the deflated 2000 value as a percentage of the 1990 value:

$$\frac{\text{2000 value}}{\text{1990 value}} = \frac{502,645.5}{280,000} \times 100 = 179.52$$

i.e. a real increase of 79.52 per cent from 1990 to 2000.

Worked example 22.6 Use of price indexes – index-linking

If a person retired in March 1999 with an index-linked pension of £10,000, and the pension is linked to the retail prices index, which increased from 100.0 in March 1999 to 101.8 in March 2000:

(a) what would the pension be in March 2000?

(b) if the index was not index-linked how much would it have been worth in March 2000 in terms of March 1999 prices?

Solution

(a) $£10,000 \times \dfrac{101.8}{100} = £10,180$

(b) $\dfrac{£10,000}{101.8} \times 100 = £9,823.18$

Key terms and concepts

Composite price index A summary measure of the average price change of a group of items.

Index numbers A set of percentage values, expressed with reference to a base of 100, used to measure differences over time or space in the magnitude of either a single variable or a group of related variables.

Laspeyres weighted aggregate price index A weighted aggregate price index where the weight for each item is its base period quantity (denoted q_{io}).

$$PI_n = \frac{\Sigma p_{in} q_{io}}{\Sigma p_{io} q_{io}} \times 100$$

Paasche weighted aggregate price index A weighted aggregate price index where the weight for each item is its current period quantity (denoted q_{in}).

$$PI_n = \frac{\Sigma p_{in} q_{in}}{\Sigma p_{io} q_{in}} \times 100$$

Price relative A simple price index which is computed by dividing a current unit price for a single item by its base-period unit price.

$$PI_n = \frac{p_n}{p_o} \times 100$$

Price index An index that is designed to measure changes in prices over time or differences in prices between different locations.

Quantity index An index that is designed to measure changes in quantities over time or differences in quantities between different locations.

Simple price index Price index which is computed by dividing a current unit price for a single item by its base-period unit price (also called a 'price relative').

$$PI_n = \frac{p_n}{p_o} \times 100$$

Weighted average of simple price indexes A composite price index computed by taking a weighted average of simple price indexes (price relatives), using either base or current weights to produce Laspreyes- or Paasche-type index numbers respectively:

$$PI_n = \frac{\Sigma (p_{in}/p_{io}) W_i}{\Sigma W_i} \times 100$$

Weighted aggregate price index A composite price index where the prices of the items in the composite are weighted by their relative importance (denoted q_i).

$$PI_n = \frac{\Sigma p_{in} q_i}{\Sigma p_{io} q_i} \times 100$$

Weighted aggregate quantity index A composite quantity index where the quantities of the items in the composite index are weighted in terms of unchanged prices (base period prices or current period prices may be used) to produce Laspeyres- or Paasche-type quantity indexes respectively.

Base-weighted (Laspeyres): $\qquad \frac{\Sigma q_{in} p_{io}}{\Sigma q_{io} p_{io}} \times 100$

Current-weighted (Paasche): $\qquad \frac{\Sigma q_{in} p_{in}}{\Sigma q_{io} p_{in}} \times 100$

Weighted averages of quantity relatives A composite quantity index computed by taking a weighted average of quantity relatives, using either base or current weights, to produce Laspeyres- or Paasche-type index numbers respectively.

Base-weighted (Laspeyres): $\dfrac{\Sigma(q_{in}/q_{io})p_{io}q_{io}}{\Sigma p_{io}q_{io}} \times 100$

Current-weighted (Paasche): $\dfrac{\Sigma(q_{in}/q_{io})p_{in}q_{in}}{\Sigma p_{in}q_{in}} \times 100$

Self-study exercises

22.1. Define, using appropriate formulae, the differences between Laspeyres and Paasche price and volume (quantity) index numbers.

22.2. The table below shows prices and expenditure on drinks and tobacco in 1990 and 1999. Calculate price relatives (simple price indexes) and use them to compile value-weighted price indexes for 1999 (1990 = 100) of the Laspeyres and Paasche types for all drink and tobacco combined.

	Price (£)		Expenditure (£000m)	
Commodity	1990	1999	1990	1999
Beer (can)	0.1	0.2	1.3	4.2
Spirits (bottle)	2.0	5.0	0.6	2.0
Wine (litre)	0.5	2.5	0.3	1.6
Tobacco (tin)	1.0	4.0	1.7	4.0

22.3. A survey of expenditure on fuel by a group of firms gave the results shown in the following table.

		Price (per unit)		Pattern of expenditure	
Fuel	Unit	1998 (£)	2000 (£)	1998 (%)	2000 (%)
Oil	drum	6	9	20	10
Coal	bag	5	6	40	60
Gas	therm	10	14	40	30

Total expenditure: 1998 = £1,000; 2000 = £1,500;

increase = 50 per cent.

(a) Construct Paasche and Laspeyres price indices.

(b) Express 2000 expenditure at 1998 prices.

(c) How much of the increase in expenditure of 50 per cent represented a real increase in quantity consumed?

22.4. The total daily labour costs of a firm, broken down by type of labour, were as given in the table below.

Type of labour	1991 No.	1991 Cost (£)	1996 No.	1996 Cost (£)	2001 No.	2001 Cost (£)
Unskilled	147	2,720	140	7,984	131	13,008
Skilled	63	1,449	64	4,857	69	7,245
Clerical	42	840	44	2,081	48	4,718
Managerial	28	1,568	29	4,557	32	7,136

(a) The firm maintains two indexes of average labour costs using Laspeyres and Paasche formulae respectively. Calculate these indexes for 2001 (1991 = 100) and comment on the results.

(b) How big is the difference between the two indexes expressed as an average annual percentage compound rate of change?

(c) Over this period the firm's volume of output increased by 20 per cent. The retail price index rose from 80 to 295:

(i) By how much did real earnings change?

(ii) By how much did labour costs per unit of output change?

22.5. In a particular country it is argued that inflation has had a much greater impact on pensioner households over the period 1993–2001 than upon the population in general, quite apart from the fact that pensioners are dependent upon fixed incomes. You are asked to apply your knowledge of index numbers to judge whether the evidence supports the argument or not.

(a) Given the information in the table below, what is your verdict?

(b) State any reservations you may have about the verdict and what additional information, if any, would be needed to resolve them.

Item	Price index in 2001 (1993 = 100)	Distribution of expenditure in 1993 Pensioner households	Distribution of expenditure in 1993 All households
Housing	173	25	10
Fuel, light and power	161	14	6
Food, drink and tobacco	152	35	46
Clothing and footwear	132	4	10
Durable goods	135	4	6
Transport and vehicles	147	3	9
Other goods and services	155	15	13
Total – all items	152	100	100

22.6. The table below shows total consumers' expenditure for three separate years measured in current prices and in constant (1998) prices.

	Consumers' expenditure	
Year	At current prices (£ million)	At constant (1998) prices (£ million)
1998	136,890	136,890
1999	152,239	137,063
2000	167,128	138,865

(a) Explain the purpose of publishing these figures both at current and constant prices.

(b) From the data obtain both price *and* volume index numbers for 2000 with a base year of 1998.

(c) Name the type of index numbers you have obtained in (b) above and explain your answers.

22.7. The table below shows the total values of sales and amounts sold for three fuels in 1998 and 2000. Taking 1998 as the base year calculate an index of sales *volume* (quantity) for all fuels combined in 2000.

Fuel	Units	Amounts sold (millions)		Value of sales (£ billion)	
		1998	2000	1998	2000
Electricity	Kwh	100 000	100 000	3	5
Gas	Therms	10 000	9000	2	3
Coal	Tonnes	70	50	1	1

22.8. The following table shows gross domestic expenditure in 2000 and the price indexes for three of its components (based on 1995 = 100).

Components	Amount (£ billion)	Price index (1995 = 100)
Consumer expenditure	135.4	190
Government expenditure	48.3	198
Investment expenditure	40.1	193

Total gross domestic expenditure in 1995 was £106.7 billion.

(a) Based on these three components derive

(i) a price index

(ii) a volume index

for 2000 (with 1995 = 100) for total gross domestic expenditure.

(b) What further data would you require to obtain an alternative set of answers?

22.9. What are the main problems involved in calculating an index of prices for a heterogeneous group of commodities, such as the index of retail prices, over a long period of time?

References for further reading

Anderson, David R. (1998) *Statistics for Business and Economics*, 7th edn, USA: South Western College Publishing, ISBN 0538875933

Anderson, David R., Sweeney, Dennis J. and Williams, Thomas A. (1999) *Essentials of Statistics for Business and Economics*, 2nd edn, USA: South Western College Publishing, ISBN 0324003285

Bierman, Harold, Bonini, Charles P. and Hausman, Warren H. (1992) *Quantitative Analysis for Business Decisions*, 8th edn, UK: McGraw Hill Publishing Company, ISBN 0256114021 (International edition)

Carver, Robert (1998) *Doing Data Analysis with Minitab 12*, Belmont, Ca: Duxbury Press, ISBN 0534359248

Curwin, Jon and Slater, Roger (1996) *Quantitative Methods for Business Decisions*, 4th edn, London, UK: International Thomson Business Press, ISBN 1861520271

Daly, F., Hand, D.J., Jones, M.C., Lunn, A.D. and McConway, K.J. (1995) *Elements of Statistics*, Wokingham, UK: Addison-Wesley Publishing Company, ISBN 0201422786

Evans, Michael J (1999) *Minitab Manual for Moore and McCabe's Introduction to the Practice of Statistics* 3rd edn., Salt Lake City, USA: W.H. Freeman, ISBN 0716727854

Fleming, M.C. and Nellis, J.G. (1996) *The Essence of Statistics for Business*, 2nd edn. Hemel Hempstead, UK: Prentice Hall Europe, ISBN 0133987779

Hackett, Graham and Caunt, David (1994) *Quantitative Methods*, Oxford, UK: Basil Blackwell Ltd, ISBN 0631195378

Keller, Gerald and Warrack, Brian (1999) *Statistics for Management and Economics*, 5th edn, Duxbury, UK: ISBN 0534368301

Kvanli, Alan H., Guynes, Stephen C. and Pavur, Robert J. (1995) *Introduction to Business Statistics: A Computer Integrated Data Analysis Approach*, 5th edn, USA: South Western College Publishing, ISBN 0324012071

Levin, Richard I. and Rubin, David S. (1997) *Statistics for Management*, 7th edn, Englewood Cliffs, NJ: Prentice Hall International, ISBN 0136067166

Levine, David M., Berenson, Mark L. and Stephan, David (1999) *Statistics for Managers Using Microsoft Excel*, Upper Saddle River, NJ, USA: Prentice Hall, ISBN 0130203122

Mann, Prem S. (1997) *Introductory Statistics*, 3rd edn, New York, USA: John Wiley & Sons, ISBN 0471165468

Mason, Robert D., Lind, Douglas A and Marchal, William G. (1998) *Statistics: An Introduction* 5th edn., Belmont, Ca: Duxbury Press, ISBN 00309691744

McClave, James T., Benson, P. George and Sincich, Terry (1998) *Statistics for Business and Economics*, 7th edn, Upper Saddle River, NJ, USA: Prentice Hall, ISBN 0139505458

Moore, David S and McCabe George P (1998) *Introduction to the Practic'e of Statistics* 3rd edn., Salt Lake City, USA: W.H. Freeman, ISBN 0716735024 (see also Evans, Michael J (1999))

Waters, Donald (1997) *Quantitative Methods for Business*, Wokingham, UK: Addison-Wesley Publishing Company, ISBN 0201403978

Weiss, Neil (1997) *Introductory Statistics* 4th edn., Reading, MA: Addison Wesley, ISBN 0201545675 (see also Zehna, Peter, W (1997))

Whigham, David (1998) *Quantitative Business Methods Using Excel*, Oxford, UK: Oxford University Press, ISBN 0198775458

Wisniewski, Mik (1997) *Quantitative Methods for Decision Makers*, London, UK: Financial Times Management, ISBN 0273624040

Zehna, Peter W (1997) *Minitab Supplement to accompany Introductory Statistics* by Weiss, Neil (1997), Reading, MA: Addison Wesley, ISBN 0201883244

Appendix A: Statistical tables

TABLE A1   BINOMIAL PROBABILITY DISTRIBUTION

The table shows the value of the probability of x successes in n trials of a binomial experiment where p is the probability of a 'success' on one trial (and $q(= 1 - p)$ is the probability of 'failure'). Thus,

$$P(x) = {}^{n}C_{x}(p^{x})(q^{n-x})$$

where

$${}^{n}C_{x} = \frac{n!}{x!(n - x)!}$$

The table gives probabilities for selected values of p from 0.05 to 0.50. For values of p exceeding 0.50, the value of $P(x)$ can be obtained by substituting q for p and by finding the probability of $n - x$ failures, i.e. $P(n - x)$.

							p				
n	x	0.05	0.10	0.15	0.20	0.25	0.30	0.35	0.40	0.45	0.50
1	0	0.9500	0.9000	0.8500	0.8000	0.7500	0.7000	0.6500	0.6000	0.5500	0.5000
	1	0.0500	0.1000	0.1500	0.2000	0.2500	0.3000	0.3500	0.4000	0.4500	0.5000
2	0	0.9025	0.8100	0.7225	0.6400	0.5625	0.4900	0.4225	0.3600	0.3025	0.2500
	1	0.0950	0.1800	0.2550	0.3200	0.3750	0.4200	0.4550	0.4800	0.4950	0.5000
	2	0.0025	0.0100	0.0225	0.0400	0.0625	0.0900	0.1225	0.1600	0.2025	0.2500
3	0	0.8574	0.7290	0.6141	0.5120	0.4219	0.3430	0.2746	0.2160	0.1664	0.1250
	1	0.1354	0.2430	0.3251	0.3840	0.4219	0.4410	0.4436	0.4320	0.4084	0.3750
	2	0.0071	0.0270	0.0574	0.0960	0.1406	0.1890	0.2389	0.2880	0.3341	0.3750
	3	0.0001	0.0010	0.0034	0.0080	0.0156	0.0270	0.0429	0.0640	0.0911	0.1250
4	0	0.8145	0.6561	0.5220	0.4096	0.3164	0.2401	0.1785	0.1296	0.0915	0.0625
	1	0.1715	0.2916	0.3685	0.4096	0.4219	0.4116	0.3845	0.3456	0.2995	0.2500
	2	0.0135	0.0486	0.0975	0.1536	0.2109	0.2646	0.3105	0.3456	0.3675	0.3750
	3	0.0005	0.0036	0.0115	0.0256	0.0469	0.0756	0.1115	0.1536	0.2005	0.2500
	4	0.0000	0.0001	0.0005	0.0016	0.0039	0.0081	0.0150	0.0256	0.0410	0.0625
5	0	0.7738	0.5905	0.4437	0.3277	0.2373	0.1681	0.1160	0.0778	0.0503	0.0312
	1	0.2036	0.3280	0.3915	0.4096	0.3955	0.3602	0.3124	0.2592	0.2059	0.1562
	2	0.0214	0.0729	0.1382	0.2048	0.2637	0.3087	0.3364	0.3456	0.3369	0.3125
	3	0.0011	0.0081	0.0244	0.0512	0.0879	0.1323	0.1811	0.2304	0.2757	0.3125
	4	0.0000	0.0004	0.0022	0.0064	0.0146	0.0284	0.0488	0.0768	0.1128	0.1562
	5	0.0000	0.0000	0.0001	0.0003	0.0010	0.0024	0.0053	0.0102	0.0185	0.0312

(continued)

							p				
n	x	0.05	0.10	0.15	0.20	0.25	0.30	0.35	0.40	0.45	0.50
6	0	0.7351	0.5314	0.3771	0.2621	0.1780	0.1176	0.0754	0.0467	0.0277	0.0156
	1	0.2321	0.3543	0.3993	0.3932	0.3560	0.3025	0.2437	0.1866	0.1359	0.0938
	2	0.0305	0.0984	0.1762	0.2458	0.2966	0.3241	0.3280	0.3110	0.2780	0.2344
	3	0.0021	0.0146	0.0415	0.0819	0.1318	0.1852	0.2355	0.2765	0.3032	0.3125
	4	0.0001	0.0012	0.0055	0.0154	0.0330	0.0595	0.0951	0.1382	0.1861	0.2344
	5	0.0000	0.0001	0.0004	0.0015	0.0044	0.0102	0.0205	0.0369	0.0609	0.0938
	6	0.0000	0.0000	0.0000	0.0001	0.0002	0.0007	0.0018	0.0041	0.0083	0.0156
7	0	0.6983	0.4783	0.3206	0.2097	0.1335	0.0824	0.0490	0.0280	0.0152	0.0078
	1	0.2573	0.3720	0.3960	0.3670	0.3115	0.2471	0.1848	0.1306	0.0872	0.0547
	2	0.0406	0.1240	0.2097	0.2753	0.3115	0.3177	0.2985	0.2613	0.2140	0.1641
	3	0.0036	0.0230	0.0617	0.1147	0.1730	0.2269	0.2679	0.2903	0.2918	0.2734
	4	0.0002	0.0026	0.0109	0.0287	0.0577	0.0972	0.1442	0.1935	0.2388	0.2734
	5	0.0009	0.0002	0.0012	0.0043	0.0115	0.0250	0.0466	0.0774	0.1172	0.1641
	6	0.0000	0.0000	0.0001	0.0004	0.0013	0.0036	0.0084	0.0172	0.0320	0.0547
	7	0.0000	0.0000	0.0000	0.0000	0.0001	0.0002	0.0006	0.0016	0.0037	0.0078
8	0	0.6634	0.4305	0.2725	0.1678	0.1001	0.0576	0.0319	0.0168	0.0084	0.0039
	1	0.2793	0.3826	0.3847	0.3355	0.2670	0.1977	0.1373	0.0896	0.0548	0.0312
	2	0.0515	0.1488	0.2376	0.2936	0.3115	0.2965	0.2587	0.2090	0.1569	0.1094
	3	0.0054	0.0331	0.0839	0.1468	0.2076	0.2541	0.2786	0.2787	0.2568	0.2188
	4	0.0004	0.0046	0.0815	0.0459	0.0865	0.1361	0.1875	0.2322	0.2627	0.2734
	5	0.0000	0.0004	0.0026	0.0092	0.0231	0.0467	0.0808	0.1239	0.1719	0.2188
	6	0.0000	0.0000	0.0002	0.0011	0.0038	0.0100	0.0217	0.0413	0.0703	0.1094
	7	0.0000	0.0000	0.0000	0.0001	0.0004	0.0012	0.0033	0.0079	0.0164	0.0312
	8	0.0000	0.0000	0.0000	0.0000	0.0000	0.0001	0.0002	0.0007	0.0017	0.0039
9	0	0.6302	0.3874	0.2316	0.1342	0.0751	0.0404	0.0207	0.0101	0.0046	0.0020
	1	0.2985	0.3874	0.3679	0.3020	0.2253	0.1556	0.1004	0.0605	0.0339	0.0176
	2	0.0629	0.1722	0.2597	0.3020	0.3003	0.2668	0.2162	0.1612	0.1110	0.0703
	3	0.0077	0.0446	0.1069	0.1762	0.2336	0.2668	0.2716	0.2508	0.2119	0.1641
	4	0.0006	0.0074	0.0283	0.0661	0.1168	0.1715	0.2194	0.2508	0.2600	0.2461
	5	0.0000	0.0008	0.0050	0.0165	0.0389	0.0735	0.1181	0.1672	0.2128	0.2461
	6	0.0000	0.0001	0.0006	0.0028	0.0087	0.0210	0.0424	0.0743	0.1160	0.1641
	7	0.0000	0.0000	0.0000	0.0003	0.0012	0.0039	0.0098	0.0212	0.0407	0.0703
	8	0.0000	0.0000	0.0000	0.0000	0.0001	0.0004	0.0013	0.0035	0.0083	0.0716
	9	0.0000	0.0000	0.0000	0.0000	0.0000	0.0000	0.0001	0.0003	0.0008	0.0020
10	0	0.5987	0.3487	0.1969	0.1074	0.0563	0.0282	0.0135	0.0060	0.0025	0.0010
	1	0.3151	0.3874	0.3474	0.2684	0.1877	0.1211	0.0725	0.0403	0.0207	0.0098
	2	0.0746	0.1937	0.2759	0.3020	0.2816	0.2335	0.1757	0.1209	0.0763	0.0439
	3	0.0105	0.0574	0.1298	0.2013	0.2503	0.2668	0.2522	0.2150	0.1665	0.1172
	4	0.0010	0.0112	0.0401	0.0881	0.1460	0.2001	0.2377	0.2508	0.2384	0.2051
	5	0.0001	0.0015	0.0085	0.0264	0.0584	0.1029	0.1536	0.2007	0.2340	0.2461
	6	0.0000	0.0001	0.0012	0.0055	0.0162	0.0368	0.0689	0.1115	0.1596	0.2051
	7	0.0000	0.0000	0.0001	0.0008	0.0031	0.0090	0.0212	0.0425	0.0746	0.1172

(*continued*)

						p					
n	x	0.05	0.10	0.15	0.20	0.25	0.30	0.35	0.40	0.45	0.50
	8	0.0000	0.0000	0.0000	0.0001	0.0004	0.0014	0.0043	0.0106	0.0229	0.0439
	9	0.0000	0.0000	0.0000	0.0000	0.0000	0.0001	0.0005	0.0016	0.0042	0.0098
	10	0.0000	0.0000	0.0000	0.0000	0.0000	0.0000	0.0000	0.0001	0.0003	0.0010
11	0	0.5688	0.3138	0.1673	0.0859	0.0422	0.0198	0.0088	0.0036	0.0014	0.0005
	1	0.3293	0.3835	0.3248	0.2362	0.1549	0.0932	0.0518	0.0266	0.0125	0.0054
	2	0.0867	0.2131	0.2866	0.2953	0.2581	0.1998	0.1395	0.0887	0.0513	0.0269
	3	0.0137	0.0710	0.1517	0.2215	0.2581	0.2568	0.2254	0.1774	0.1259	0.0806
	4	0.0014	0.0158	0.0536	0.1107	0.1721	0.2201	0.2428	0.2365	0.2060	0.1611
	5	0.0001	0.0025	0.0132	0.0388	0.0803	0.1321	0.1830	0.2207	0.2360	0.2256
	6	0.0000	0.0003	0.0023	0.0097	0.0268	0.0566	0.0985	0.1471	0.1931	0.2256
	7	0.0000	0.0000	0.0003	0.0017	0.0064	0.0173	0.0379	0.0701	0.1128	0.1611
	8	0.0000	0.0000	0.0000	0.0002	0.0011	0.0037	0.0102	0.0234	0.0462	0.0806
	9	0.0000	0.0000	0.0000	0.0000	0.0001	0.0005	0.0018	0.0052	0.0126	0.0269
	10	0.0000	0.0000	0.0000	0.0000	0.0000	0.0000	0.0002	0.0007	0.0021	0.0054
	11	0.0000	0.0000	0.0000	0.0000	0.0000	0.0000	0.0000	0.0000	0.0002	0.0005
12	0	0.5404	0.2824	0.1422	0.0687	0.0317	0.0138	0.0057	0.0022	0.0008	0.0002
	1	0.3413	0.3766	0.3012	0.2062	0.1267	0.0712	0.0368	0.0174	0.0075	0.0029
	2	0.0988	0.2301	0.2924	0.2835	0.2323	0.1678	0.1088	0.0639	0.0339	0.0161
	3	0.0173	0.0852	0.1720	0.2362	0.2581	0.2397	0.1954	0.1419	0.0923	0.0537
	4	0.0021	0.0213	0.0683	0.1329	0.1936	0.2311	0.2367	0.2128	0.1700	0.1208
	5	0.0002	0.0038	0.0193	0.0532	0.1032	0.1585	0.2039	0.2270	0.2225	0.1934
	6	0.0000	0.0005	0.0040	0.0155	0.0401	0.0792	0.1281	0.1766	0.2124	0.2256
	7	0.0000	0.0000	0.0006	0.0033	0.0115	0.0291	0.0591	0.1009	0.1489	0.1934
	8	0.0000	0.0000	0.0001	0.0005	0.0024	0.0078	0.0199	0.0420	0.0762	0.1208
	9	0.0000	0.0000	0.0000	0.0001	0.0004	0.0015	0.0048	0.0125	0.0277	0.0537
	10	0.0000	0.0000	0.0000	0.0000	0.0000	0.0002	0.0008	0.0025	0.0068	0.0161
	11	0.0000	0.0000	0.0000	0.0000	0.0000	0.0000	0.0001	0.0003	0.0010	0.0029
	12	0.0000	0.0000	0.0000	0.0000	0.0000	0.0000	0.0000	0.0000	0.0001	0.0002
13	0	0.5133	0.2542	0.1209	0.0550	0.0238	0.0097	0.0037	0.0013	0.0004	0.0001
	1	0.3512	0.3672	0.2774	0.1787	0.1029	0.0540	0.0259	0.0113	0.0045	0.0016
	2	0.1109	0.2448	0.2937	0.2680	0.2059	0.1388	0.0836	0.0453	0.0220	0.0095
	3	0.0214	0.0997	0.1900	0.2457	0.2517	0.2181	0.1651	0.1107	0.0660	0.0349
	4	0.0028	0.0277	0.0838	0.1535	0.2097	0.2337	0.2222	0.1845	0.1350	0.0873
	5	0.0003	0.0055	0.0266	0.0691	0.1258	0.1803	0.2154	0.2214	0.1989	0.1571
	6	0.0000	0.0008	0.0063	0.0230	0.0559	0.1030	0.1546	0.1968	0.2169	0.2095
	7	0.0000	0.0001	0.0011	0.0058	0.0186	0.0442	0.0833	0.1312	0.1775	0.2095
	8	0.0000	0.0000	0.0001	0.0011	0.0047	0.0142	0.0336	0.0656	0.1089	0.1571
	9	0.0000	0.0000	0.0000	0.0001	0.0009	0.0034	0.0101	0.0243	0.0495	0.0873
	10	0.0000	0.0000	0.0000	0.0000	0.0001	0.0006	0.0022	0.0065	0.0162	0.0349
	11	0.0000	0.0000	0.0000	0.0000	0.0000	0.0001	0.0003	0.0012	0.0036	0.0095
	12	0.0000	0.0000	0.0000	0.0000	0.0000	0.0000	0.0000	0.0001	0.0005	0.0016
	13	0.0000	0.0000	0.0000	0.0000	0.0000	0.0000	0.0000	0.0000	0.0000	0.0001

(continued)

						p					
n	x	0.05	0.10	0.15	0.20	0.25	0.30	0.35	0.40	0.45	0.50
14	0	0.4877	0.2288	0.1028	0.0440	0.0178	0.0068	0.0024	0.0008	0.0002	0.0001
	1	0.3593	0.3559	0.2539	0.1539	0.0832	0.0407	0.0181	0.0073	0.0027	0.0009
	2	0.1229	0.2570	0.2912	0.2501	0.1802	0.1134	0.0634	0.0317	0.0141	0.0056
	3	0.0259	0.1142	0.2056	0.2501	0.2402	0.1943	0.1366	0.0845	0.0462	0.0222
	4	0.0037	0.0348	0.0998	0.1720	0.2202	0.2290	0.2022	0.1549	0.1040	0.0611
	5	0.0004	0.0078	0.0352	0.0860	0.1468	0.1963	0.2178	0.2066	0.1701	0.1222
	6	0.0000	0.0013	0.0093	0.0322	0.0734	0.1262	0.1759	0.2066	0.2088	0.1833
	7	0.0000	0.0002	0.0019	0.0092	0.0280	0.0618	0.1082	0.1574	0.1952	0.2095
	8	0.0000	0.0000	0.0003	0.0020	0.0082	0.0232	0.0510	0.0918	0.1398	0.1833
	9	0.0000	0.0000	0.0000	0.0003	0.0018	0.0066	0.0183	0.0408	0.0762	0.0122
	10	0.0000	0.0000	0.0000	0.0000	0.0003	0.0014	0.0049	0.0136	0.0312	0.0611
	11	0.0000	0.0000	0.0000	0.0000	0.0000	0.0002	0.0010	0.0033	0.0093	0.0222
	12	0.0000	0.0000	0.0000	0.0000	0.0000	0.0000	0.0001	0.0005	0.0019	0.0056
	13	0.0000	0.0000	0.0000	0.0000	0.0000	0.0000	0.0000	0.0001	0.0002	0.0009
	14	0.0000	0.0000	0.0000	0.0000	0.0000	0.0000	0.0000	0.0000	0.0000	0.0001
15	0	0.4633	0.2059	0.0874	0.0352	0.0134	0.0047	0.0016	0.0005	0.0001	0.0000
	1	0.3658	0.3432	0.2312	0.1319	0.0668	0.0305	0.0126	0.0047	0.0016	0.0005
	2	0.1348	0.2669	0.2856	0.2309	0.1559	0.0916	0.0476	0.0219	0.0090	0.0032
	3	0.0307	0.1285	0.2184	0.2501	0.2252	0.1700	0.1110	0.0634	0.0318	0.0139
	4	0.0049	0.0428	0.1156	0.1876	0.2252	0.2186	0.1792	0.1268	0.0780	0.0417
	5	0.0006	0.0105	0.0449	0.1032	0.1651	0.2061	0.2123	0.1859	0.1404	0.0916
	6	0.0000	0.0019	0.0132	0.0430	0.0917	0.1472	0.1906	0.2066	0.1914	0.1527
	7	0.0000	0.0003	0.0030	0.0138	0.0393	0.0811	0.1319	0.1771	0.2013	0.1964
	8	0.0000	0.0000	0.0005	0.0035	0.0131	0.0348	0.0710	0.1181	0.1647	0.1964
	9	0.0000	0.0000	0.0001	0.0007	0.0034	0.0116	0.0298	0.0612	0.1048	0.1527
	10	0.0000	0.0000	0.0000	0.0001	0.0007	0.0030	0.0096	0.0245	0.0515	0.0916
	11	0.0000	0.0000	0.0000	0.0000	0.0001	0.0006	0.0024	0.0074	0.0191	0.0417
	12	0.0000	0.0000	0.0000	0.0000	0.0000	0.0001	0.0004	0.0016	0.0052	0.0139
	13	0.0000	0.0000	0.0000	0.0000	0.0000	0.0000	0.0001	0.0003	0.0010	0.0032
	14	0.0000	0.0000	0.0000	0.0000	0.0000	0.0000	0.0000	0.0000	0.0001	0.0005
	15	0.0000	0.0000	0.0000	0.0000	0.0000	0.0000	0.0000	0.0000	0.0000	0.0000
16	0	0.4401	0.1853	0.0743	0.0281	0.0100	0.0033	0.0010	0.0003	0.0001	0.0000
	1	0.3706	0.3294	0.2097	0.1126	0.0535	0.0228	0.0087	0.0030	0.0009	0.0002
	2	0.1463	0.2745	0.2775	0.2111	0.1336	0.0732	0.0353	0.0150	0.0056	0.0018
	3	0.0359	0.1423	0.2285	0.2463	0.2079	0.1465	0.0888	0.0468	0.0215	0.0085
	4	0.0061	0.0514	0.1311	0.2001	0.2252	0.2040	0.1553	0.1014	0.0572	0.0278
	5	0.0008	0.0137	0.0555	0.1201	0.1802	0.2099	0.2008	0.1623	0.1123	0.0667
	6	0.0001	0.0028	0.0180	0.0550	0.1101	0.1649	0.1982	0.1983	0.1684	0.1222
	7	0.0000	0.0004	0.0045	0.0197	0.0524	0.1010	0.1524	0.1889	0.1969	0.1746
	8	0.0000	0.0001	0.0009	0.0055	0.0197	0.0487	0.0923	0.1417	0.1812	0.1964
	9	0.0000	0.0000	0.0001	0.0012	0.0058	0.0185	0.0442	0.0840	0.1318	0.1746
	10	0.0000	0.0000	0.0000	0.0002	0.0014	0.0056	0.0167	0.0392	0.0755	0.1222
	11	0.0000	0.0000	0.0000	0.0000	0.0002	0.0013	0.0049	0.0142	0.0337	0.0667

(*continued*)

							p				
n	x	0.05	0.10	0.15	0.20	0.25	0.30	0.35	0.40	0.45	0.50
	12	0.0000	0.0000	0.0000	0.0000	0.0000	0.0002	0.0011	0.0040	0.0115	0.0278
	13	0.0000	0.0000	0.0000	0.0000	0.0000	0.0000	0.0002	0.0008	0.0029	0.0085
	14	0.0000	0.0000	0.0000	0.0000	0.0000	0.0000	0.0000	0.0001	0.0005	0.0018
	15	0.0000	0.0000	0.0000	0.0000	0.0000	0.0000	0.0000	0.0000	0.0001	0.0002
	16	0.0000	0.0000	0.0000	0.0000	0.0000	0.0000	0.0000	0.0000	0.0000	0.0000
17	0	0.4181	0.1668	0.0631	0.0225	0.0075	0.0023	0.0007	0.0002	0.0000	0.0000
	1	0.3741	0.3150	0.1893	0.0957	0.0426	0.0169	0.0060	0.0019	0.0005	0.0001
	2	0.1575	0.2800	0.2673	0.1914	0.1136	0.0581	0.0260	0.0102	0.0035	0.0010
	3	0.0415	0.1556	0.2359	0.2393	0.1893	0.1245	0.0701	0.0341	0.0144	0.0052
	4	0.0076	0.0605	0.1457	0.2093	0.2209	0.1868	0.1320	0.0796	0.0411	0.0182
	5	0.0010	0.0175	0.0668	0.1361	0.1914	0.2081	0.1849	0.1379	0.0875	0.0472
	6	0.0001	0.0039	0.0236	0.0680	0.1276	0.1784	0.1991	0.1839	0.1432	0.0944
	7	0.0000	0.0007	0.0065	0.0267	0.0668	0.1201	0.1685	0.1927	0.1841	0.1484
	8	0.0000	0.0001	0.0014	0.0084	0.0279	0.0644	0.1134	0.1606	0.1883	0.1855
	9	0.0000	0.0000	0.0003	0.0021	0.0093	0.0276	0.0611	0.1070	0.1540	0.1855
	10	0.0000	0.0000	0.0000	0.0004	0.0025	0.0095	0.0263	0.0571	0.1008	0.1484
	11	0.0000	0.0000	0.0000	0.0001	0.0005	0.0026	0.0090	0.0242	0.0525	0.0944
	12	0.0000	0.0000	0.0000	0.0000	0.0001	0.0006	0.0024	0.0021	0.0215	0.0472
	13	0.0000	0.0000	0.0000	0.0000	0.0000	0.0001	0.0005	0.0021	0.0068	0.0182
	14	0.0000	0.0000	0.0000	0.0000	0.0000	0.0000	0.0001	0.0004	0.0016	0.0052
	15	0.0000	0.0000	0.0000	0.0000	0.0000	0.0000	0.0000	0.0001	0.0003	0.0010
	16	0.0000	0.0000	0.0000	0.0000	0.0000	0.0000	0.0000	0.0000	0.0000	0.0001
	17	0.0000	0.0000	0.0000	0.0000	0.0000	0.0000	0.0000	0.0000	0.0000	0.0000
18	0	0.3972	0.1501	0.0536	0.0180	0.0056	0.0016	0.0004	0.0001	0.0000	0.0000
	1	0.3763	0.3002	0.1704	0.0811	0.0338	0.0126	0.0042	0.0012	0.0003	0.0001
	2	0.1683	0.2835	0.2556	0.1723	0.0958	0.0458	0.0190	0.0069	0.0022	0.0006
	3	0.0473	0.1680	0.2406	0.2297	0.1704	0.1046	0.0547	0.0246	0.0095	0.0031
	4	0.0093	0.0700	0.1592	0.2153	0.2130	0.1681	0.1104	0.0614	0.0291	0.0117
	5	0.0014	0.0218	0.0787	0.1507	0.1988	0.2017	0.1664	0.1146	0.0666	0.0327
	6	0.0002	0.0052	0.0301	0.0816	0.1436	0.1873	0.1941	0.1655	0.1181	0.0708
	7	0.0000	0.0010	0.0091	0.0350	0.0820	0.1376	0.1792	0.1892	0.1657	0.1214
	8	0.0000	0.0002	0.0022	0.0120	0.0376	0.0811	0.1327	0.1734	0.1864	0.1669
	9	0.0000	0.0000	0.0004	0.0033	0.0139	0.0386	0.0794	0.1284	0.1694	0.1855
	10	0.0000	0.0000	0.0001	0.0008	0.0042	0.0149	0.0385	0.0771	0.1248	0.1669
	11	0.0000	0.0000	0.0000	0.0001	0.0010	0.0046	0.0151	0.0374	0.0742	0.1214
	12	0.0000	0.0000	0.0000	0.0000	0.0002	0.0012	0.0047	0.0145	0.0354	0.0708
	13	0.0000	0.0000	0.0000	0.0000	0.0000	0.0002	0.0012	0.0044	0.0134	0.0327
	14	0.0000	0.0000	0.0000	0.0000	0.0000	0.0000	0.0002	0.0011	0.0039	0.0117
	15	0.0000	0.0000	0.0000	0.0000	0.0000	0.0000	0.0000	0.0002	0.0009	0.0031
	16	0.0000	0.0000	0.0000	0.0000	0.0000	0.0000	0.0000	0.0000	0.0001	0.0006
	17	0.0000	0.0000	0.0000	0.0000	0.0000	0.0000	0.0000	0.0000	0.0000	0.0001
	18	0.0000	0.0000	0.0000	0.0000	0.0000	0.0000	0.0000	0.0000	0.0000	0.0000
19	0	0.3774	0.1351	0.0456	0.0144	0.0042	0.0011	0.0003	0.0001	0.0000	0.0000

(continued)

419

n	x	0.05	0.10	0.15	0.20	0.25	0.30	0.35	0.40	0.45	0.50
	1	0.3774	0.2852	0.1529	0.0685	0.0268	0.0093	0.0029	0.0008	0.0002	0.0000
	2	0.1787	0.2852	0.2428	0.1540	0.0803	0.0358	0.0138	0.0046	0.0013	0.0003
	3	0.0533	0.1796	0.2428	0.2182	0.1517	0.0869	0.0422	0.0175	0.0062	0.0018
	4	0.0112	0.0798	0.1714	0.2182	0.2023	0.1491	0.0909	0.0467	0.0203	0.0074
	5	0.0018	0.0266	0.0907	0.1636	0.2023	0.1916	0.1468	0.0933	0.0497	0.0222
	6	0.0002	0.0069	0.0374	0.0955	0.1574	0.1916	0.1844	0.1451	0.0949	0.0518
	7	0.0000	0.0014	0.0122	0.0443	0.0974	0.1525	0.1844	0.1797	0.1443	0.0961
	8	0.0000	0.0002	0.0032	0.0166	0.0487	0.0981	0.1489	0.1797	0.1771	0.1442
	9	0.0000	0.0000	0.0007	0.0051	0.0198	0.0514	0.0980	0.1464	0.1771	0.1762
	10	0.0000	0.0000	0.0001	0.0013	0.0066	0.0220	0.0528	0.0976	0.1449	0.1762
	11	0.0000	0.0000	0.0000	0.0003	0.0018	0.0077	0.0233	0.0532	0.0970	0.1442
	12	0.0000	0.0000	0.0000	0.0000	0.0004	0.0022	0.0083	0.0237	0.0529	0.0961
	13	0.0000	0.0000	0.0000	0.0000	0.0001	0.0005	0.0024	0.0085	0.0233	0.0518
	14	0.0000	0.0000	0.0000	0.0000	0.0000	0.0001	0.0006	0.0024	0.0082	0.0222
	15	0.0000	0.0000	0.0000	0.0000	0.0000	0.0000	0.0001	0.0005	0.0022	0.0074
	16	0.0000	0.0000	0.0000	0.0000	0.0000	0.0000	0.0000	0.0001	0.0005	0.0018
	17	0.0000	0.0000	0.0000	0.0000	0.0000	0.0000	0.0000	0.0000	0.0001	0.0003
	18	0.0000	0.0000	0.0000	0.0000	0.0000	0.0000	0.0000	0.0000	0.0000	0.0000
	19	0.0000	0.0000	0.0000	0.0000	0.0000	0.0000	0.0000	0.0000	0.0000	0.0000
20	0	0.3585	0.1216	0.0388	0.0115	0.0032	0.0008	0.0002	0.0000	0.0000	0.0000
	1	0.3774	0.2702	0.1368	0.0576	0.0211	0.0068	0.0020	0.0005	0.0001	0.0000
	2	0.1887	0.2852	0.2293	0.1369	0.0669	0.0278	0.0100	0.0031	0.0008	0.0002
	3	0.0596	0.1901	0.2428	0.2054	0.1339	0.0716	0.0323	0.0123	0.0040	0.0011
	4	0.0133	0.0898	0.1821	0.2182	0.1897	0.1304	0.0738	0.0350	0.0139	0.0046
	5	0.0022	0.0319	0.1028	0.1746	0.2023	0.1789	0.1272	0.0746	0.0365	0.0148
	6	0.0003	0.0089	0.0454	0.1091	0.1686	0.1916	0.1712	0.1244	0.0746	0.0370
	7	0.0000	0.0020	0.0160	0.0545	0.1124	0.1643	0.1844	0.1659	0.1221	0.0739
	8	0.0000	0.0004	0.0046	0.0222	0.0609	0.1144	0.1614	0.1797	0.1623	0.1201
	9	0.0000	0.0001	0.0011	0.0074	0.0271	0.0654	0.1158	0.1597	0.1771	0.1602
	10	0.0000	0.0000	0.0002	0.0020	0.0099	0.0308	0.0686	0.1171	0.1593	0.1762
	11	0.0000	0.0000	0.0000	0.0005	0.0030	0.0120	0.0336	0.0710	0.1185	0.1602
	12	0.0000	0.0000	0.0000	0.0001	0.0008	0.0039	0.0136	0.0355	0.0727	0.1201
	13	0.0000	0.0000	0.0000	0.0000	0.0002	0.0010	0.0045	0.0146	0.0366	0.0739
	14	0.0000	0.0000	0.0000	0.0000	0.0000	0.0002	0.0012	0.0049	0.0150	0.0370
	15	0.0000	0.0000	0.0000	0.0000	0.0000	0.0000	0.0003	0.0013	0.0049	0.0148
	16	0.0000	0.0000	0.0000	0.0000	0.0000	0.0000	0.0000	0.0003	0.0013	0.0046
	17	0.0000	0.0000	0.0000	0.0000	0.0000	0.0000	0.0000	0.0000	0.0002	0.0011
	18	0.0000	0.0000	0.0000	0.0000	0.0000	0.0000	0.0000	0.0000	0.0000	0.0002
	19	0.0000	0.0000	0.0000	0.0000	0.0000	0.0000	0.0000	0.0000	0.0000	0.0000
	20	0.0000	0.0000	0.0000	0.0000	0.0000	0.0000	0.0000	0.0000	0.0000	0.0000

Source: Taken from *Tables of the Binomial Probability Distribution*, Applied Mathematics Series, US Department of Commerce, 1950, by courtesy of the National Institute of Standards and Technology (formerly the National Bureau of Standards).

TABLE A2 POISSON PROBABILITY DISTRIBUTION

The table shows the value of the probability of x occurrences, $P(x)$, given by

$$P(x) = \frac{e^{-\mu}\mu^x}{x!}$$

for selected values of x and for $\mu = 0.005 - 8.0$.

x					μ					
	0.005	0.01	0.02	0.03	0.04	0.05	0.06	0.07	0.08	0.09
0	0.9950	0.9900	0.9802	0.9704	0.9608	0.9512	0.9418	0.9324	0.9231	0.9139
1	0.0050	0.0099	0.0192	0.0291	0.0384	0.0476	0.0565	0.0653	0.0738	0.0823
2	0.0000	0.0000	0.0002	0.0004	0.0008	0.0012	0.0017	0.0023	0.0030	0.0037
3	0.0000	0.0000	0.0000	0.0000	0.0000	0.0000	0.0000	0.0001	0.0001	0.0001

	0.1	0.2	0.3	0.4	0.5	0.6	0.7	0.8	0.9	1.0
0	0.9048	0.8187	0.7408	0.6703	0.6065	0.5488	0.4966	0.4493	0.4066	0.3679
1	0.0905	0.1637	0.2222	0.2681	0.3033	0.3293	0.3476	0.3595	0.3659	0.3679
2	0.0045	0.0164	0.0333	0.0536	0.0758	0.0988	0.1217	0.1438	0.1647	0.1839
3	0.0002	0.0011	0.0033	0.0072	0.0126	0.0198	0.0284	0.0383	0.0494	0.0613
4	0.0000	0.0001	0.0002	0.0007	0.0016	0.0030	0.0050	0.0077	0.0111	0.0153
5	0.0000	0.0000	0.0000	0.0001	0.0002	0.0004	0.0007	0.0012	0.0020	0.0031
6	0.0000	0.0000	0.0000	0.0000	0.0000	0.0000	0.0001	0.0002	0.0003	0.0005
7	0.0000	0.0000	0.0000	0.0000	0.0000	0.0000	0.0000	0.0000	0.0000	0.0001

	1.1	1.2	1.3	1.4	1.5	1.6	1.7	1.8	1.9	2.0
0	0.3329	0.3012	0.2725	0.2466	0.2231	0.2019	0.1827	0.1653	0.1496	0.1353
1	0.3662	0.3614	0.3543	0.3452	0.3347	0.3230	0.3106	0.2975	0.2842	0.2707
2	0.2014	0.2169	0.2303	0.2417	0.2510	0.2584	0.2640	0.2678	0.2700	0.2707
3	0.0738	0.0867	0.0998	0.1128	0.1255	0.1378	0.1496	0.1607	0.1710	0.1804
4	0.0203	0.0260	0.0324	0.0395	0.0471	0.0551	0.0636	0.0723	0.0812	0.0902
5	0.0045	0.0062	0.0084	0.0111	0.0141	0.0176	0.0216	0.0260	0.0309	0.0361
6	0.0008	0.0012	0.0018	0.0026	0.0035	0.0047	0.0061	0.0078	0.0098	0.0120
7	0.0001	0.0002	0.0003	0.0005	0.0008	0.0011	0.0015	0.0020	0.0027	0.0034
8	0.0000	0.0000	0.0001	0.0001	0.0001	0.0002	0.0003	0.0005	0.0006	0.0009
9	0.0000	0.0000	0.0000	0.0000	0.0000	0.0000	0.0001	0.0001	0.0001	0.0002

	2.1	2.2	2.3	2.4	2.5	2.6	2.7	2.8	2.9	3.0
0	0.1225	0.1108	0.1003	0.0907	0.0821	0.0743	0.0672	0.0608	0.0550	0.0498
1	0.2572	0.2438	0.2306	0.2177	0.2052	0.1931	0.1815	0.1703	0.1596	0.1494
2	0.2700	0.2681	0.2652	0.2613	0.2565	0.2510	0.2450	0.2384	0.2314	0.2240
3	0.1890	0.1966	0.2033	0.2090	0.2138	0.2176	0.2205	0.2225	0.2237	0.2240
4	0.0992	0.1092	0.1169	0.1254	0.1336	0.1414	0.1488	0.1557	0.1622	0.1680
5	0.0417	0.0476	0.0538	0.0602	0.0668	0.0735	0.0804	0.0872	0.0940	0.1008
6	0.0146	0.0174	0.0206	0.0241	0.0278	0.0319	0.0362	0.0407	0.0455	0.0504
7	0.0044	0.0055	0.0068	0.0083	0.0099	0.0118	0.0139	0.0163	0.0188	0.0216
8	0.0011	0.0015	0.0019	0.0025	0.0031	0.0038	0.0047	0.0057	0.0068	0.0081

(continued)

x						μ				
	2.1	2.2	2.3	2.4	2.5	2.6	2.7	2.8	2.9	3.0
9	0.0003	0.0004	0.0005	0.0007	0.0009	0.0011	0.0014	0.0018	0.0022	0.0027
10	0.0001	0.0001	0.0001	0.0002	0.0002	0.0003	0.0004	0.0005	0.0006	0.0008
11	0.0000	0.0000	0.0000	0.0000	0.0000	0.0001	0.0001	0.0001	0.0002	0.0002
12	0.0000	0.0000	0.0000	0.0000	0.0000	0.0000	0.0000	0.0000	0.0000	0.0001
	3.1	3.2	3.3	3.4	3.5	3.6	3.7	3.8	3.9	4.0
0	0.0450	0.0408	0.0369	0.0334	0.0302	0.0273	0.0247	0.0224	0.0202	0.0183
1	0.1397	0.1304	0.1217	0.1135	0.1057	0.0984	0.0915	0.0850	0.0789	0.0733
2	0.2165	0.2087	0.2008	0.1929	0.1850	0.1771	0.1692	0.1615	0.1539	0.1465
3	0.2237	0.2226	0.2209	0.2186	0.2158	0.2125	0.2087	0.2046	0.2001	0.1954
4	0.1734	0.1781	0.1823	0.1858	0.1888	0.1912	0.1931	0.1944	0.1951	0.1954
5	0.1075	0.1140	0.1203	0.1264	0.1322	0.1377	0.1429	0.1477	0.1522	0.1563
6	0.0555	0.0608	0.0662	0.0716	0.0771	0.0826	0.0881	0.0936	0.0989	0.1402
7	0.0246	0.0278	0.0312	0.0348	0.0385	0.0425	0.0466	0.0508	0.0551	0.0595
8	0.0095	0.0111	0.0129	0.0148	0.0169	0.0191	0.0215	0.0241	0.0269	0.0298
9	0.0033	0.0040	0.0047	0.0056	0.0066	0.0076	0.0089	0.0102	0.0116	0.0132
10	0.0010	0.0013	0.0016	0.0019	0.0023	0.0028	0.0033	0.0039	0.0045	0.0053
11	0.0003	0.0004	0.0005	0.0006	0.0007	0.0009	0.0011	0.0013	0.0016	0.0019
12	0.0001	0.0001	0.0001	0.0002	0.0002	0.0003	0.0003	0.0004	0.0005	0.0006
13	0.0000	0.0000	0.0000	0.0000	0.0001	0.0001	0.0001	0.0001	0.0002	0.0002
14	0.0000	0.0000	0.0000	0.0000	0.0000	0.0000	0.0000	0.0000	0.0000	0.0001
	4.1	4.2	4.3	4.4	4.5	4.6	4.7	4.8	4.9	5.0
0	0.0166	0.0150	0.0136	0.0123	0.0111	0.0101	0.0091	0.0082	0.0074	0.0067
1	0.0679	0.0630	0.0583	0.0540	0.0500	0.0462	0.0427	0.0395	0.0365	0.0337
2	0.1393	0.1323	0.1254	0.1188	0.1125	0.1063	0.1005	0.0948	0.0894	0.0842
3	0.1904	0.1852	0.1798	0.1743	0.1687	0.1631	0.1574	0.1517	0.1460	0.1404
4	0.1951	0.1944	0.1933	0.1917	0.1898	0.1875	0.1849	0.1820	0.1789	0.1755
5	0.1600	0.1633	0.1662	0.1687	0.1708	0.1725	0.1738	0.1747	0.1753	0.1755
6	0.1093	0.1143	0.1191	0.1237	0.1281	0.1323	0.1362	0.1398	0.1432	0.1462
7	0.0640	0.0686	0.0732	0.0778	0.0824	0.0869	0.0914	0.0959	0.1002	0.1044
8	0.0328	0.0360	0.0393	0.0428	0.0463	0.0500	0.0537	0.0575	0.0614	0.0653
9	0.0150	0.0168	0.0188	0.0209	0.0232	0.0255	0.0280	0.0307	0.0334	0.0363
10	0.0061	0.0071	0.0081	0.0092	0.0104	0.0118	0.0132	0.0147	0.0164	0.0181
11	0.0023	0.0027	0.0032	0.0037	0.0043	0.0049	0.0056	0.0064	0.0073	0.0082
12	0.0008	0.0009	0.0011	0.0014	0.0016	0.0019	0.0022	0.0026	0.0030	0.0034
13	0.0002	0.0003	0.0004	0.0005	0.0006	0.0007	0.0008	0.0009	0.0011	0.0013
14	0.0001	0.0001	0.0001	0.0001	0.0002	0.0002	0.0003	0.0003	0.0004	0.0005
15	0.0000	0.0000	0.0000	0.0000	0.0001	0.0001	0.0001	0.0001	0.0001	0.0002
	5.1	5.2	5.3	5.4	5.5	5.6	5.7	5.8	5.9	6.0
0	0.0061	0.0055	0.0050	0.0045	0.0041	0.0037	0.0033	0.0030	0.0027	0.0025
1	0.0311	0.0287	0.0265	0.0244	0.0225	0.0207	0.0191	0.0176	0.0162	0.0149
2	0.0793	0.0746	0.0701	0.0659	0.0618	0.0580	0.0544	0.0509	0.0477	0.0446
3	0.1348	0.1293	0.1239	0.1185	0.1133	0.1082	0.1033	0.0985	0.0938	0.0892

(*continued*)

x					μ					
	5.1	5.2	5.3	5.4	5.5	5.6	5.7	5.8	5.9	6.0
4	0.1719	0.1681	0.1641	0.1600	0.1558	0.1515	0.1472	0.1428	0.1383	0.1339
5	0.1753	0.1748	0.1740	0.1728	0.1714	0.1697	0.1678	0.1656	0.1632	0.1606
6	0.1490	0.1515	0.1537	0.1555	0.1571	0.1584	0.1594	0.1601	0.1605	0.1606
7	0.1086	0.1125	0.1163	0.1200	0.1234	0.1267	0.1298	0.1326	0.1353	0.1377
8	0.0692	0.0731	0.0771	0.0810	0.0849	0.0887	0.0925	0.0962	0.0998	0.1033
9	0.0392	0.0423	0.0454	0.0486	0.0519	0.0552	0.0586	0.0620	0.0654	0.0688
10	0.0200	0.0220	0.0241	0.0262	0.0285	0.0309	0.0334	0.0359	0.0386	0.0413
11	0.0093	0.0104	0.0116	0.0129	0.0143	0.0157	0.0173	0.0190	0.0207	0.0225
12	0.0039	0.0045	0.0051	0.0058	0.0065	0.0073	0.0082	0.0092	0.0102	0.0113
13	0.0015	0.0018	0.0021	0.0024	0.0028	0.0032	0.0036	0.0041	0.0046	0.0052
14	0.0006	0.0007	0.0008	0.0009	0.0011	0.0013	0.0015	0.0017	0.0019	0.0022
15	0.0002	0.0002	0.0003	0.0003	0.0004	0.0005	0.0006	0.0007	0.0008	0.0009
16	0.0001	0.0001	0.0001	0.0001	0.0001	0.0002	0.0002	0.0002	0.0003	0.0003
17	0.0000	0.0000	0.0000	0.0000	0.0000	0.0001	0.0001	0.0001	0.0001	0.0001

	6.1	6.2	6.3	6.4	6.5	6.6	6.7	6.8	6.9	7.0
0	0.0022	0.0020	0.0018	0.0017	0.0015	0.0014	0.0012	0.0011	0.0010	0.0009
1	0.0137	0.0126	0.0116	0.0106	0.0098	0.0090	0.0082	0.0076	0.0070	0.0064
2	0.0417	0.0390	0.0364	0.0340	0.0318	0.0296	0.0276	0.0258	0.0240	0.0223
3	0.0848	0.0806	0.0765	0.0726	0.0688	0.0652	0.0617	0.0584	0.0552	0.0521
4	0.1294	0.1249	0.1205	0.1162	0.1118	0.1076	0.1034	0.0992	0.0952	0.0912
5	0.1579	0.1549	0.1519	0.1487	0.1454	0.1420	0.1385	0.1349	0.1314	0.1277
6	0.1605	0.1601	0.1595	0.1586	0.1575	0.1562	0.1546	0.1529	0.1511	0.1490
7	0.1399	0.1418	0.1435	0.1450	0.1462	0.1472	0.1480	0.1486	0.1489	0.1490
8	0.1066	0.1099	0.1130	0.1160	0.1188	0.1215	0.1240	0.1263	0.1284	0.1304
9	0.0723	0.0757	0.0791	0.0825	0.0858	0.0891	0.0923	0.0954	0.0985	0.1014
10	0.0441	0.0469	0.0498	0.0528	0.0558	0.0588	0.0618	0.0649	0.0679	0.0710
11	0.0245	0.0265	0.0285	0.0307	0.0330	0.0353	0.0377	0.0401	0.0426	0.0452
12	0.0124	0.0137	0.0150	0.0164	0.0179	0.0194	0.0210	0.0227	0.0245	0.0264
13	0.0058	0.0065	0.0073	0.0081	0.0089	0.0098	0.0108	0.0119	0.0130	0.0142
14	0.0025	0.0029	0.0033	0.0037	0.0041	0.0046	0.0052	0.0058	0.0064	0.0071
15	0.0010	0.0012	0.0014	0.0016	0.0018	0.0020	0.0023	0.0026	0.0029	0.0033
16	0.0004	0.0005	0.0005	0.0006	0.0007	0.0008	0.0010	0.0011	0.0013	0.0014
17	0.0001	0.0002	0.0002	0.0002	0.0003	0.0003	0.0004	0.0004	0.0005	0.0006
18	0.0000	0.0001	0.0001	0.0001	0.0001	0.0001	0.0001	0.0002	0.0002	0.0002
19	0.0000	0.0000	0.0000	0.0000	0.0000	0.0000	0.0000	0.0001	0.0001	0.0001

	7.1	7.2	7.3	7.4	7.5	7.6	7.7	7.8	7.9	8.0
0	0.0008	0.0007	0.0007	0.0006	0.0006	0.0005	0.0005	0.0004	0.0004	0.0003
1	0.0059	0.0054	0.0049	0.0045	0.0041	0.0038	0.0035	0.0032	0.0029	0.0027
2	0.0208	0.0194	0.0180	0.0167	0.0156	0.0145	0.0134	0.0125	0.0116	0.0107
3	0.0492	0.0464	0.0438	0.0413	0.0389	0.0366	0.0345	0.0324	0.0305	0.0286
4	0.0874	0.0836	0.0799	0.0764	0.0729	0.0696	0.0663	0.0632	0.0602	0.0573
5	0.1241	0.1204	0.1167	0.1130	0.1094	0.1057	0.1021	0.0986	0.0951	0.0916
6	0.1468	0.1445	0.1420	0.1394	0.1367	0.1339	0.1311	0.1282	0.1252	0.1221

(continued)

x	μ									
	7.1	7.2	7.3	7.4	7.5	7.6	7.7	7.8	7.9	8.0
7	0.1489	0.1486	0.1481	0.1474	0.1465	0.1454	0.1442	0.1428	0.1413	0.1396
8	0.1321	0.1337	0.1351	0.1363	0.1373	0.1382	0.1388	0.1392	0.1395	0.1396
9	0.1042	0.1070	0.1096	0.1121	0.1144	0.1167	0.1187	0.1207	0.1224	0.1241
10	0.0740	0.0770	0.0800	0.0829	0.0858	0.0887	0.0914	0.0941	0.0967	0.0993
11	0.0478	0.0504	0.0531	0.0558	0.0585	0.0613	0.0640	0.0667	0.0695	0.0722
12	0.0283	0.0303	0.0323	0.0344	0.0366	0.0388	0.0411	0.0434	0.0457	0.0481
13	0.0154	0.0168	0.0181	0.0196	0.0211	0.0227	0.0243	0.0260	0.0278	0.0296
14	0.0078	0.0086	0.0095	0.0104	0.0113	0.0123	0.0134	0.0145	0.0157	0.0169
15	0.0037	0.0041	0.0046	0.0051	0.0057	0.0062	0.0069	0.0075	0.0083	0.0090
16	0.0016	0.0019	0.0021	0.0024	0.0026	0.0030	0.0033	0.0037	0.0041	0.0045
17	0.0007	0.0008	0.0009	0.0010	0.0012	0.0013	0.0015	0.0017	0.0019	0.0021
18	0.0003	0.0003	0.0004	0.0004	0.0005	0.0006	0.0006	0.0007	0.0008	0.0009
19	0.0001	0.0001	0.0001	0.0002	0.0002	0.0002	0.0003	0.0003	0.0003	0.0004
20	0.0000	0.0000	0.0001	0.0001	0.0001	0.0001	0.0001	0.0001	0.0001	0.0002
21	0.0000	0.0000	0.0000	0.0000	0.0000	0.0000	0.0000	0.0000	0.0001	0.0001

TABLE A3 STANDARD NORMAL DISTRIBUTION

The entries in this table are the probabilities that a random variable having the standard normal distribution assumes a value between 0 and Z_1; the probability is represented by the area under the curve (the shaded area). Areas for negative values of z are obtained by symmetry.

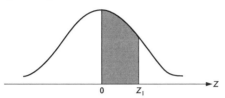

Z	Second decimal place in Z									
	0.00	0.01	0.02	0.03	0.04	0.05	0.06	0.07	0.08	0.09
0.0	0.0000	0.0040	0.0080	0.0120	0.0160	0.0199	0.0239	0.0279	0.0319	0.0359
0.1	0.0398	0.0438	0.0478	0.0517	0.0557	0.0596	0.0636	0.0675	0.0714	0.0753
0.2	0.0793	0.0832	0.0871	0.0910	0.0948	0.0987	0.1026	0.1064	0.1103	0.1141
0.3	0.1179	0.1217	0.1255	0.1293	0.1331	0.1368	0.1406	0.1443	0.1480	0.1517
0.4	0.1554	0.1591	0.1628	0.1664	0.1700	0.1736	0.1772	0.1808	0.1844	0.1879
0.5	0.1915	0.1950	0.1985	0.2019	0.2054	0.2088	0.2123	0.2157	0.2190	0.2224
0.6	0.2257	0.2291	0.2324	0.2357	0.2389	0.2422	0.2454	0.2486	0.2517	0.2549
0.7	0.2580	0.2611	0.2642	0.2673	0.2704	0.2734	0.2764	0.2794	0.2823	0.2852
0.8	0.2881	0.2910	0.2939	0.2967	0.2995	0.3023	0.3051	0.3078	0.3106	0.3133
0.9	0.3159	0.3186	0.3212	0.3238	0.3264	0.3289	0.3315	0.3340	0.3365	0.3389
1.0	0.3413	0.3438	0.3461	0.3485	0.3508	0.3531	0.3554	0.3577	0.3599	0.3621
1.1	0.3643	0.3665	0.3686	0.3708	0.3729	0.3749	0.3770	0.3790	0.3810	0.3830
1.2	0.3849	0.3869	0.3888	0.3907	0.3925	0.3944	0.3962	0.3980	0.3997	0.4015
1.3	0.4032	0.4049	0.4066	0.4082	0.4099	0.4115	0.4131	0.4147	0.4162	0.4177
1.4	0.4192	0.4207	0.4222	0.4236	0.4251	0.4265	0.4279	0.4292	0.4306	0.4319
1.5	0.4332	0.4345	0.4357	0.4370	0.4382	0.4394	0.4406	0.4418	0.4429	0.4441
1.6	0.4452	0.4463	0.4474	0.4484	0.4495	0.4505	0.4515	0.4525	0.4535	0.4545
1.7	0.4554	0.4564	0.4573	0.4582	0.4591	0.4599	0.4608	0.4616	0.4625	0.4633
1.8	0.4641	0.4649	0.4656	0.4664	0.4671	0.4678	0.4686	0.4693	0.4699	0.4706
1.9	0.4713	0.4719	0.4726	0.4732	0.4738	0.4744	0.4750	0.4756	0.4761	0.4767
2.0	0.4772	0.4778	0.4783	0.4788	0.4793	0.4796	0.4803	0.4808	0.4812	0.4817
2.1	0.4821	0.4826	0.4830	0.4834	0.4838	0.4842	0.4846	0.4850	0.4854	0.4857
2.2	0.4861	0.4864	0.4868	0.4871	0.4875	0.4878	0.4881	0.4884	0.4887	0.4890
2.3	0.4893	0.4896	0.4898	0.4901	0.4904	0.4906	0.4909	0.4911	0.4913	0.4916
2.4	0.4918	0.4920	0.4922	0.4925	0.4927	0.4929	0.4931	0.4932	0.4934	0.4936
2.5	0.4938	0.4940	0.4941	0.4943	0.4945	0.4946	0.4948	0.4949	0.4951	0.4952
2.6	0.4953	0.4955	0.4956	0.4957	0.4959	0.4960	0.4961	0.4962	0.4963	0.4974
2.7	0.4965	0.4966	0.4967	0.4968	0.4969	0.4970	0.4971	0.4972	0.4973	0.4974
2.8	0.4974	0.4975	0.4976	0.4977	0.4977	0.4978	0.4979	0.4979	0.4980	0.4981
2.9	0.4981	0.4982	0.4982	0.4983	0.4984	0.4984	0.4985	0.4985	0.4986	0.4986
3.0	0.4987	0.4987	0.4987	0.4988	0.4988	0.4989	0.4989	0.4989	0.4990	0.4990
3.1	0.4990	0.4991	0.4991	0.4991	0.4992	0.4992	0.4992	0.4992	0.4993	0.4993
3.2	0.4993	0.4993	0.4994	0.4994	0.4994	0.4994	0.4994	0.4995	0.4995	0.4995
3.3	0.4995	0.4995	0.4995	0.4996	0.4996	0.4996	0.4996	0.4996	0.4996	0.4997
3.4	0.4997	0.4997	0.4997	0.4997	0.4997	0.4997	0.4997	0.4997	0.4997	0.4998
3.5	0.4998									
4.0	0.49997									
4.5	0.499997									
5.0	0.4999997									

Source: Reprinted with permission from *Standard Mathematical Tables*, 15th edn, © CRC Press Inc., Boca Raton, FL.

TABLE A4   *t* DISTRIBUTION

Entries in the table give t_α values, where α is the area or probability in the upper tail of the *t* distribution. For example, with ten degrees of freedom and an area of 0.05 in the upper tail, $t_{0.05} = 1.812$.

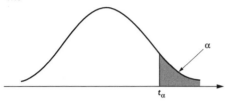

Degrees of freedom	$t_{0.100}$	$t_{0.050}$	$t_{0.025}$	$t_{0.010}$	$t_{0.005}$
1	3.078	6.314	12.706	31.821	63.657
2	1.886	2.920	4.303	6.965	9.925
3	1.638	2.353	3.182	4.541	5.841
4	1.533	2.132	2.776	3.747	4.604
5	1.476	2.015	2.571	3.365	4.032
6	1.440	1.943	2.447	3.143	3.707
7	1.415	1.895	2.365	2.998	3.499
8	1.397	1.860	2.306	2.896	3.355
9	1.383	1.833	2.262	2.821	3.250
10	1.372	1.812	2.228	2.764	3.169
11	1.363	1.796	2.201	2.718	3.106
12	1.356	1.782	2.179	2.681	3.055
13	1.350	1.771	2.160	2.650	3.012
14	1.345	1.761	2.145	2.624	2.977
15	1.341	1.753	2.131	2.602	2.947
16	1.337	1.746	2.120	2.583	2.921
17	1.333	1.740	2.110	2.567	2.898
18	1.330	1.734	2.101	2.552	2.878
19	1.328	1.729	2.093	2.539	2.861
20	1.325	1.725	2.086	2.528	2.845
21	1.323	1.721	2.080	2.518	2.831
22	1.321	1.717	2.074	2.508	2.819
23	1.319	1.714	2.069	2.500	2.808
24	1.318	1.711	2.064	2.492	2.797
25	1.316	1.708	2.060	2.485	2.787
26	1.315	1.706	2.056	2.479	2.779
27	1.314	1.703	2.052	2.473	2.771
28	1.313	1.701	2.048	2.467	2.763
29	1.311	1.699	2.045	2.462	2.756
30	1.310	1.697	2.042	2.457	2.750
40	1.303	1.684	2.021	2.423	2.704
60	1.296	1.671	2.000	2.390	2.660
120	1.289	1.658	1.980	2.358	2.617
∞	1.282	1.645	1.960	2.326	2.576

TABLE A5   CRITICAL VALUES OF SPEARMAN'S RANK CORRELATION COEFFICIENT

The table shows critical values of Spearman's rank correlation coefficient for selected values of n and α.

n	$\alpha = 0.05$	$\alpha = 0.025$	$\alpha = 0.01$
5	0.900	—	—
6	0.829	0.886	0.943
7	0.714	0.786	0.893
8	0.643	0.738	0.833
9	0.600	0.683	0.783
10	0.564	0.648	0.745
11	0.523	0.623	0.736
12	0.497	0.591	0.703
13	0.475	0.566	0.673
14	0.457	0.545	0.646
15	0.441	0.525	0.623
16	0.425	0.507	0.601
17	0.412	0.490	0.582
18	0.399	0.476	0.564
19	0.388	0.462	0.549
20	0.377	0.450	0.534
21	0.368	0.438	0.521
22	0.359	0.428	0.508
23	0.351	0.418	0.496
24	0.343	0.409	0.485
25	0.336	0.400	0.475
26	0.329	0.392	0.465
27	0.323	0.385	0.456
28	0.317	0.377	0.448
29	0.311	0.370	0.440
30	0.305	0.364	0.432

Source: Reproduced by permission from E. G. Olds (1938) 'Distribution of sums of squares of rank differences for small samples', *Annals of Mathematical Statistics 9*.

TABLE A6   F DISTRIBUTION

The table shows values of F for various degrees of freedom, v_1 and v_2, from 1 to ∞, and for $\alpha = 0.05$ and $\alpha = 0.01$.

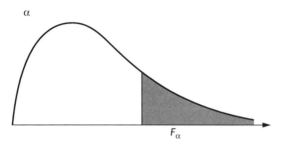

Denominator degrees of freedom (v_2)	Numerator degrees of freedom (v_1)								
	1	2	3	4	5	6	7	8	9
1	161.4	199.5	215.7	224.6	230.2	234.0	236.8	238.9	240.5
2	18.51	19.00	19.16	19.25	19.30	19.33	19.35	19.37	19.38
3	10.13	9.55	9.28	9.12	9.01	8.94	8.89	8.85	8.81
4	7.71	6.94	6.59	6.39	6.26	6.16	6.09	6.04	6.00
5	6.61	5.79	5.41	5.19	5.05	4.95	4.88	4.82	4.77
6	5.99	5.14	4.76	4.53	4.39	4.28	4.21	4.15	4.10
7	5.59	4.74	4.35	4.12	3.97	3.87	3.79	3.73	3.68
8	5.32	4.46	4.07	3.84	3.69	3.58	3.50	3.44	3.39
9	5.12	4.26	3.86	3.63	3.48	3.37	3.29	3.23	3.18
10	4.96	4.10	3.71	3.48	3.33	3.22	3.14	3.07	3.02
11	4.84	3.98	3.59	3.36	3.20	3.09	3.01	2.95	2.90
12	4.75	3.89	3.49	3.26	3.11	3.00	2.91	2.85	2.80
13	4.67	3.81	3.41	3.18	3.03	2.92	2.83	2.77	2.71
14	4.60	3.74	3.34	3.11	2.96	2.85	2.76	2.70	2.65
15	4.54	3.68	3.29	3.06	2.90	2.79	2.71	2.64	2.59
16	4.49	3.63	3.24	3.01	2.85	2.74	2.66	2.59	2.54
17	4.45	3.59	3.20	2.96	2.81	2.70	2.61	2.55	2.49
18	4.41	3.55	3.16	2.93	2.77	2.66	2.58	2.51	2.46
19	4.38	3.52	3.13	2.90	2.74	2.63	2.54	2.48	2.42
20	4.35	3.49	3.10	2.87	2.71	2.60	2.51	2.45	2.39
21	4.32	3.47	3.07	2.84	2.68	2.57	2.49	2.42	2.37
22	4.30	3.44	3.05	2.82	2.66	2.55	2.46	2.40	2.34
23	4.28	3.42	3.03	2.80	2.64	2.53	2.44	2.37	2.32
24	4.26	3.40	3.01	2.78	2.62	2.51	2.42	2.36	2.30
25	4.24	3.39	2.99	2.76	2.60	2.49	2.40	2.34	2.28
26	4.23	3.37	2.98	2.74	2.59	2.47	2.39	2.32	2.27
27	4.21	3.35	2.96	2.73	2.57	2.46	2.37	2.31	2.25
28	4.20	3.34	2.95	2.71	2.56	2.45	2.36	2.29	2.24
29	4.18	3.33	2.93	2.70	2.55	2.43	2.35	2.28	2.22
30	4.17	3.32	2.92	2.69	2.53	2.42	2.33	2.27	2.21
40	4.08	3.23	2.84	2.61	2.45	2.34	2.25	2.18	2.12
60	4.00	3.15	2.76	2.53	2.37	2.25	2.17	2.10	2.04
120	3.92	3.07	2.68	2.45	2.29	2.17	2.09	2.02	1.96
∞	3.84	3.00	2.60	2.37	2.21	2.10	2.01	1.94	1.88

Numerator degrees of freedom (v_1)										Denominator degrees of freedom (v_2)
10	12	15	20	24	30	40	60	120	∞	
241.9	243.9	245.9	248.0	249.1	250.1	251.1	252.2	253.3	254.3	1
19.40	19.41	19.43	19.45	19.45	19.46	19.47	19.48	19.49	19.50	2
8.79	8.74	8.70	8.66	8.64	8.62	8.59	8.57	8.55	8.53	3
5.96	5.91	5.86	5.80	5.77	5.75	5.72	5.69	5.66	5.63	4
4.74	4.68	4.62	4.56	4.53	4.50	4.46	4.43	4.40	4.36	5
4.06	4.00	3.94	3.87	3.84	3.81	3.77	3.74	3.70	3.67	6
3.64	3.57	3.51	3.44	3.41	3.38	3.34	3.30	3.27	3.23	7
3.35	3.28	3.22	3.15	3.12	3.08	3.04	3.01	2.97	2.93	8
3.14	3.07	3.01	2.94	2.90	2.86	2.83	2.79	2.75	2.71	9
2.98	2.91	2.85	2.77	2.74	2.70	2.66	2.62	2.58	2.54	10
2.85	2.79	2.72	2.65	2.61	2.57	2.53	2.49	2.45	2.40	11
2.75	2.69	2.62	2.54	2.51	2.47	2.43	2.38	2.34	2.30	12
2.67	2.60	2.53	2.46	2.42	2.38	2.34	2.30	2.25	2.21	13
2.60	2.53	2.46	2.39	2.35	2.31	2.27	2.22	2.18	2.13	14
2.54	2.48	2.40	2.33	2.29	2.25	2.20	2.16	2.11	2.07	15
2.49	2.42	2.35	2.28	2.24	2.19	2.15	2.11	2.06	2.01	16
2.45	2.38	2.31	2.23	2.19	2.15	2.10	2.06	2.01	1.96	17
2.41	2.34	2.27	2.19	2.15	2.11	2.06	2.02	1.97	1.92	18
2.38	2.31	2.23	2.16	2.11	2.07	2.03	1.98	1.93	1.88	19
2.35	2.28	2.20	2.12	2.08	2.04	1.99	1.95	1.90	1.84	20
2.32	2.25	2.18	2.10	2.05	2.01	1.96	1.92	1.87	1.81	21
2.30	2.23	2.15	2.07	2.03	1.98	1.94	1.89	1.84	1.78	22
2.27	2.20	2.13	2.05	2.01	1.96	1.91	1.86	1.81	1.76	23
2.25	2.18	2.11	2.03	1.98	1.94	1.89	1.84	1.79	1.73	24
2.24	2.16	2.09	2.01	1.96	1.92	1.87	1.82	1.77	1.71	25
2.22	2.15	2.07	1.99	1.95	1.90	1.85	1.80	1.75	1.69	26
2.20	2.13	2.06	1.97	1.93	1.88	1.84	1.79	1.73	1.67	27
2.19	2.12	2.04	1.96	1.91	1.87	1.82	1.77	1.71	1.65	28
2.18	2.10	2.03	1.94	1.90	1.85	1.81	1.75	1.70	1.64	29
2.16	2.09	2.01	1.93	1.89	1.84	1.79	1.74	1.68	1.62	30
2.08	2.00	1.92	1.84	1.79	1.74	1.69	1.64	1.58	1.51	40
1.99	1.92	1.84	1.75	1.70	1.65	1.59	1.53	1.47	1.39	60
1.91	1.83	1.75	1.66	1.61	1.55	1.50	1.43	1.35	1.25	120
1.83	1.75	1.67	1.57	1.52	1.46	1.39	1.32	1.22	1.00	∞

(b) α=0.01

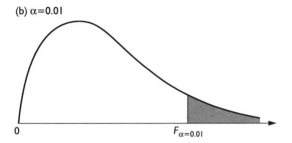

$F_{\alpha=0.01}$

Denominator degrees of freedom (v_2)	Numerator degrees of freedom (v_1)								
	1	2	3	4	5	6	7	8	9
1	4,052	4,999.5	5,403	5,625	5,764	5,859	5,928	5,982	6,022
2	98.50	99.00	99.17	99.25	99.30	99.33	99.36	99.37	99.39
3	34.12	30.82	29.46	28.71	28.24	27.91	27.67	27.49	27.35
4	21.20	18.00	16.69	15.98	15.52	15.21	14.98	14.80	14.66
5	16.26	13.27	12.06	11.39	10.97	10.67	10.46	10.29	10.16
6	13.75	10.92	9.78	9.15	8.75	8.47	8.26	8.10	7.98
7	12.25	9.55	8.45	7.85	7.46	7.19	6.99	6.84	6.72
8	11.26	8.65	7.59	7.01	6.63	6.37	6.18	6.03	5.91
9	10.56	8.02	6.99	6.42	6.06	5.80	5.61	5.47	5.35
10	10.04	7.56	6.55	5.99	5.64	5.39	5.20	5.06	4.94
11	9.65	7.21	6.22	5.67	5.32	5.07	4.89	4.74	4.63
12	9.33	6.93	5.95	5.41	5.06	4.82	4.64	4.50	4.39
13	9.07	6.70	5.74	5.21	4.86	4.62	4.44	4.30	4.19
14	8.86	6.51	5.56	5.04	4.69	4.46	4.28	4.14	4.03
15	8.68	6.36	5.42	4.89	4.56	4.32	4.14	4.00	3.89
16	8.53	6.23	5.29	4.77	4.44	4.20	4.03	3.89	3.78
17	8.40	6.11	5.18	4.67	4.34	4.10	3.93	3.79	3.68
18	8.29	6.01	5.09	4.58	4.25	4.01	3.84	3.71	3.60
19	8.18	5.93	5.01	4.50	4.17	3.94	3.77	3.63	3.52
20	8.10	5.85	4.94	4.43	4.10	3.87	3.70	3.56	3.46
21	8.02	5.78	4.87	4.37	4.04	3.81	3.64	3.51	3.40
22	7.95	5.72	4.82	4.31	3.99	3.76	3.59	3.45	3.35
23	7.88	5.66	4.76	4.26	3.94	3.71	3.54	3.41	3.30
24	7.82	5.61	4.72	4.22	3.90	3.67	3.50	3.36	3.26
25	7.77	5.57	4.68	4.18	3.85	3.63	3.46	3.32	3.22
26	7.72	5.53	4.64	4.14	3.82	3.59	3.42	3.29	3.18
27	7.68	5.49	4.60	4.11	3.78	3.56	3.39	3.26	3.15
28	7.64	5.45	4.57	4.07	3.75	3.53	3.36	3.23	3.12
29	7.60	5.42	4.54	4.04	3.73	3.50	3.33	3.20	3.09
30	7.56	5.39	4.51	4.02	3.70	3.47	3.30	3.17	3.07
40	7.31	5.18	4.31	3.83	3.51	3.29	3.12	2.99	2.89
60	7.08	4.98	4.13	3.65	3.34	3.12	2.95	2.82	2.72
120	6.85	4.79	3.95	3.48	3.17	2.96	2.79	2.66	2.56
∞	6.63	4.61	3.78	3.32	3.02	2.80	2.64	2.51	2.41

Numerator degrees of freedom (v_1)										Denominator degrees of freedom (v_2)
10	12	15	20	24	30	40	60	120	∞	
6,056	6,106	6,157	6,209	6,235	6,261	6,287	6,313	6,339	6,366	1
99.40	99.42	99.43	99.45	99.46	99.47	99.47	99.48	99.49	99.50	2
27.23	27.05	26.87	26.69	26.60	26.50	26.41	26.32	26.22	26.13	3
14.55	14.37	14.20	14.02	13.93	13.84	13.75	13.65	13.56	13.46	4
10.05	9.89	9.72	9.55	9.47	9.38	9.29	9.20	9.11	9.02	5
7.87	7.72	7.56	7.40	7.31	7.23	7.14	7.06	6.97	6.88	6
6.62	6.47	6.31	6.16	6.07	5.99	5.91	5.82	5.74	5.65	7
5.81	5.67	5.52	5.36	5.28	5.20	5.12	5.03	4.95	4.86	8
5.26	5.11	4.96	4.81	4.73	4.65	4.57	4.48	4.40	4.31	9
4.85	4.71	4.56	4.41	4.33	4.25	4.17	4.08	4.00	3.91	10
4.54	4.40	4.25	4.10	4.02	3.94	3.86	3.78	3.69	3.60	11
4.30	4.16	4.01	3.86	3.78	3.70	3.62	3.54	3.45	3.36	12
4.10	3.96	3.82	3.66	3.59	3.51	3.43	3.34	3.25	3.17	13
3.94	3.80	3.66	3.51	3.43	3.35	3.27	3.18	3.09	3.00	14
3.80	3.67	3.52	3.37	3.29	3.21	3.13	3.05	2.96	2.87	15
3.69	3.55	3.41	3.26	3.18	3.10	3.02	2.93	2.84	2.75	16
3.59	3.46	3.31	3.16	3.08	3.00	2.92	2.83	2.75	2.65	17
3.51	3.37	3.23	3.08	3.00	2.92	2.84	2.75	2.66	2.57	18
3.43	3.30	3.15	3.00	2.92	2.84	2.76	2.67	2.58	2.49	19
3.37	3.23	3.09	2.94	2.86	2.78	2.69	2.61	2.52	2.42	20
3.31	3.17	3.03	2.88	2.80	2.72	2.64	2.55	2.46	2.36	21
3.26	3.12	2.98	2.83	2.75	2.67	2.58	2.50	2.40	2.31	22
3.21	3.07	2.93	2.78	2.70	2.62	2.54	2.45	2.35	2.26	23
3.17	3.03	2.89	2.74	2.66	2.58	2.49	2.40	2.31	2.21	24
3.13	2.99	2.85	2.70	2.62	2.54	2.45	2.36	2.27	2.17	25
3.09	2.96	2.81	2.66	2.58	2.50	2.42	2.33	2.23	2.13	26
3.06	2.93	2.78	2.63	2.55	2.47	2.38	2.29	2.20	2.10	27
3.03	2.90	2.75	2.60	2.52	2.44	2.35	2.26	2.17	2.06	28
3.00	2.87	2.73	2.57	2.49	2.41	2.33	2.23	2.14	2.03	29
2.98	2.84	2.70	2.55	2.47	2.39	2.30	2.21	2.11	2.01	30
2.80	2.66	2.52	2.37	2.29	2.20	2.11	2.02	1.92	1.80	40
2.63	2.50	2.35	2.20	2.12	2.03	1.94	1.84	1.73	1.60	60
2.47	2.34	2.19	2.03	1.95	1.86	1.76	1.66	1.53	1.38	120
2.32	2.18	2.04	1.88	1.79	1.70	1.59	1.47	1.32	1.00	∞

Source: From M. Merrington and C. M. Thompson (1943) 'Tables of percentage points of the inverted beta (*F*)-distribution', *Biometrika* 33, 73–88. Reproduced by permission of the Biometrika trustees.

TABLE A7 χ^2 DISTRIBUTION

Entries in the table give χ^2_α, where α is the area or probability in the tail of the distribution (the shaded area). For example, with ten degrees of freedom and $\alpha = 0.05$, $\chi^2_\alpha = 18.307$.

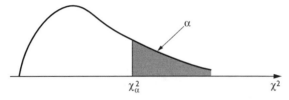

Degrees of freedom	$\chi^2_{0.995}$	$\chi^2_{0.990}$	$\chi^2_{0.975}$	$\chi^2_{0.950}$	$\chi^2_{0.900}$
1	0.0000393	0.0001571	0.0009821	0.0039321	0.0157908
2	0.0100251	0.0201007	0.0506356	0.102587	0.210720
3	0.0717212	0.114832	0.215795	0.351846	0.584375
4	0.206990	0.297110	0.484419	0.710721	1.063623
5	0.411740	0.554300	0.831211	1.145476	1.61031
6	0.675727	0.872085	1.237347	1.63539	2.20413
7	0.989265	1.239043	1.68987	2.16735	2.83311
8	1.344419	1.646482	2.17973	2.73264	3.48954
9	1.734926	2.087912	2.70039	3.32511	4.16816
10	2.15585	2.55821	3.24697	3.94030	4.86518
11	2.60321	3.05347	3.81575	4.57481	5.57779
12	3.07382	3.57056	4.40379	5.22603	6.30380
13	3.56503	4.10691	5.00874	5.89186	7.04150
14	4.07468	4.66043	5.62872	6.57063	7.78953
15	4.60094	5.22935	6.26214	7.26094	8.54675
16	5.14224	5.81221	6.90766	7.96164	9.31223
17	5.69724	6.40776	7.56418	8.67176	10.0852
18	6.26481	7.01491	8.23075	9.39046	10.8649
19	6.84398	7.63273	8.90655	10.1170	11.6509
20	7.43386	8.26040	9.59083	10.8508	12.4426
21	8.03366	8.89720	10.28293	11.5913	13.2396
22	8.64272	9.54249	10.9823	12.3380	14.0415
23	9.26042	10.19567	11.6885	13.0905	14.8479
24	9.88623	10.8564	12.4011	13.8484	15.6587
25	10.5197	11.5240	13.1197	14.6114	16.4734
26	11.1603	12.1981	13.8439	15.3791	17.2919
27	11.8076	12.8786	14.5733	16.1513	18.1138
28	12.4613	13.5648	15.3079	16.9279	18.9392
29	13.1211	14.2565	16.0471	17.7083	19.7677
30	13.7867	14.9535	16.7908	18.4926	20.5992
40	20.7065	22.1643	24.4331	26.5093	29.0505
50	27.9907	29.7067	32.3574	34.7642	37.6886
60	35.5346	37.4848	40.4817	43.1879	46.4589
70	43.2752	45.4418	48.7576	51.7393	55.3290
80	51.1720	53.5400	57.1532	60.3915	64.2778
90	59.1963	61.7541	65.6466	69.1260	73.2912
100	67.3276	70.0648	74.2219	77.9295	82.3581

Degrees of freedom	$\chi^2_{0.100}$	$\chi^2_{0.050}$	$\chi^2_{0.025}$	$\chi^2_{0.010}$	$\chi^2_{0.005}$
1	2.70554	3.84146	5.02389	6.63490	7.87944
2	4.60517	5.99147	7.37776	9.21034	10.5966
3	6.25139	7.81473	9.34840	11.3449	12.8381
4	7.77944	9.48773	11.1433	13.2767	14.8602
5	9.23635	11.0705	12.8325	15.0863	16.7496
6	10.6446	12.5916	14.4494	16.8119	18.5476
7	12.0170	14.0671	16.0128	18.4753	20.2777
8	13.3616	15.5073	17.5346	20.0902	21.9550
9	14.6837	16.9190	19.0228	21.6660	23.5893
10	15.9871	18.3070	20.4831	23.2093	25.1882
11	17.2750	19.6751	21.9200	24.7250	26.7569
12	18.5494	21.0261	23.3367	26.2170	28.2995
13	19.8119	22.3621	24.7356	27.6883	29.8194
14	21.0642	23.6848	26.1190	29.1413	31.3193
15	22.3072	24.9958	27.4884	30.5779	32.8013
16	23.5418	26.2962	28.8454	31.9999	34.2672
17	24.7690	27.5871	30.1910	33.4087	35.7185
18	25.9894	28.8693	31.5264	34.8053	37.1564
19	27.2036	30.1435	32.8523	36.1908	38.5822
20	28.4120	31.4104	34.1696	37.5662	39.9968
21	29.6151	32.6705	35.4789	38.9321	41.4010
22	30.8133	33.9244	36.7807	40.2894	42.7956
23	32.0069	35.1725	38.0757	41.6384	44.1813
24	33.1963	36.4151	39.3641	42.9798	45.5585
25	34.3816	37.6525	40.6465	44.3141	46.9278
26	35.5631	38.8852	41.9232	45.6417	48.2899
27	36.7412	40.1133	43.1944	46.9630	49.6449
28	37.9159	41.3372	44.4607	48.2782	50.9933
29	39.0875	42.5569	45.7222	49.5879	52.3356
30	40.2560	43.7729	46.9792	50.8922	53.6720
40	51.8050	55.7585	59.3417	63.6907	66.7659
50	63.1671	67.5048	71.4202	76.1539	79.4900
60	74.3970	79.0819	83.2976	88.3794	91.9517
70	85.5271	90.5312	95.0231	100.425	104.215
80	96.5782	101.879	106.629	112.329	116.321
90	107.565	113.145	118.136	124.116	128.229
100	118.498	124.342	129.561	135.807	140.169

Source: Reproduced from *Biometrika Tables for Statisticians*, Cambridge: Cambridge University Press, 1954, by permission of the Biometrika trustees.

TABLE A8 CRITICAL VALUES FOR THE DURBIN-WATSON STATISTIC

Entries in the table give the critical values for a one-tailed Durbin-Watson test for autocorrelation. For a two-tailed test, the level of significance is doubled.

	k	1		2		3		4		5	
n		d_L	d_U	d_L	d_U	d_L	d_U	d_L	d_U	d_L	d_U
15		1.08	1.36	0.95	1.54	0.82	1.75	0.69	1.97	0.56	2.21
16		1.10	1.37	0.98	1.54	0.86	1.73	0.74	1.93	0.62	2.15
17		1.13	1.38	1.02	1.54	0.90	1.71	0.78	1.90	0.67	2.10
18		1.16	1.39	1.05	1.53	0.93	1.69	0.82	1.87	0.71	2.06
19		1.18	1.40	1.08	1.53	0.97	1.68	0.86	1.85	0.75	2.02
20		1.20	1.41	1.10	1.54	1.00	1.68	0.90	1.83	0.79	1.99
21		1.22	1.42	1.13	1.54	1.03	1.67	0.93	1.81	0.83	1.96
22		1.24	1.43	1.15	1.54	1.05	1.66	0.96	1.80	0.86	1.94
23		1.26	1.44	1.17	1.54	1.08	1.66	0.99	1.79	0.90	1.92
24		1.27	1.45	1.19	1.55	1.10	1.66	1.01	1.78	0.93	1.90
25		1.29	1.45	1.21	1.55	1.12	1.66	1.04	1.77	0.95	1.89
26		1.30	1.46	1.22	1.55	1.14	1.65	1.06	1.76	0.98	1.88
27		1.32	1.47	1.24	1.56	1.16	1.65	1.08	1.76	1.01	1.86
28		1.33	1.48	1.26	1.56	1.18	1.65	1.10	1.75	1.03	1.85
29		1.34	1.48	1.27	1.56	1.20	1.65	1.12	1.74	1.05	1.84
30		1.35	1.49	1.28	1.57	1.21	1.65	1.14	1.74	1.07	1.83
31		1.36	1.50	1.30	1.57	1.23	1.65	1.16	1.74	1.09	1.83
32		1.37	1.50	1.31	1.57	1.24	1.65	1.18	1.73	1.11	1.82
33		1.38	1.51	1.32	1.58	1.26	1.65	1.19	1.73	1.13	1.81
34		1.39	1.51	1.33	1.58	1.27	1.65	1.21	1.73	1.15	1.81
35		1.40	1.52	1.34	1.58	1.28	1.65	1.22	1.73	1.16	1.80
36		1.41	1.52	1.35	1.59	1.29	1.65	1.24	1.73	1.18	1.80
37		1.42	1.53	1.36	1.59	1.31	1.66	1.25	1.72	1.19	1.80
38		1.43	1.54	1.37	1.59	1.32	1.66	1.26	1.72	1.21	1.79
39		1.43	1.54	1.38	1.60	1.33	1.66	1.27	1.72	1.22	1.79
40		1.44	1.54	1.39	1.60	1.34	1.66	1.29	1.72	1.23	1.79
45		1.48	1.57	1.43	1.62	1.38	1.67	1.34	1.72	1.29	1.78
50		1.50	1.59	1.46	1.63	1.42	1.67	1.38	1.72	1.34	1.77
55		1.53	1.60	1.49	1.64	1.45	1.68	1.41	1.72	1.38	1.77
60		1.55	1.62	1.51	1.65	1.48	1.69	1.44	1.73	1.41	1.77
65		1.57	1.63	1.54	1.66	1.50	1.70	1.47	1.73	1.44	1.77
70		1.58	1.64	1.55	1.67	1.52	1.70	1.49	1.74	1.46	1.77
75		1.60	1.65	1.57	1.68	1.54	1.71	1.51	1.74	1.49	1.77
80		1.61	1.66	1.59	1.69	1.56	1.72	1.53	1.74	1.51	1.77
85		1.62	1.67	1.60	1.70	1.57	1.72	1.55	1.75	1.52	1.77
90		1.63	1.68	1.61	1.70	1.59	1.73	1.57	1.75	1.54	1.78
95		1.64	1.69	1.62	1.71	1.60	1.73	1.58	1.75	1.56	1.78
100		1.65	1.69	1.63	1.72	1.61	1.74	1.59	1.76	1.57	1.78

Significance Points of d_L and d_U: $\alpha = 0.05$
Number of Independent Variables

(*continued*)

	k	1		2		3		4		5	
n		d_L	d_U	d_L	d_U	d_L	d_U	d_L	d_U	d_L	d_U
15		0.95	1.23	0.83	1.40	0.71	1.61	0.59	1.84	0.48	2.09
16		0.98	1.24	0.86	1.40	0.75	1.59	0.64	1.80	0.53	2.03
17		1.01	1.25	0.90	1.40	0.79	1.58	0.68	1.77	0.57	1.98
18		1.03	1.26	0.93	1.40	0.82	1.56	0.72	1.74	0.62	1.93
19		1.06	1.28	0.96	1.41	0.86	1.55	0.76	1.72	0.66	1.90
20		1.08	1.28	0.99	1.41	0.89	1.55	0.79	1.70	0.70	1.87
21		1.10	1.30	1.01	1.41	0.92	1.54	0.83	1.69	0.73	1.84
22		1.12	1.31	1.04	1.42	0.95	1.54	0.86	1.68	0.77	1.82
23		1.14	1.32	1.06	1.42	0.97	1.54	0.89	1.67	0.80	1.80
24		1.16	1.33	1.08	1.43	1.00	1.54	0.91	1.66	0.83	1.79
25		1.18	1.34	1.10	1.43	1.02	1.54	0.94	1.65	0.86	1.77
26		1.19	1.35	1.12	1.44	1.04	1.54	0.96	1.65	0.88	1.76
27		1.21	1.36	1.13	1.44	1.06	1.54	0.99	1.64	0.91	1.75
28		1.22	1.37	1.15	1.45	1.08	1.54	1.01	1.64	0.93	1.74
29		1.24	1.38	1.17	1.45	1.10	1.54	1.03	1.63	0.96	1.73
30		1.25	1.38	1.18	1.46	1.12	1.54	1.05	1.63	0.98	1.73
31		1.26	1.39	1.20	1.47	1.13	1.55	1.07	1.63	1.00	1.72
32		1.27	1.40	1.21	1.47	1.15	1.55	1.08	1.63	1.02	1.71
33		1.28	1.41	1.22	1.48	1.16	1.55	1.10	1.63	1.04	1.71
34		1.29	1.41	1.24	1.48	1.17	1.55	1.12	1.63	1.06	1.70
35		1.30	1.42	1.25	1.48	1.19	1.55	1.13	1.63	1.07	1.70
36		1.31	1.43	1.26	1.49	1.20	1.56	1.15	1.63	1.09	1.70
37		1.32	1.43	1.27	1.49	1.21	1.56	1.16	1.62	1.10	1.70
38		1.33	1.44	1.28	1.50	1.23	1.56	1.17	1.62	1.12	1.70
39		1.34	1.44	1.29	1.50	1.24	1.56	1.19	1.63	1.13	1.69
40		1.35	1.45	1.30	1.51	1.25	1.57	1.20	1.63	1.15	1.69
45		1.39	1.48	1.34	1.53	1.30	1.58	1.25	1.63	1.21	1.69
50		1.42	1.50	1.38	1.54	1.34	1.59	1.30	1.64	1.26	1.69
55		1.45	1.52	1.41	1.56	1.37	1.60	1.33	1.64	1.30	1.69
60		1.47	1.54	1.44	1.57	1.40	1.61	1.37	1.65	1.33	1.69
65		1.49	1.55	1.46	1.59	1.43	1.62	1.40	1.66	1.36	1.69
70		1.51	1.57	1.48	1.60	1.45	1.63	1.42	1.66	1.39	1.70
75		1.53	1.58	1.50	1.61	1.47	1.64	1.45	1.67	1.42	1.70
80		1.54	1.59	1.52	1.62	1.49	1.65	1.47	1.67	1.44	1.70
85		1.56	1.60	1.53	1.63	1.51	1.65	1.49	1.68	1.46	1.71
90		1.57	1.61	1.55	1.64	1.53	1.66	1.50	1.69	1.48	1.71
95		1.58	1.62	1.56	1.65	1.54	1.67	1.52	1.69	1.50	1.71
100		1.59	1.63	1.57	1.65	1.55	1.67	1.53	1.70	1.51	1.72

Significance Points of d_L and d_U: $\alpha = 0.025$
Number of Independent Variables

(*continued*)

		Significance Points of d_L and d_U: $\alpha = 0.01$ Number of Independent Variables								
	k	1		2		3		4		5	
n		d_L	d_U	d_L	d_U	d_L	d_U	d_L	d_U	d_L	d_U
15		0.81	1.07	0.70	1.25	0.59	1.46	0.49	1.70	0.39	1.96
16		0.84	1.09	0.74	1.25	0.63	1.44	0.53	1.66	0.44	1.90
17		0.87	1.10	0.77	1.25	0.67	1.43	0.57	1.63	0.48	1.85
18		0.90	1.12	0.80	1.26	0.71	1.42	0.61	1.60	0.52	1.80
19		0.93	1.13	0.83	1.26	0.74	1.41	0.65	1.58	0.56	1.77
20		0.95	1.15	0.86	1.27	0.77	1.41	0.68	1.57	0.60	1.74
21		0.97	1.16	0.89	1.27	0.80	1.41	0.72	1.55	0.63	1.71
22		1.00	1.17	0.91	1.28	0.83	1.40	0.75	1.54	0.66	1.69
23		1.02	1.19	0.94	1.29	0.86	1.40	0.77	1.53	0.70	1.67
24		1.04	1.20	0.96	1.30	0.88	1.41	0.80	1.53	0.72	1.66
25		1.05	1.21	0.98	1.30	0.90	1.41	0.83	1.52	0.75	1.65
26		1.07	1.22	1.00	1.31	0.93	1.41	0.85	1.52	0.78	1.64
27		1.09	1.23	1.02	1.32	0.95	1.41	0.88	1.51	0.81	1.63
28		1.10	1.24	1.04	1.32	0.97	1.41	0.90	1.51	0.83	1.62
29		1.12	1.25	1.05	1.33	0.99	1.42	0.92	1.51	0.85	1.61
30		1.13	1.26	1.07	1.34	1.01	1.42	0.94	1.51	0.88	1.61
31		1.15	1.27	1.08	1.34	1.02	1.42	0.96	1.51	0.90	1.60
32		1.16	1.28	1.10	1.35	1.04	1.43	0.98	1.51	0.92	1.60
33		1.17	1.29	1.11	1.36	1.05	1.43	1.00	1.51	0.94	1.59
34		1.18	1.30	1.13	1.36	1.07	1.43	1.01	1.51	0.95	1.59
35		1.19	1.31	1.14	1.37	1.08	1.44	1.03	1.51	0.97	1.59
36		1.21	1.32	1.15	1.38	1.10	1.44	1.04	1.51	0.99	1.59
37		1.22	1.32	1.16	1.38	1.11	1.45	1.06	1.51	1.00	1.59
38		1.23	1.33	1.18	1.39	1.12	1.45	1.07	1.52	1.02	1.58
39		1.24	1.34	1.19	1.39	1.14	1.45	1.09	1.52	1.03	1.58
40		1.25	1.34	1.20	1.40	1.15	1.46	1.10	1.52	1.05	1.58
45		1.29	1.38	1.24	1.42	1.20	1.48	1.16	1.53	1.11	1.58
50		1.32	1.40	1.28	1.45	1.24	1.49	1.20	1.54	1.16	1.59
55		1.36	1.43	1.32	1.47	1.28	1.51	1.25	1.55	1.21	1.59
60		1.38	1.45	1.35	1.48	1.32	1.52	1.28	1.56	1.25	1.60
65		1.41	1.47	1.38	1.50	1.35	1.53	1.31	1.57	1.28	1.61
70		1.43	1.49	1.40	1.52	1.37	1.55	1.34	1.58	1.31	1.61
75		1.45	1.50	1.42	1.53	1.39	1.56	1.37	1.59	1.34	1.62
80		1.47	1.52	1.44	1.54	1.42	1.57	1.39	1.60	1.36	1.62
85		1.48	1.53	1.46	1.55	1.43	1.58	1.41	1.60	1.39	1.63
90		1.50	1.54	1.47	1.56	1.45	1.59	1.43	1.61	1.41	1.64
95		1.51	1.55	1.49	1.57	1.47	1.60	1.45	1.62	1.42	1.64
100		1.52	1.56	1.50	1.58	1.48	1.60	1.46	1.63	1.44	1.65

Source: J. Durbin and G. S. Watson, 'Testing for serial correlation in least square regression II,' *Biometrika*, **38**, 1951, 159–178.

Appendix B: Description of MINITAB and its application

We provide here a brief outline of the MINITAB statistical package which is used to provide computer solutions to many of the Worked examples along with Microsoft Excel (see Appendix C below). The version of MINITAB used here is the latest available at the time of writing – Release 12 for Windows 95 and Windows NT. References to MINITAB publications, the MINITAB website and contact address are given at the end of this Appendix.

MINITAB Release 12 is a very popular statistical package which is widely used in industry, research and teaching. It provides an extensive range of data analysis and graphics capabilities which are easy to apply in a user-friendly Windows environment. The package is also available for Windows 3.1, Macintosh/Power Macintosh, DOS micro-computers and most of the leading workstations, minicomputers and mainframe computers.

MINITAB Release 12 covers all of the techniques considered in this book – except for index numbers – and many more besides which go beyond the scope of the core principles considered here. In brief, the package includes:

- Comprehensive statistical capabilities, including descriptive statistics, exploratory data analysis, regression and correlation, analysis of variance, sample size and power calculations, multivariate analysis, non-parametric techniques, time series, cross-tabulations as well as simulations and distributions.

- High-resolution graphics capabilities which are fully editable, including a brushing capability for identifying points on plots and pinpointing the actual data point in the Data window, as well as the facility of pasting graphs into other applications.

- Powerful data management capabilities allowing the import of data from other versions of MINITAB, spreadsheets, databases and text files into a project, as well the facility of easily creating data subsets.

- A macro facility which allows the user to write a program of MINITAB commands to automate repetitive tasks or to extend MINITAB's functionality.

- A graphical interface that provides an easy-to-use and efficient work environment.

The MINITAB environment

The main parts of the MINITAB environment are illustrated in Figures B1 and B2 below.

437

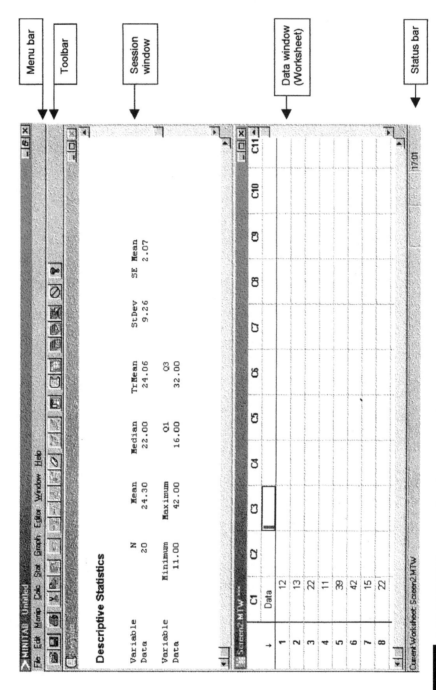

Figure B1 Initial MINITAB screen showing Data and Session windows with results

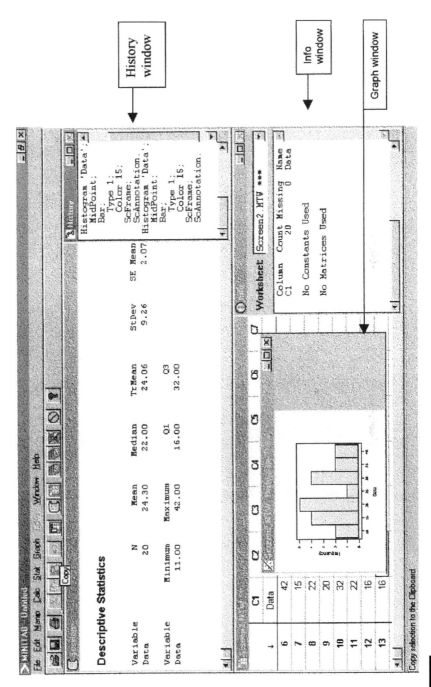

Figure B2 MINITAB screen showing Info, History and Graph windows overlaid on initial screen

The initial MINITAB screen, shown in Figure B1, illustrates the following:

■ The **Menu bar** which is used to choose commands.

■ The **Toolbar** which displays buttons for commonly-used functions – the buttons change depending on which MINITAB window is active.

■ The **Session window** displays text output such as tables of statistics.

■ The **Data window** for entering, editing and viewing columns of data for each worksheet.

■ The **Status bar** which displays explanatory text whenever the mouse-pointer is pointing to a menu item or toolbar button.

In addition, as shown in Figure B2, facilities are included to overlay the initial screen with various additional windows as follows:

■ The **History window** which records all of the commands used so far in the course of a MINITAB session.

■ The **Info window** which summarizes each open worksheet.

■ The **Graph window** which displays any graphical outputs.

In MINITAB, there are three ways to access commands: with menus, the Toolbar and session commands. Most commands use data in some way: they draw graphs based on the data, change existing data or create data. Data are entered and stored in worksheets and a *Minitab Project* can contain many worksheets. When issuing a command (by any method) that uses data, the command acts on the current worksheet. The current worksheet is the one associated with the active Data window. A window is made active by clicking on it or choosing it from the Window menu. If no Data window is active, the command acts on the Data window that was most recently active.

Illustrations of the use of MINITAB are included throughout this book in the Applications and Worked Examples sections alongside manual calculations and Excel applications. We now explain below the application of MINITAB to create graphs and charts and then its use for statistical analysis.

Using MINITAB for creating graphs and charts

In MINITAB, choosing **Graph** on the menu bar accesses all the graphics capabilities. The options then available fall into four groups:

Core graphs

These cover the basic types of graphs and charts and include plots, time series plots, charts such as bar charts and line charts etc., histograms, boxplots and others which are not specifically covered in this book (matrix plots, draftsman plots and contour plots). Core graphs also allow the creation of multiple graphs that allow different types of graphs to be placed in the same graph window.

3D graphs

These include scatter plots and other types which are not considered here (wireframe plots and surface plots) based on three variables.

Speciality graphs

These are graphs which combine elements of core graphs and include: dotplots, pie charts, marginal plots and probability plots.

Character graphs

These are 'typewriter' style graphs such as stem and leaf plots that appear as text in the output.

Some of the graphs lie outside the scope of this book, but most of the core graphs and some of the special and character graphs are covered in the Worked examples in Chapter 2.

Using MINITAB for statistical analysis

The analysis tools in MINITAB are extensive. Several of them go beyond the scope of this book and are therefore not covered here. All of the tools are activated through three buttons on the menu bar. These are:

- **Stat** which gives access to the following tools:
  - Basic statistics (including descriptive statistics, confidence intervals, significance tests, correlation)
  - Regression
  - Analysis of variance (ANOVA)
  - Multivariate analysis
  - Nonparametrics
  - Tables
  - Time series
  - Exploratory data analysis
  - Power and sample size
- **Calc** which includes the following tools (among others which go beyond the scope of this book):
  - Calculator (this allows use of the standard mathematical operators and a large number of pre-defined functions)
  - Column statistics
  - Row statistics
- **Manip** which covers the following:

  - Sub-set worksheet
  - Split worksheet
  - Sort
  - Rank
  - Delete rows
  - Erase variables

- Copy columns
- Stack/unstack
- Concatenate
- Code
- Change data type
- Display data

Most of the tools listed above are used in the Worked examples.

References

Meet MINITAB, Release 12 for Windows. Minitab Inc, PA, USA: Revised 1999. ISBN 0-925636-38-X.

MINITAB User's Guide 1: Data, Graphics and Macros, Release 12 for Windows. Minitab Inc, PA, USA: February 1998. ISBN 0-925636-39-8.

MINITAB User's Guide 2: Data Analysis and Quality Tools, Release 12 for Windows. Minitab Inc, PA, USA: February 1998. ISBN 0-925636-40-1.

Ryan, B.F. and Joiner, B.L. (1994) *MINITAB Handbook*, 3rd edition, Belmont, CA: Duxbury Press. ISBN 0-534-21240-9.

Contact addresses

For additional information contact:

Minitab Inc.
3081 Enterprise Drive
State College
PA 16801
USA

Telephone numbers:
800-448-3555 from within USA or Canada
(+1) 814-238-3280 from outside USA or Canada
FAX (+1) 814-238-4383

email:
sales@minitab.com (from within USA or Canada)
intlsales@minitab.com (outside the USA or Canada)

Worldwide web:
http://www.minitab.com

Appendix C: Description of Excel and its application

Microsoft Excel is one of the most widely used computer programs and belongs to the category of programs known as *spreadsheets*. It is the power and versatility of Excel that makes it so popular. Accountants, sales people, managers and nearly everyone working in business today use it. Spreadsheets have not traditionally been used for statistical analysis but as the use of computers has become more widespread so spreadsheet programs have developed to include the capability of carrying out a wide range of statistical procedures. They also include a wide range of graphical capabilities. In combination, these facilities enable the user not only to undertake analyses but also to prepare reports and present data in meaningful ways.

The version of Excel used in this book is the latest available at the time of writing – Excel 2000 for Windows 95 and Windows NT. References to Excel publications, the Microsoft website and contact address are given at the end of this Appendix.

Excel 2000 covers most of the techniques considered in this book. In brief, the package includes a wide range of procedures for:

- Entering text, numbers, values, dates and times in worksheets using menus and toolbars

- Editing and formatting worksheets

- Extracting data from a list – sorting and summarizing data as well as finding, adding, deleting and filtering records

- Graphical analysis of data

- Data analysis in statistical and other fields – using basic mathematical operators and pre-defined functions as well as linking worksheets with formulae

The Excel Environment

A complete document in Excel is referred to as a *Workbook*. A workbook can be made up of one or many pages, called *worksheets*. Each worksheet is named (default = sheet1, sheet2 etc.) and the current worksheet appears in the *active window*. The workbook can be saved in a file.

When Excel is launched the Excel application window is opened and displays a blank workbook as shown in Figure C1 on the following page.

The initial Excel screen illustrates the following:

- The **Title bar** which indicates what is in the window. For the main Excel window

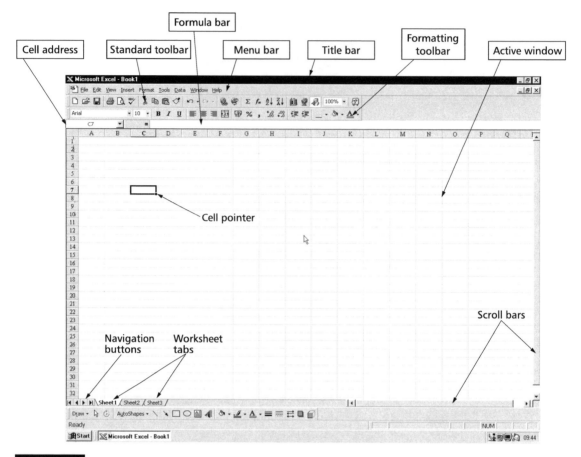

Figure C1 Initial Excel application screen showing a blank workbook

this includes the name of the application (Microsoft Excel) and the name of the workbook.

■ The **Menu bar** which is used to issue commands. Behind each menu is a drop-down list of commands. Excel automatically customises menus based on how often a command is used. In use, Excel automatically adjusts the menus to display the most frequently used commands. Options not used very often can be displayed by clicking the down arrows at the bottom of the menu.

■ The **Toolbars** – **Standard toolbar** and **Formatting toolbar** – each of which consist of a group of buttons which provide an easy way of accessing the most frequently-used operations. Excel provides extensive facilities for the user to control what buttons are shown and even to define new buttons.

■ The **Cell address** which shows the location at any one time of the cell pointer denoted by an alphabetical column reference and a numerical row reference (e.g. C7).

■ The **Formula bar** which shows any formulae or standard functions used to generate values in particular cells. Excel includes numerous standard functions and a complete list, together with a short description, may be seen by clicking on the 'equals sign' (=) symbol in the formula bar which will display the formula palette.

- The **Active window** which shows the particular worksheet (highlighted by the worksheet tabs) which is currently active.

- The **Cell pointer** which points to the **active** cell, surrounded by a bold border, in which the user can enter data.

- **Navigation buttons** which allow the user to move from one worksheet to another.

- **Worksheet tabs** that identify the worksheets contained in the workbook. These worksheets can each be given their own name.

- **Scroll bars** that enable the user to move the screen window on the worksheet until it is over the cells the user wishes to see. There are two scroll bars which allow movement horizontally and vertically.

We now consider the use of the program for creating graphs and charts and then its use for statistical analysis.

Using Excel for creating graphs and charts

Excel contains many pre-defined commands for creating graphical presentations which are referred to as *charts*. These cover all the standard types of charts including area charts, column charts, bar charts, line charts, pie charts etc. and others described as: doughnut charts, stock charts, XY scatter charts, bubble charts, radar charts, surface charts and cone, cylinder and pyramid charts. All charts can be edited, formatted, re-sized, dragged to a different location and printed as well as being downloaded for use in other applications.

The procedure in Excel for creating graphs and charts is more automatic than in MINITAB but the range of charting options is not as large. There are also some slight differences in terminology. The easiest way to create charts in Excel is to use the **Chart Wizard**. This guides the user through a standard sequence of steps which remains the same for all charts.

Before choosing **Chart Wizard** select the data to be used in the chart (make sure the data you select are symmetrical and not selected randomly). Do not include blank cells as these create empty spaces in the chart. Non-consecutive ranges of cells may be selected by holding down the CTRL key and dragging over cells you wish to select.

When the data are selected click the **Chart Wizard button** to activate it:

Step 1

From the **Chart Type** list, choose a **chart type** and click on a **chart sub-type** in the right panel (to view a sample chart of the data and the chart type selected, click the **Press and Hold to view Sample** button). Click **NEXT** to continue.

Step 2

This provides a preview of your chart using the data you had selected. To select a different range of data click the **Collapse Dialog Box** (on the right of the **Data** range box) and select new data. Click **NEXT** to continue.

Step 3

This provides additional chart information. Use the tabs to provide information about various elements of the chart:

Titles	provide a chart title and axes titles
Axes	display or hide axes
Gridlines	display gridlines for the chart and display or hide the third dimension of a 3D chart.
Legend	position the legend or uncheck the **show legend** button to remove it
Data labels	display data labels to provide additional information about a value
Data table	display the selected range as part of the chart

The chart reflects your changes. Click **NEXT** to continue.

Step 4

This positions the chart either as an object on the current worksheet or on a new blank sheet. Click **FINISH** to place the chart.

The various elements of a chart (e.g. chart title, legend, axes titles etc.) may be edited and formatted by clicking on the chart to select it (the chart is surrounded by a border with handles on each corner). The **Chart Formatting toolbar** appears. Click on any element of the chart and use the toolbar to reformat. Alternatively, double click on a chart element and the formatting dialog box opens. The handles on each corner of the chart allow the chart to be moved and resized by dragging.

Using Excel for statistical analysis

In using Excel for statistical analysis it is necessary to make a distinction between **Data analysis tools** and **Functions**.

Data analysis tools are accessed through the **Tools** button on the toolbar and cover the following topics (among others which fall outside the scope of this book):

- Analysis of variance (ANOVA)
- Correlation
- Descriptive statistics
- *F*-test
- Histogram
- Moving average
- Rank and percentile
- Regression
- *t*-test
- *Z*-test

Functions are pre-defined formulas. Excel includes a large number of functions and a complete list, together with a short description, can be displayed by clicking on the 'equals sign' (=) on the formula bar. This displays the **formula palette** and the name of a particular function. Clicking on the down arrow displays functions other than the one displayed. Click on 'More functions' and the **Paste function** dialogue box appears – as shown below in Figure C2.

It will be seen that this gives a list of **function categories** on the left and a list of the **function names** that apply to each selected category on the right. Click on a function name and a brief description appears at the foot of the box. The function category most relevant

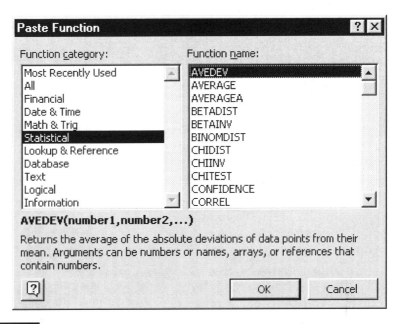

Figure C2   List of Excel function categories and function names

to this book is one named **Statistical**. Some 80 functions are listed against this category. The main ones are as follows:

AVEDEV   Returns the average of the absolute deviations of data points from their mean

AVERAGE   Returns the average of its arguments

BINOMDIST   Returns the individual term binomial distribution probability

CHIDIST   Returns the one-tailed probability of the chi-squared distribution

CHITEST   Returns the test for independence

CONFIDENCE   Returns the confidence interval for a population mean

CORREL   Returns the correlation coefficient between two data sets

COUNT   Counts how many numbers are in the list of arguments

COUNTA   Counts how many values are in the list of arguments

COVAR   Returns covariance, the average of the products of paired deviations

CRITBINOM   Returns the smallest value for which the cumulative binomial distribution is less than or equal to a criterion value

DEVSQ   Returns the sum of squares of deviations

FDIST   Returns the F probability distribution

FINV   Returns the inverse of the F probability distribution

FORECAST   Returns a value along a linear trend

FREQUENCY   Returns a frequency distribution as a vertical array

FTEST   Returns the result of an F-test

GEOMEAN   Returns the geometric mean

LARGE   Returns the k-th largest value in a data set

LINEST   Returns the parameters of a linear trend

MEDIAN   Returns the median of the given numbers

MODE   Returns the most common value in a data set

NORMDIST   Returns the normal cumulative distribution

NORMINV   Returns the inverse of the normal cumulative distribution

NORMSDIST   Returns the standard normal cumulative distribution

NORMSINV   Returns the inverse of the standard normal cumulative distribution

PEARSON   Returns the Pearson product moment correlation coefficient

PERCENTILE   Returns the k-th percentile of values in a range

PERCENTRANK   Returns the percentage rank of a value in a data set

PERMUT   Returns the number of permutations for a given number of objects

POISSON   Returns the Poisson distribution

PROB   Returns the probability that values in a range are between two limits

QUARTILE   Returns the quartile of a data set

RANK   Returns the rank of a number in a list of numbers

RSQ   Returns the square of the Pearson product moment correlation coefficient

SLOPE   Returns the slope of the linear regression line

SMALL   Returns the k-th smallest value in a data set

STANDARDIZE   Returns a normalized value

STDEV   Estimates standard deviation based on a sample

STDEVP   Calculates standard deviation based on the entire population

STEYX   Returns the standard error of the predicted y-value for each x-value in the regression

TDIST   Returns the Student's t-distribution

TINV   Returns the inverse of the Student's t-distribution

TREND   Returns values along a linear trend

TRIMMEAN   Returns the mean of the interior of a data set

TTEST   Returns the probability associated with a Student's t-test

VAR   Estimates variance based on a sample

VARP   Calculates variance based on the entire population

ZTEST   Returns the two-tailed P-value of a Z-test

As a function is a formula, its use is always preceded by an $=$ symbol followed by the function name and then one or more 'arguments' separated by commas and enclosed in parentheses. The following is an example of a function and its syntax:

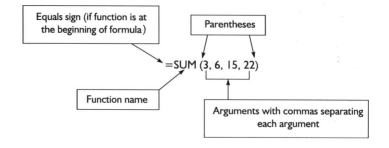

Wherever appropriate the application of the functions is illustrated in giving solutions to the Worked examples in each chapter.

References

Berk, K.N. and Carey, P. (1997) *Data Analysis with Microsoft Excel*, USA: International Thomson Publishing Company. ISBN 0-53452-929-1.

Catapult Inc. (ed). (1999) *Microsoft Excel 2000 Step by Step*, USA: Microsoft Press. ISBN 1-57231-974-7.

Dodge, M. and Stinson, C. (1999) *Running Microsoft Excel 2000*, USA: Microsoft Press. ISBN 1-57231-935-6.

Nelson, S.L. (1999) *Microsoft Pocket Guide to Microsoft Excel 2000*, USA: Microsoft Press. ISBN 1-57231-971-2.

Online Press Inc. (ed). (1999) *Quick Course in Microsoft Excel 2000*, USA: Microsoft Press. ISBN 1-57231-977-1.

Perspection (ed). (1999) *Microsoft Excel 2000 at a Glance*, USA: Microsoft Press. ISBN 1-57231-942-9.

Contact addresses

For additional information contact:

Microsoft Corporation
International Customer Service
One Microsoft Way
Redmond
WA 98052-6399
USA

Telephone number:
(+1) 425-882-8080

Worldwide web:
http://www.microsoft.com

Answers to self-study exercises

Lecturers can visit the accompanying website at http://www.cengage.co.uk for full worked solutions to all these exercises.

Chapter 3

3.1 Mean $= 38.17$ median $= 40$ mode $= 45$

3.2 Mean $= 7.6$ median $= 8$ mode $= 10$

3.3 For this to occur, the distribution of incomes in the engineering sector must be negatively skewed while those in the retailing sector must be positively skewed and the skewness in each case must be sufficient to ensure that the locations of these measures are as required by the question.

3.4 (a) Mean $= 41.1$ hours   median $= 42.5$ hours   mode $= 45$ hours
(b) Median (skewed distribution)

3.5 (a) £103.8
(b) 5.7%

3.6 Cola

3.7 2.42%

3.8 14.2 years

3.9 26.8%

3.10 (a) 0.41592%
(b) 57.6 m

Chapter 4

4.1 (a) 81   (b) Q1 $= 8.25$ Q3 $= 30.75$ (c) 22.5
(d) 11.25   (e) 79.8   (f) 24.135   (g) 582.5
(h) 1.027 or 102.7%

4.2 (a) (i) Values: 1280   Weights: 3.2
(ii) Values: 1626.34   Weights: 4.3
(iii) Values: 2,645,000   Weights: 18.5
(iv) Values: 29.04%   Weights: 23.89%
(b) Value (29.04 > 23.89)

4.3 (a)
Classes	Frequencies
2 and less than 3	2
3 and less than 4	10
4 and less than 5	8
5 and less than 6	17
6 and less than 7	9
7 and less than 8	3
8 and less than 9	1

(b) (i)   Q1 = 4.06   Q2 = 6.06
(ii) 3.3
(c) Q1 = 3.975   Q3 = 6.025   D1 = 3.51
(d) Semi-interquartile range = 1.0

4.4   (a) Height (coefficient of variation for height (0.22) > coefficient of variation for weight (0.07))
(b) Height (coefficient of variation for height (0.27) > coefficient of variation for weight (0.11))
(c) No (0.22 < 0.27)
(d) Yes (0.11 > 0.07)

Chapter 5

5.1

1	1	7
3	2	45
3	3	
7	4	2356
(7)	5	0346777
11	6	017
8	7	068
5	8	268
2	9	24

5.2

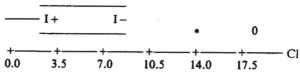

(a) Median of 4.0 with lower and upper quartile values around 2.0 and 6.0 respectively. Most observations concentrated in the range between 0.0 and 7.0.
(b) Two. One 'possible' outlier (14.0) and one 'extreme' outlier (18.5).

5.3

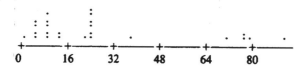

Chapter 6

6.1   (a) Classical   (b) Empirical
(c) Subjective

6.2   18

6.3   (a) 0.0   (b) 0.8

6.4   (a) 0.05   (b) 0.0325   (c) 0.6675

6.5   (a) (i) 0.60   (ii) 0.04   (iii) 0.36
(iv) 0.06   (v) 0.12   (vi) 0.08
(b) —
(c) 0.0696
(d) 0.069

6.6   (a) 120   (b) 30 240

6.7 (a) 260 000   (b) 10 000

6.8 6

6.9 252

6.10 10 000. Combinations involving the same four digits are not equivalent to one another.

6.11 (a) $^{25}C_{15} = 3\,268\,760$
(b) $^{25}C_{15} - (^{21}C_{15} + {}^{19}C_{15} + {}^{17}C_{15} + {}^{18}C_{15} + {}^{15}C_{15}) = 3\,209\,667$

Chapter 7

7.1 (a) 0.205   (b) 0.172

7.2 (a) 0.1296   (b) 0.1792

7.3 0.5987, 0.3151, 0.0746, 0.0105, 0.0010 respectively

7.4 0.0539

7.5 (a) 16   (b) 3.58

7.6 p = 0.342, n = 43.7, say 44

7.7 (a) 0.07   (b) 0.2   (c) 0.464   (d) £135

7.8 11

Chapter 8

8.1 0.039

8.2 (a) 0.122   (b) 2 boxes

8.3 (a) 0.0498   (b) 0.1465   (c) 0.3528   (d) 0.0009

8.4 (a) 0.2231   (b) 0.1913   (c) 0.1673   (d) 18.6%

8.5 0.0892

8.6 (a) 0.7981   (b) 0.2584

8.7 0.013

8.8 0.6703

Chapter 9

9.1 Symmetrical 'bell-shaped', asymptotic curve with defined mathematical properties which allow the measurement of probabilities (areas under the curve), given the mean and standard deviation.

9.2 A variable, x, is transformed into a 'standard normal' form by measuring in terms of deviations from its mean relative to the standard deviation.

9.3 (a) 0.1587   (b) 0.3413   (c) 0.1587   (d) 0.6826
(e) 0.025   (f) 0.0668   (g) 0.84

9.4 (a) 0.1190   (b) 0.3192   (c) 0.4448

9.5 Q3 = 3 569.5   Q1 = 2 430.5

9.6 0.3085

9.7 Mean = 1.8121, standard deviation = 0.7092

9.8 (a) Mean = 61.396, standard deviation = 0.901
(b) 0.03754   (c) 62.549%

9.9 (a) (i) 0.9224   (ii) 0.0228   (iii) 0.0548
(b) £56.03

9.10 (a) 10.56%   (b) 73.4%   (c) 0.5132   (d) 0.0061

9.11 $np \geq 5$ and $nq \geq 5$

9.12 (a) 0.1124   (b) 0.0803   (c) 0.2912
     (d) 0.9015   (e) 0.8212   (f) 0.9015

9.13 0.8962

9.14 (a) 0.0559   (b) 0.0297   (c) 0.5452

9.15 (a) 0.0065   (b) 0.0143   (c) 30.5

Chapter 11

11.1 (a) 6.0   (b) 5.7   (c) 2.0

11.2 (a) 25   (b) $16/\sqrt{n}$ (c) 0.0003
     (d) 0.0082   (e) 0.0002   (f) 0.0505

11.3 0.9938

11.4 (a) 0.0668   (b) 0.0013

11.5 (a) 0.0062   (b) 0.3372

11.6 (a) mean $= 45.875$ standard deviation $= 2.25$
     (Normal distribution assumed and continuity correction applied.)
     (b) 0.2709   (c) 0.1616   (d) Yes: $40.59 < \bar{x} < 49.41$,
     where 40.59 and 49.41 denote the lower and upper 95 per cent confidence limits.
     (e) 0.9732   (f) 48.19

11.7 (a) 385   (b) 83.54%

11.8 (a) £0.23   (b) £2.09   (c) 13.36%

11.9 433.5 hours

Chapter 12

12.1 (a) 0.7
     (b) (i) 105 ± 1.4 (ii) 105 ± 1.8
     (c) (i) increased   (ii) reduced   (iii) increased

12.2 (a) 62.3%   (b) 62.3 ± 0.392

12.3 (a) 0.1587
     (b) 1011.65 ml

12.4 (a) $\bar{x} \pm 5$ (b) 99.9%   (c) 62

12.5 (a) 97   (b) 221

12.6 (a) —
     (b) (i) $\mu = £300$. $\sigma^2 = £11\,025$
        (ii) 90%: 300 ± 8.64
           95%: 300 ± 10.29
        (iii) 99.58%
     (c) (i) Narrower
        (ii) Not affected (unless finite population correction factor applied).

12.7 (a) 1.531   (b) $£(10.5 \pm 0.4)$mn

Chapter 13

13.1 (a) 1.753   (b) -2.602 (c) -1.341

13.2 (a) 0.025   (b) 0.01   (c) 0.94

13.3 9.58 ± 0.28

13.4 2.35 ± 0.31

Chapter 14

14.1 Unknown, but with large samples ($n > 30$) Z may be used as an approximation of t.

14.2 (a) 78.675 and 81.325
(b) 78.355 or 81.645
(c) 74.89 or 85.11
(d) 74.833 and 85.167
(e) 76.511

14.3 (a) Z (b) t (normal population)
(c) Z (approximation of t) (d) Z

14.4 H_0 rejected

14.5 Yes (H_0 rejected.)

14.6 Yes (H_0 rejected.)

14.7 Claim rejected (H_0 rejected.)

14.8 Guarantee met (H_0 accepted.)

Chapter 15

15.1 (a) H_0 rejected (A > B)
(b) H_0 rejected (A $\neq$ B)

15.2 Yes. H_0 rejected ($\mu_1 \neq \mu_2$).

15.3 (a) Yes. H_0 rejected ($\mu_1 \neq u_2$).
(b) SL: H_0 rejected ($\mu > 1000$).
   LL: H_0 accepted ($\mu = 1500$).
(c) 136

15.4 Yes H_0 ($\mu_1 - \mu_2 = 0.08$) rejected.

15.5 Significant difference. $H_0(\mu_1 - \mu_2 = 0)$ rejected.

15.6 $H_0(\mu_A - \mu_B = 0)$ rejected.

15.7 No. Accept H_0 ($\mu_D = 0$).

15.8 (a) Yes. Reject $H_0(\mu_D = 0)$.
(b) Yes.

15.9 Yes. Reject H_0 ($\mu_D = 0$).

Chapter 16

16.1 (a) 0.33
(b) No. H_0 ($\pi_1 = \pi_2$) accepted.

16.2 23.7% to 38.7%

16.3 (a) 0.0485 (b) 0.0597 (c) 0.0017

16.4 Reject H_0 ($\pi = 0.75$) at 5% level – manager's claim not accepted.
Accept H_0 ($\pi = 0.75$) at 1% level – manager's claim accepted.

16.5 (a) 385 (b) 0.3 ± 0.107
(c) Yes. H_0 ($\pi_1 - \pi_2 = 0$) rejected.

16.6 (a) No. H_0 ($\pi_1 - \pi_2 = 0$) accepted at 5% level.
(b) Yes. H_0 ($\pi = 0.44$) rejected at 5% and 1% levels.
(c) 16 641

Chapter 17

17.1 Independent random samples from normally distributed populations with equal variances.

17.2 (a) 440

(b) 21

(c) 47.27%

(d) 3

(e) 2

(f) 104

(g) 4.46

(h) $v_1 = 2$ and $v_2 = 18$

(i) 3.55

(j) Yes $(4.46 > 3.55)$

17.3 Null hypothesis cannot be rejected [F $(= 0.413) <$ F$_{0.05}$ with $v_1 = 2$ and $v_2 = 15$ df $(= 3.68)$].

17.4 (a) 3

(b) 5

(c) 4

(d) 47

(e) Whether or not there is a statistically significant difference between the population means of the variable classified by column and, likewise, between the population means of the variable classified by row.

(f) 2 and 4 for columns and rows respectively.

(g) Columns $= 4.64$; rows $= 4.27$.

(h) Cannot reject null hypotheses of no difference between population means of column variable and of row variable (F$_\alpha =$ F$_{0.05}$, with 2 and 4 df $= 6.94$; in each case F $<$ F$_\alpha$).

17.5 No. [For columns: F $(= 0.35) <$ F$_{0.05}$ with 2 and 12 df $(= 3.89)$; for rows: F $(= 0.79) <$ F$_{0.05}$ with 6 and 12 df $(= 3.00)$].

Chapter 18

18.1 (a) Independent: U. Dependent: S.

(b) S $= 3.35 - 0.209$U

H$_0$ (B $= 0$) rejected: B is significantly less than zero.

18.2 (a) Yes. H$_0$ ($\rho = 0$) rejected.

(b) 99.13% of the variation in cost is explained.

(c) 6.1733 ± 0.8023

(d) £11.88 ($=$ intercept)

(e) £61.7

18.3 (a) $Y_c = -1.17 + 0.115x$

(b) 0.9984. Significantly greater than zero at 5% level of significance.

(c) 99.68%

(d) £4.58, say £4.6

(e) £4.6 $\pm$ 0.1

18.4 (a) Negative

(b) £0.3575m

(c) No. H$_0$: B $= 0$ accepted at 5% level of significance.

(d) 26%

18.5 (a) 2.12

(b) (i) d$_L$ $= 1.08$ and d$_U$ $= 1.36$

(ii) d$_L$ $= 0.81$ and d$_U$ $= 1.07$

(c) No evidence of significant positive or negative autocorrelation at either the 5% level of significance $[d_U(= 1.36) < d (= 2.12) < 4 - d_U (= 2.64)]$ or the at the 1% level of significance $[d_U(= 1.07) < d (= 2.12) < 4 - d_U (= 2.93)]$.

18.6 The evidence is inconclusive $(d = 1.34)$: $d_L (= 1.22) < d (= 1.34) < d_U (= 1.42)$.

18.7 (a) (i) 136.1, say 136　　(ii) 136.1 ± 40.3.

(b) $t = \dfrac{b}{s_b} = \dfrac{-72.17}{14.69} = -4.91$

$$t = r\sqrt{\frac{n-2}{1-r^2}} = -0.8666\sqrt{\frac{10-2}{1-0.751}} = -4.91$$

(c) 24.9%. Multiple regression including additional independent variables.

18.8 (a) Aptitude test most useful: r_s for S & A is positive $(+0.56)$ whereas r_s for S & Y is negative (-0.44).

(b) Correlation between A & S (0.539) greater than that between Y & S (0.385) and both positive. Therefore A better than Y but neither is statistically significant at the 5% level

(c) Rely on aptitude test rather than interview.

18.9 (a) No difference when data are ranked integers. Otherwise r_s is a less precise measure of co-variability in the sense that ranking disregards the absolute magnitude of variables.

(b) -0.606

(c) Both r and r_s are negative but only r_s is statistically significant at 5% level. Clear tendency for C to decline with increased TV but not a closely predictable relationship.

18.10 Positive rank correlation $(r_s = +0.158)$ but not statistically significant.

18.11 (a) Expectations: a: positive; b: negative. Both statistically significant. Goodness-of-fit significant but low $(r^2 = 0.321)$. No evidence of positive or negative autocorrelation [at 5% level of signicance, $d_L = 1.13, d_U = 1.38$. $d_U < d (= 1.6) < (4 - d_U)$].

(b) No. High unexplained variation. Fit curvilinear regression and/or add additional explanatory variables.

18.12 (a) Yes

(b) Yes (at 5% level)

(c) 2.46%

(d) (i) £1033　　(ii) $-£133$

18.13 t ratios $= 5.673$ for constant and 4.179 for weight.

$s = 7.616$　　$R - sq = 0.686$

Chapter 19

19.1 (a) n $= 51$; 43 cases used.

(b) Price

(c) 4

(d) Price decreases by £0.308 for each additional year of age.

(e) H_0 rejected. H_0: $B_1 = B_2 = B_3 = B_4 = 0$; H_1: at least one $B \neq 0$.

(f) Significant. H_0: BEDS $= 0$; H_1: BEDS > 0.

(g) 55.8% of variation in price is explained by variation in the group of explanatory variables.

(h) £79,630

19.2 (a) Yes. GNP and BAL positive, POL negative.

(b) Reject H_0 (F $= 189.4$).

(c) GNP $(t = 15.01)$ and POL $(t = -8.30)$ significant; BAL $(t = 1.743)$ not significant.

(d) Fall by £1.743m.

(e) £279.47m.

19.3 (a) Time $= 84.20 - 0.8872$ score $+ 0.2686$ age.

(b) Time decreases as score increases and age decreases.

(c) Constant $= 5.312$; score $= -5.612$; age $= 1.801$.

(d) b_1 significant ($t = -5.612$); b_2 not significant ($t = 1.801$).

(e) Yes (F $= 122.4$).

(f) SSR $= 2$ df; SSE $= 9$ df.

(g) $R^2 = 0.964$. Adj $R^2 = 0.956$.

(h) 96.4% of variation in time is explained.

(i) 42.6 minutes.

(j) Age. Not statistically significant.

Chapter 20

20.1 Yes. $\chi^2 = 7.8 < \chi^2_{0.05}$, 5 df $= 11.07$.

20.2 No. $\chi^2 = 0.3 < \chi^2_{0.05}$, 1 df $= 3.84$.

20.3 Accept null hypothesis: distribution not significantly different from normal
($\chi^2 = 3.276 < \chi^2_{0.05}$, 3 df $= 7.8.2$).

20.4 Yes. $\chi^2 = 9.398 < \chi^2_{0.05}$, 4 df $= 9.49$.

20.5 (a) No. $\chi^2 = 1.5 < \chi^2_{0.05}$, 2 df $= 5.99$.

(b) No. $\chi^2 = 13.75 > \chi^2_{0.05}$, 4 df $= 9.49$.

(c) No. $\chi^2 = 6.724 > \chi^2_{0.05}$, 2 df $= 5.99$.

20.6 (a) (i) No. $\chi^2 = 1.6 < \chi^2_{0.05}$, 2 df $= 5.99$.

(ii) Yes. $\chi^2 = 7.5 > \chi^2_{0.05}$, 2 df $= 5.99$.

(b) No. $\chi^2 = 2.93 < \chi^2_{0.05}$, 4 df $= 9.49$.

Chapter 21

21.1 Breaking down into constituent elements: trend, seasonal, cyclical and irregular or residual components.

21.2 (a) trend   (b) cyclical

(c) irregular   (d) seasonal

(e) seasonal   (f) irregular

21.3

	Q1	Q2	Q3	Q4
1998	8.3	8.6	8.9	9.3
1999	9.6	9.9	10.2	10.6
2000	10.9	11.2	11.6	11.9

Equation $= 7.92 + 0.332t$, where t is time (1998 Q1 $= 1$)

21.4 (a)

	1997	1998	1999	2000
Jan–Mar (Q1)		7.125	11.375	14.125
Apr–Jun (Q2)		8.750	11.875	14.875
Jul–Sep (Q3)	4.125	9.875	12.500	
Oct–Dec (Q4)	5.500	10.625	13.375	

(b) Less. Actual rise: 34%; seasonal index: +54%.

Seasonal index:

Q1	Q2	Q3	Q4
1.01	1.54	0.83	0.62

(c) Trend: strong positive trend – response to rising demand.
Seasonal factors: marked rise in Q2 each year – higher yields of milk in Q2 and/or dairy herds increased in size in this season to take advantage of higher yields and produce butter for storage.

21.5
Spring	Summer	Autumn	Winter
98.94	94.48	102.36	104.22

Rise. Winter 2000 s.a. = $(53/104.22) \times 100 = 50.85$
Spring 2001 s.a. = $(50.4/98.94) \times 100 = 50.94$

21.6
	Spring	Summer	Autumn	Winter
1997	46.3	45.9	44.4	45.6
1998	47.7	47.0	45.2	44.1
1999	45.9	42.5	45.0	45.6
2000	44.4	45.7	48.9	50.9

21.7 (a)
Q1	Q2	Q3	Q4
101.7	89.6	93.3	115.4

(b) + 1.7% over trend; − 11.9% over previous quarter.

(c)
	Q1	Q2	Q3	Q4
1997	5.9	5.4	5.3	4.9
1998	4.4	4.6	4.9	4.9
1999	4.8	5.6	5.2	6.3
2000	6.3	6.9	7.9	8.1

(d)
	Q1	Q2	Q3	Q4
2001	7.4	6.7	7.2	9.1

21.8 Peaks: 1987 1993 Interval: 6 years
Troughs: 1990 1997 Inrerval: 7 years

Chapter 22

22.1

	Price	Quantity
Laspeyres:	$\dfrac{\Sigma p_{in}q_{io}}{\Sigma p_{io}q_{io}} \times 100$	$\dfrac{\Sigma q_{in}p_{io}}{\Sigma q_{io}p_{io}} \times 100$
Paasche:	$\dfrac{\Sigma p_{in}q_{in}}{\Sigma p_{io}q_{in}} \times 100$	$\dfrac{\Sigma q_{in}p_{in}}{\Sigma q_{io}p_{in}} \times 100$

22.2
Beer	Spirits	Wine	Tobacco
200	250	500	400

Laspeyres index = 317.94   Paasche index = 316.95

22.3 (a) Laspeyres = 134   Paasche = 129
(b) £1,119 using Laspeyres index; £1,163 using Paasche index
(c) 11.9% using the Laspeyres price index; 16.4% using the Paasche price index.

22.4 (a) Laspeyres = 480.23   Paasche = 474.78.
(b) Absolute difference = 0.14% (Laspeyres = 16.98%, Paasche = 16.84%)
Relative difference (480.23/474.78) = 0.114% per annum compound

(c) (i) 30.23% on Laspeyres basis, 28.75% on Paasche basis.

(ii) 4.002 times using Laspeyres labour cost index, 3.957 times using Paasche labour cost index.

22.5 (a) Greater impact on pensioners. Pensioner price index = 157 > 152

(b) Reservations:

(i) appropriateness of 1993 weights. (ii) appropriateness of single index number for each expenditure category for both pensioner and other households.

Additional information:

To resolve (i) above – information about current consumption patterns. To resolve (ii) above – information about price movements and consumption patterns at less aggregated level.

22.6 (a) To measure real (i.e. volume) changes in expenditure.

(b) Price: 120.4   Volume: 101.4

(c) Price: Paasche index; Volume: Laspeyres index

22.7 Average of relatives indices:

Either base-weighted = 91.9 or current weighted = 93.4

Aggregative indices:

Either base-weighted = 91.9 or current weighted = 92.8

22.8 (a) (i) 192.26   (ii) 109.12

(b) (i)   Alternative price indices for 2000 (Laspeyres or Paasche) for use as deflators.

(ii)   The constituents of gross domestic expenditure in 1995 which could be revalued to 2000 price levels.

(iii) Actual prices and quantities.

22.9 (i) Changing relative importance of commodities – needed to revise weights.

(ii) Introduction of new commodities and withdrawal of old commodities.

(iii) Qualitative change.

Glossary

A

correlation.

Acceptance region   Region in which the null hypothesis cannot be rejected.

Addition rule for mutually exclusive events   If A and B are two mutually exclusive events, then the probability of obtaining *either* A *or* B is equal to the probability of obtaining A *plus* the probability of obtaining B:

$$P(A\ or\ B) = P(A) + P(B)$$

Addition rule for non-mutually exclusive events   If A and B are not mutually exclusive events, then we must subtract the probability of the joint occurrence of A and B from the sum of their probabilities:

$$P(A\ or\ B) = P(A) + P(B) - P(A\ and\ B)$$

Additive model of time series

$$y_t = T_t + S_t + C_t + R_t$$

Adjusted multiple coefficient of determination $(\bar{R}^2)$ Allows comparison of the overall explanatory power of regression models which have the same dependent variable but a different number of explanatory variables.

Alternative hypothesis   Specifies the hypothesis assumed true if the null hypothesis is rejected.

Analysis of variance (ANOVA)   A statistical technique for testing whether or not the means of two or more populations are equal.

Analysis of variance table   A tabular summary of the results of an analysis of variance procedure with entries showing the sources of variation, the degrees of freedom, the sums of squares and the mean squares (variances).

ANOVA   Analysis of variance.

Arithmetic mean   A measure of the central tendency of a data set. It is computed by summing all the values in the data set and dividing by the number of items.

Array   A set of data in which the individual values are arranged in either ascending or descending order.

Autocorrelation   Situation where values of the dependent variable, y are related to the values of y in previous time periods. Also referred to as *serial*

B

Bar chart   Similar to a line chart but used when discrete data are grouped into classes with each class corresponding to a rectangle, the base of which is the class interval and the height of which is the class frequency.

Bayes' rule for conditional probabilities   A general method for revising prior probabilities in the light of new information to provide posterior probabilities.

Between samples sum of squares (SSB)   The sum of squared deviations between each sample mean and the overall (grand) mean.

Bi-modal distribution   A distribution which has two modes.

Binomial distribution properties:

$$\text{Mean} = np; \ \text{Variance} = npq$$

Binomial experiment   A series of identical random trials in which each trial has only two possible mutually exclusive and complementary outcomes and in which all trial outcomes are statistically independent of one another.

Binomial probability distribution   A probability distribution showing the probability of X successes in n trials of a binomial experiment:

$$P(X\ successes) = {}^nC_x p^x q^{n-x}$$

Boxplot   A graphical display of the location of the quartiles of a data set and the overall spread of the data (also referred to as *box and whisker plot*).

C

Central limit theorem   A theorem that allows us to use the normal distribution to approximate the sampling distribution whenever the sample is large, even if the distribution of the parent population is not normal.

Chi-squared (χ^2) distribution A derived family of distributions, which depend on the number of degrees of freedom available, used in tests of goodness-of-fit and independence as well as other applications.

Chi-squared (χ^2) statistic Statistic based on the differences between the observed and expected frequencies in tests of goodness-of-fit and independence.

Class frequencies The number of observations contained in each class.

Class intervals The range of values which define the width of each class (group).

Class limits (class boundaries) The upper and lower limits for the classes of a frequency distribution.

Classes The groups into which a set of data may be classified.

Classical approach A method of assigning probabilities which assumes equally likely outcomes.

Coefficient of determination (r^2) A measure of the variation explained by the estimated regression equation. It is a measure of how well the estimated regression equation fits the data.

Coefficient of multiple correlation (R) A statistical measure of the degree of association between the dependent variable and all the explanatory variables jointly:

$$R = \sqrt{R^2}$$

Coefficient of variation A measure of relative dispersion for a data set, found by dividing the standard deviation by the mean and multiplying by 100 to express the coefficient as a percentage.

Combinations These refer to the number of ways in which a set of objects can be arranged without regard to the order of their selection.

Component (stacked) bar chart A bar chart in which the component sub-categories of a set of data are represented by bars which are stacked one on top of the other.

Composite price index A summary measure of the average price change of a group of items.

Compound formula A formula for calculating compound rates of change or the outcome of compound rates of change.

Conditional probability The probability of one event (say B) occurring given that another event (say A) has already occurred, denoted P(B|A).

Confidence interval An interval estimate of the population parameter defined according to a specified level of confidence (probability).

Confidence levels Degree of confidence associated with an interval estimate of the population parameter.

Confidence limits The upper and lower limits that together define an interval estimate.

Consistency A sample estimator is said to be consistent if its value approaches that of the population parameter being estimated as the sample size increases.

Contingency table A table in which data are cross-classified according to two characteristics simultaneously.

Continuity correction factor A correction factor applied to the values of a discrete variable when using the normal distribution as an approximation to the binomial distribution (equal to half the distance between the discrete values which is added to, or subtracted from, each of them).

Continuous variable A quantitative variable that may be measured on a continuous scale and can take any value within an interval.

Control chart A diagram used for quality control purposes in which the results of repeated sampling are recorded.

Control limits Limits on a control chart defined as $\mu_0 \pm 3\sigma_{\bar{x}}$.

Correction for tied ranks Correction required in ranking data (in the context of calculating Spearman's rank correlation coefficient) when there are tied ranks.

Correlation analysis Concerned with providing a statistical measure of the strength of any relationship between variables.

Correlation coefficient (r) Provides a numerical summary measure of the degree of correlation between two variables.

Critical region Region in which the null hypothesis cannot be accepted (sometimes referred to as rejection region).

Critical value A value that is compared with the test statistic to determine whether H_0 is to be accepted or rejected.

Cumulative density function (CDF) This measures areas under the normal curve and gives the cumulative probability associated with a PDF up to a specified value.

Cumulative frequency curve (ogive) A graph of the cumulative frequency distribution.

Cumulative frequency distribution A tabular summary of a set of data that shows the total number of data items with values *either* less than or equal to the upper limit of the class *or* greater than or equal to the lower limit of each class.

Cyclical component (C) The component of a time series that results in periodic above-trend and below-trend movements lasting more than one year.

D

Decision tree Tree diagram used in the context of decision-making.

Dependent events Two events are dependent when the occurrence (or non-occurrence) of one event affects the probability of occurrence of the other event.

Dependent variable The variable that is being predicted or explained by the independent variable. It is denoted by y in the regression equation.

Discrete variable A quantitative variable in which the scale of measurement varies in discrete steps.

Dotplot A horizontal display of the individual values of a data set.

Durbin-Watson statistic Statistic used to test for the presence of first-order autocorrelation.

E

Efficiency For a given sample size, efficiency requires that the standard error of the sample statistic is as small as possible.

Equation of a straight line: $y = a + bx$

Error sum of squares (SSE) The variation which remains after subtracting from the total variation (SST) any variation arising from measured factors (also referred to as *residual sum of squares* or *sum of squares error*).

Errors in hypothesis testing Errors relating to the possibility of rejecting a null hypothesis when it is in fact true and, conversely, to accept it when it is in fact false.

Errors in variables Situation in which variables in a regression model include measurement errors.

Estimated regression equation The estimate of the regression equation obtained by the method of least squares: $y_c = a + bx$

Event The outcome of an experiment.

Expected frequencies Frequencies expected, in tests of goodness-of-fit and independence, under the null hypothesis that the observed frequencies follow a certain pattern or theoretical distribution.

Experiment Any statistical process which generates well-defined outcomes (events).

Exploratory data analysis A technique which provides a graphical means of exploring the underlying structure of a data set.

Extrapolation Predicting values of y from a regression equation using values of x which lie outside the range of the recorded values of x from which the regression equation itself was derived.

F

F distribution A family of probability distributions based on the ratio between two sample variances.

F statistic The ratio between two estimates of variance in random samples from normal distributions: if both samples are drawn from normal distributions with the *same* variance then the ratio of the two estimates follows the F distribution.

F test In the context of multiple regression, a test of the statistical significance of the regression model as a whole, i.e. its overall explanatory power, given by R^2.

Factor The variable of interest in an ANOVA procedure.

Finite population correction factor (fpc) Used to modify the expressions for $\sigma_{\bar{x}}$ when sampling without replacement from a finite population.

First-order autocorrelation Situation where the value of y in time period t is related to its value in time period $t - 1$.

Frequency density The class frequency per unit class interval.

Frequency distribution table A tabular summary of a set of data showing the frequency (or number) of items in each of several non-overlapping classes.

Frequency polygon A graph formed by plotting class frequencies against the mid-points of the corresponding classes, connecting the points to form a graph and extending it to each end of the distribution to meet the horizontal axis at the mid-point at what would have been the next class below or above respectively.

Functional form Form of the relationship between variables.

G

Geometric mean A measure of central tendency of a data set used when it is desired to produce an average of rates of change:

$$g = \sqrt[n]{(x_1 \times x_2 \times x_3 \times \cdots \times x_n)}$$

Goodness-of-fit test   A statistical test conducted to determine whether to accept or reject a hypothesized probability distribution for a population.

Grand mean   In the context of ANOVA, the mean of all the sample observations.

H

Heteroscedasticity   Term used to describe the situation where the error terms (e_i) do not have a constant variance.

Histogram   A chart similar to a bar chart and may be used in the cases of continuous or discrete grouped data but with no gaps left between the bars.

Homoscedasticity   Term used to describe the situation where the error terms (e_i) have a constant variance.

Hypothesis testing   Procedures using sample statistics to test hypothesized values of population parameters.

Hypothesis testing for a single population proportions   Procedure using sample data to test hypotheses about a population proportion.

Hypothesis testing for two population proportions   Procedure to test the difference between two population proportions based on two independent samples.

Hypothesis testing of the difference between two population means   Procedures using sample statistics to test hypotheses about the difference between two population means.

I

Independent (or explanatory) variable   The variable used to predict the value of the dependent variable. It is denoted by x in the regression equation.

Independent events   Two events are independent when the occurrence (or non-occurrence) of one event has no effect on the probability of occurrence of the other event.

Independent samples   Samples in which the observations are taken from different objects.

Index numbers   A set of percentage values, expressed with reference to a base of 100, used to measure differences over time or space in the magnitude of either a single variable or a group of related variables.

Intercept   The point at which a regression line intersects the vertical axis on which the dependent variable is measured. It is the value of y when x is zero.

Inter-decile range   A measure showing the difference between the upper and lower decile values.

Inter-percentile range   A measure showing the difference between the upper and lower percentile values.

Interpolation   Predicting values of y from a regression equation using values of x which lie inside the range of recorded values of x from which the regression equation itself was derived.

Inter-quartile range   A measure of dispersion which covers the central 50 per cent of the observations in a data set – i.e. the difference between the lower quartile (the value below which the lowest 25 per cent of the observations lie), denoted Q_1, and the upper quartile (the value above which the highest 25 per cent of the observation lie), denoted Q_3.

Interval estimate   An estimate of a population parameter that provides an interval of values believed to contain the value of the parameter being estimated. The interval estimate is also referred to as a confidence interval.

Irregular (random) component (R)   The component of a time series that reflects the variation of the series other than that which can be explained by trend, cyclical and seasonal components.

L

Laspeyres weighted aggregate price index   A weighted aggregate price index where the weight for each item is its base period quantity.

Level of significance   The maximum probability of a Type I error that the user will tolerate in the hypothesis-testing procedure.

Line graph   A graphical presentation of data often used for showing movements in a variable over time.

Linear trend method   A method of measuring the trend which applies regression analysis to the time series data with time taken as the independent (i.e. explanatory) variable:

$$T_y = a + bx$$

Logarithmic scale   A scale in which the actual values of a variable are measured as the logarithms of the values.

Lorenz curve   A curve showing the degree of inequality in a frequency distribution.

M

Matched samples   Samples in which the observations are taken from the same objects.

Mean deviation   A measure of dispersion of a data set expressed as the average of the absolute differences between each observation and the arithmetic mean of all observations.

Mean square between columns (MSC)   The sum of squares between columns (SSC) divided by degrees of freedom – a measure of the variation among means of samples (arranged in columns) taken from different populations.

Mean square between rows (MSR)   The sum of squares between rows (SSR) divided by degrees of freedom – a measure of the variation among means of samples (arranged in rows) taken from different populations.

Mean square between samples (MSB)   Between samples sum of squares (SSB) divided by its degrees of freedom – a measure of the variation among means of samples taken from different populations.

Mean square error (MSE)   Error sum of squares divided by its degrees of freedom.

Mean square within samples (MSW)   Within sample sum of squares (SSW) divided by its degrees of freedom – a measure of the variation within samples of all samples taken from different populations.

Median   A measure of central tendency of a data set. It is the value which splits the data set into two equal groups – one with values greater than or equal to the median, and one with values less than or equal to the median.

Method of least squares   The approach used to develop the estimated regression equation which minimizes the sum of squares of the deviations between the actual and computed (i.e. estimated) values of the dependent variable, i.e.: minimize Σe_i^2 (also referred to as the *principle of least squares*).

Mode   A measure of central tendency of a data set, defined as the most frequently occurring data value.

Modified range   A range which focuses only a central portion of a data set (i.e. excluding extreme values).

Moving average method   A method of smoothing a time series by averaging successive groups of the observations.

MSB   Mean square between samples.

MSC   Mean square between columns.

MSE   Mean square error.

MSR   Mean square between rows.

MSW   Mean square within samples.

Multicollinearity   Situation in which some of the independent variables are correlated with one another.

Multi-modal distribution   A distribution which has more than two modes.

Multiple coefficient of determination (R^2)   Provides a statistical measure of the overall goodness-of-fit for the regression model as a whole.

Multiple correlation analysis   Refers to the measuring and testing of the overall explanatory power of a multiple regression model.

Multiple regression model   A regression equation in which more than one independent variable is used to explain the dependent variable.

Multiplication rule for independent events:

$$P(A \text{ and } B) = P(A) \times P(B)$$

Multiplication rules for dependent events:

$$P(A \text{ and } B) = P(A) \times P(B|A)$$

Multiplicative model of time series

$$y_t = T_t \times S_t \times C_t \times R_t$$

Mutually exclusive events   Events which cannot occur simultaneously.

N

Normal probability distribution   A continuous probability distribution whose form is a symmetrical, bell-shaped curve and determined by the mean μ and standard deviation σ.

Null hypothesis   Specifies the hypothesized value of the parameter to be tested.

O

Observed frequencies   The actual frequencies observed in a set of data used in a test of goodness-of-fit or independence.

One-tailed test   An hypothesis test in which rejection of the null hypothesis occurs for values of the test statistic in one tail of the sampling distribution.

One-way analysis of variance   The application of the ANOVA technique to test the equality of population means when classification is by one variable (also called *single-factor analysis of variance*).

Open-ended classes   The first and/or the last classes in a frequency distribution in which the lower or upper limits respectively are not specified.

Outliers   Extreme values in a set of data.

P

p value   The probability of observing a sample outcome even more extreme than the observed value when the null hypothesis is true.

Paasche weighted aggregate price index   A weighted aggregate price index where the weight for each item is its current period quantity.

Partial regression coefficients   The b coefficients in a multiple regression model.

Permutations   These refer to the number of ways in which a set of objects can be arranged in order (the order being crucial).

Pie chart   A pictorial device for presenting categorical data in which a circle is divided into wedges (like slices of a pie), each wedge corresponding to the relative frequency of each class.

Point estimate   A single numerical value used as an estimate of a population parameter.

Poisson distribution properties:

Mean $= \mu$; Variance $= \mu$.

Poisson probability distribution   A probability distribution showing the probability of x occurrences of an event over a specified interval of time or space.

Pooled sample variance s_p^2   The weighted average of the separate variances of two samples where the weights used are the respective degrees of freedom.

Population parameters   Numerical summary measures used to describe a statistical population (such as the population mean).

Prediction interval   The interval estimate of an individual value of y for a given value of x.

Price index   An index that is designed to measure changes in prices over time or differences in prices between different locations.

Price relative   A simple price index which is computed by dividing a current unit price for a single item by its base-period unit price.

Principle of least squares   The approach used to develop the estimated regression equation which minimizes the sum of squares of the deviations between the actual and computed (i.e. estimated) values of the dependent variable: minimize Σe_i^2 (also referred to as the *method of least squares*).

Probability   A numerical measure of the likelihood that an event will occur.

Probability density function (PDF)   This determines the shape of a normally distributed variable x.

Probability distribution   An ordered listing of all possible values of a random variable and their associated probabilities.

Properties of the normal distribution   These are such that the proportion of all the observations of a normally distributed variable x that fall within a range of n standard deviations on both sides of the mean is the same for *any* normal distribution.

Q

Quantiles   Measures which divide a data set into a number of equal parts.

Quantity index   An index that is designed to measure changes in quantities over time or differences in quantities between different locations.

Quartiles, deciles and percentiles   Measures which divide a data set into four, ten and 100 parts respectively, each part containing the same number of observations.

R

Random variable   A numerical description of the outcome of an experiment (an *event*).

Range   A measure of dispersion for a data set, defined to be the difference between the highest and lowest values.

Rank correlation   Correlation between variables which are measured in rank order.

Rate of change (logarithmic graph)   A graphical presentation in which absolute values of a variable are plotted against a logarithmic scale or the logarithms of the variable (log values) are plotted on a conventional scale.

Regression analysis   Concerned with measuring the way in which one variable is related to another.

Regression coefficient   The coefficient b in the regression equation. It measures the slope of the regression line.

Regression line   Line of best fit between two related variables.

Regression sum of squares   In regression, the part of the total sum of squares (SST) that is explained by the regression model.

Rejection region   Region in which the null hypothesis cannot be accepted (sometimes referred to as the *critical region*).

Relative frequencies   Each class frequency expressed as a percentage of the total number of observations.

Relative frequency (empirical) approach   A method of assigning probabilities based on experimental or historical data.

Residual The difference between the actual value of the dependent variable and the value computed using the estimated regression equation: $e_i = y_i - y_c$.

Residual sum of squares (SSE) The variation which remains after subtracting from the total variation any variation arising from measured factors (also referred to as *error sum of squares* or *sum of squares error*).

S

Sample proportion The proportion in a set of sample data possessing a certain attribute.

Sample statistics Numerical summary measures used to describe a sample such as the sample average (whose values vary according to the random sample collected).

Sampling distribution A probability distribution of all possible values of a sample statistic.

Sampling distribution of a sample proportion A probability distribution of all possible values of a sample proportion.

Sampling distribution of the difference between two population means The distribution representing the differences between all possible sample means which could be obtained by sampling from two populations.

Sampling error The error which arises when any individual sample mean is used as an estimate of the population mean.

Sampling with replacement A sampling procedure in which a value, having been selected, is replaced and therefore can then be selected again.

Sampling without replacement A sampling procedure in which a value, having been selected, is not replaced and therefore cannot then be selected again.

Scatter diagram A diagram which consists of a set of points of paired observations on two variables.

Seasonal component (S) The component of a time series that shows a regular pattern within one calendar year.

Seasonal index An index which measures the average magnitude of seasonal effects in a particular time series.

Seasonally adjusted series Time series in which the average effect of the seasons has been removed.

Semi-interquartile range Simply defined as the inter-quartile range divided by two (also referred to as *quartile deviation*).

Serial correlation Situation where values of the dependent variable, y are related to the values of y in previous time periods. Also referred to as autocorrelation.

Simple price index Price index which is computed by dividing a current unit price for a single item by its base-period unit price (also called a *price relative*).

Single-factor analysis of variance The application of the ANOVA technique to test the equality of population means when classification is by one variable (also called *one-way analysis of variance*).

Skewed distribution A distribution in which one tail is skewed either to the right (positively skewed) or to the left (negatively skewed).

Spearman's rank correlation coefficient (r_s) Provides a numerical summary measure of the degree of correlation between two variables measured in rank order.

SSB Sum of squares between samples in ANOVA.

SSC Sum of squares between columns in ANOVA.

SSE Error sum of squares (also referred to as *residual sum of squares*). In regression, the sum of the squared differences between the actual and predicted values of y – the part of the total sum of squares (SST) that is not explained by the regression model.

SSR Sum of squares between rows in ANOVA and regression sum of squares in regression analysis (the part of the total sum of squares (SST) that is explained by the regression model).

SST Total sum of squares in ANOVA and in regression analysis (the sum of the squared differences between the actual y values and the mean value of y).

SSW Within sample sum of squares in ANOVA.

Standard deviation A measure of dispersion for a data set, found by taking the positive square root of the population or sample variance.

Standard error of a sample proportion The standard deviation of the sampling distribution of a sample proportion, p.

Standard error of difference between two proportions The standard deviation of the sampling distribution of the difference between two proportions, p_1 and p_2.

Standard error of the sample mean The standard deviation of the sampling distribution of the mean.

Standard normal distribution Distribution of the standard normal variable Z which follows a normal distribution with a mean of 0 and a standard deviation of 1.

Standard normal variable The transformation of the normal variable x, expressed as Z.

Statistical inference Drawing conclusions about statistical populations based on information obtained from sample data.

Statistical significance Test of statistical hypotheses at a pre-defined level of significance (probability).

Stem and leaf diagram A form of frequency histogram which not only shows the shape of a distribution but allows one to read off every value directly.

Subjective approach A method of assigning probabilities based upon judgement.

Sum of squares between columns (SSC) In the context of two-factor ANOVA where the observations for the samples relating to one factor are arranged in columns and the corresponding observations for the other factor are arranged in rows, the sum of squared deviations between the observations in a column and the column mean summed over all columns.

Sum of squares between rows (SSR) In the context of two-factor ANOVA where the observations for the samples relating to one factor are arranged in columns and the corresponding observations for the other factor are arranged in rows, the sum of squared deviations between the observations in a row and the row mean summed over all rows.

Sum of squares error (SSE) The variation which remains after subtracting from total variation (SST) any variation arising from measured factors (also referred to as *error sum of squares or sum of squares error*).

T

t **distribution** A family of probability distributions which can be used to develop interval estimates of a population mean and test statistical hypotheses whenever the population standard deviation is unknown and the population has a normal or near-normal probability distribution.

t **statistic** The difference between a sample mean and a population mean divided by the standard error computed from a small sample.

Test of independence A statistical test of the independence of classificatory variables.

Test statistic Statistic, such as Z or t, which defines the acceptance and rejection regions according to defined levels of probability.

Time series A set of observations measured at successive points in time or over successive periods of time.

Total sum of squares In regression, the sum of the squared differences between the actual y values and the mean value of y.

Total sum of squares (SST) Sum of squared deviations between the observations in a data set and the mean of the data set.

Tree diagram A graphical device helpful in defining sample points of an experiment involving multiple steps.

Trend (T) The long-run movement in a time series.

Trial A random experiment which results in a random outcome.

Trimmed mean A measure of central tendency which disregards a proportion of the highest and lowest values in a data set.

Two-factor analysis of variance The application of the ANOVA technique to test the equality of population means when classification is by two variables or factors (also called *two-way analysis of variance*).

Two-tailed test An hypothesis test in which rejection of the null hypothesis occurs for values of the test statistic in either tail of the sampling distribution.

Two-way analysis of variance The application of the ANOVA technique to test the equality of population means when classification is by two variables or factors (also called *two-factor analysis of variance*).

Type I error The error of rejecting H_0 when it is true.

Type II error The error of accepting H_0 when it is false.

U

Unbiasedness An estimator is said to be unbiased if the mean of the sample statistic is equal to the population parameter being estimated.

Uni-modal distribution A distribution which has only one mode.

V

Variance A measure of dispersion for a data set, found by summing the squared deviations of the data values about the mean and then dividing the total by N if the data set is a population or by $n - 1$ if the data set is for a sample.

Venn diagram A pictorial device to illustrate the concepts of exclusive and non-mutually exclusive events.

W

Warning limits Limits on a control chart defined as $\mu_0 \pm 2\sigma_{\bar{x}}$.

Weighted aggregate price index A composite price index where the prices of the items in the composite are weighted by their relative importance.

Weighted aggregate quantity index A composite quantity index where the quantities of the items in the composite index are weighted in terms of unchanged prices (base period prices or current period prices may be used) to produce Laspeyres- or Paasche-type quantity indexes respectively.

Weighted arithmetic mean A measure of the central tendency of a data set used when observations on particular values may occur more than once or when we need to reflect relative frequencies or the relative importance of certain values.

Weighted average of simple price indexes A composite price index computed by taking a weighted average of simple price indexes (price relatives), using either base or current weights to produce Laspeyres- or Paasche-type index numbers respectively.

Weighted averages of quantity relatives A composite quantity index computed by taking a weighted average of quantity relatives, using either base or current weights, to produce Laspeyres- or Paasche-type index numbers respectively.

Within sample sum of squares (SSW) The sum of squared deviations between the observations in a sample and the mean of the sample summed over all samples.

Y

Yates' continuity correction factor Used to adjust the value of the χ^2 statistic when there is only one degree of freedom present.

Z

Z distribution Distribution of the standard normal variable Z which follows a normal distribution with a mean of 0 and a standard deviation of 1 (also referred to as the *standard normal distribution*).

Z statistic The transformation of the normal variable x, expressed as Z (also referred to as the *standard normal variable*).

Index